# CONSTRUCTION AND DESIGN OF CABLE-STAYED BRIDGES

*Wiley Series of Practical Construction Guides*

M. D. MORRIS, P. E., EDITOR

Jacob Feld
CONSTRUCTION FAILURE

William G. Rapp
CONSTRUCTION OF STRUCTURAL STEEL BUILDING
FRAMES

John Philip Cook
CONSTRUCTION SEALANTS AND ADHESIVES

Ben C. Gerwick, Jr.
CONSTRUCTION OF PRESTRESSED CONCRETE STRUCTURES

S. Peter Volpe
CONSTRUCTION MANAGEMENT PRACTICE

Robert Crimmins, Reuben Samuels, and Bernard Monahan
CONSTRUCTION ROCK WORK GUIDE

B. Austin Barry
CONSTRUCTION MEASUREMENTS

D. A. Day
CONSTRUCTION EQUIPMENT GUIDE

Harold J. Rosen
CONSTRUCTION SPECIFICATION WRITING

Gordon A. Fletcher and Vernon A. Smoots
CONSTRUCTION GUIDE FOR SOILS AND FOUNDATIONS

Don A. Halperin
CONSTRUCTION FUNDING: WHERE THE MONEY COMES
FROM

Walter Podolny, Jr. and John B. Scalzi
CONSTRUCTION AND DESIGN OF CABLE-STAYED BRIDGES

# CONSTRUCTION AND DESIGN OF CABLE-STAYED BRIDGES

## WALTER PODOLNY, JR., Ph.D.

Bridge Division
Office of Engineering
Federal Highway Administration
U.S. Department of Transportation
Washington, D.C.

## JOHN B. SCALZI, Sc.D.

Lecturer
Department of Civil, Mechanical
and Environmental Engineering
George Washington University,
Washington, D.C.
Formerly Director
Marketing Technical Services
United States Steel Corporation
Pittsburgh, Pennsylvania

A Wiley-Interscience Publication

JOHN WILEY & SONS

New York    London    Sydney    Toronto

Copyright © 1976 by John Wiley & Sons, Inc.

All rights reserved. Published simultaneously in Canada.

*Library of Congress Cataloging in Publication Data:*

Scalzi, John B.
  Construction and design of cable-stayed bridges.

  (Wiley series of practical construction guides)
  "A Wiley-Interscience publication."
  Bibliography: p.
  Includes index.
  1. Bridges, Cable-stayed—Design and construction.
I. Podolny, Walter.    II. Title.

TG405.S28      624.5′5      75-46578
ISBN 0-471-75625-3

Printed in the United States of America

10 9 8 7 6 5 4 3 2 1

# Series Preface

The construction industry in the United States and other advanced nations continues to grow at a phenomenal rate. In the United States alone construction in the near future will exceed ninety billion dollars a year. With the population explosion and continued demand for new building of all kinds, the need will be for more professional practitioners.

In the past, before science and technology seriously affected the concepts, approaches, methods, and financing of structures, most practitioners developed their know-how by direct experience in the field. Now that the construction industry has become more complex there is a clear need for a more professional approach to new tools for learning and practice.

This series is intended to provide the construction practitioner with up-to-date guides which cover theory, design, and practice to help him approach his problems with more confidence. These books should be useful to all people working in construction: engineers, architects, specification experts, materials and equipment manufacturers, project superintendents, and all who contribute to the construction or engineering firm's success.

Although these books will offer a fuller explanation of the practical problems which face the construction industry, they will also serve the professional educator and student.

M. D. Morris, P.E.

# Preface

The reconstruction of bridges in Europe destroyed during World War II provided engineers with the opportunity to apply new technology to an old concept in bridge design, the cable-stayed bridge. The impetus came in the 1950s, in Germany, when many of the bridges spanning the Rhine River were replaced with various types of cable-stayed bridges.

The original concept of the cable-stayed bridge dates back to 1784 but was "shelved" by engineers because of the many collapses of the early bridges. The need to build bridges more economically combined with modern methods of analysis, construction methods, and more reliable construction materials provided bridge engineers with the impetus to develop the present-day cable-stayed bridge.

Economic studies have indicated that the cable-stayed bridge may fill the void between long-span girder bridges and suspension bridges. Some European engineers feel that the cable-stayed bridge may also replace the suspension bridge in many applications. In addition to the potential economies, some engineers believe the cable-stayed bridge adds a new dimension to the aesthetics of bridge design.

Engineers in the United States are planning and designing cable-stayed structures for pedestrian overpasses, highway bridges, and bridges for pipe lines, despite the paucity of design and construction data in the American technical literature.

The objective of this book is to bring together in one volume the current state of the art of design and construction methods for all types of cable-stayed bridges so that engineering faculties, practicing engineers, local, state, and federal bridge engineers can have a ready reference source of construction details and design data.

The book discusses the general principles of cable-stayed bridges, relating to all facets of technical design, construction details and methods, and potential economies.

The book delves into the historical development of the cable-stayed bridge from its first application to the widespread use in Germany after the war and the extensions into other countries around the world. The principal features used in modern bridges receive a thorough description, including geometrical configurations, the types and styles of the towers, and the various types of roadway decks made of different materials and methods.

Illustrations of bridges from various countries are discussed and accompanied by appropriate detail sketches and photographs of the special features of each bridge.

For the uninitiated design and construction engineer, a discussion of the manufacturing and production processes of making structural wires, rope, and strand is presented.

Among the most important aspects of cable-stayed bridges are the types and methods of making the connections between the cables and the deck and/or the towers. A discussion of the various methods is presented to enable construction and design engineers to evaluate the techniques in terms of American practices and, we hope, to improve upon them.

The theory of cables and structures and methods of analysis are contained in other textbooks, and only a discussion of the special considerations for analysis and design are included here. Such items include a summary of the general behavior of cables, a detailed explanation of the use of an equivalent modulus of elasticity for the cable as a substitute member, and a discussion of wind and aerodynamic effects. All of these factors affect the design of the cable-stayed bridge.

Because the methods of fabrication and erection influence design characteristics and construction methods, a discussion of several possible techniques is presented. These discussions may assist engineers in developing their concepts and may lead to more efficient and economical methods.

Most of the material presented in this book is not original, and though individual acknowledgment of the many sources is not possible, full credit is noted wherever the specific source can be identified.

Every effort has been made to eliminate errors, but should errors be found, the authors would appreciate such notification from the readers.

The authors hope that this book will enable other engineers to design and construct cable-stayed bridges in their own country with economy and confidence.

WALTER PODOLNY, JR.
JOHN B. SCALZI

*Arlington, Virginia*
*Burke, Virginia*
*February 1976*

# Contents

# 7   Cable Data                                       189

# 8   Cable Connections                               208

# CONSTRUCTION AND DESIGN OF CABLE-STAYED BRIDGES

# 1

# Cable-Stayed Bridges

## 1.1  MODERN CABLE-STAYED BRIDGE

The concept and practical application of the cable-stayed bridge date back to the 1600s, when a Venetian engineer named Verantius built a bridge with several diagonal chain stays.[1,2] The concept was attractive to engineers and builders for many centuries, and experimentation and development continued until its modern-day version evolved in 1950 in Germany.

The modern cable-stayed bridge consists of a superstructure of steel or reinforced concrete members that is supported at one or more points by cables extending from one or more towers, Figs. 1.1 and 1.2.

The early stayed bridges used chains or bars for the stays. The advent of the various types of structural cables, with their inherent high carrying capacity and ease of installation, led engineers and contractors to replace the chains and bars. As a result, the more specific descriptive term, "cable-stayed bridges," entered the literature.

## 1.2  RECENT WORLD APPLICATIONS

Following World War II, West Germany determined that approximately 15,000 bridges had been destroyed during the conflict. Therefore, the post-war period of rebuilding these crossings provided the opportunity for

**1**

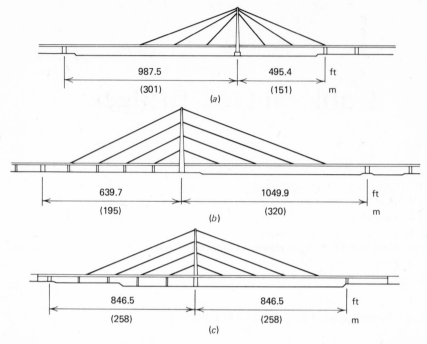

**FIGURE 1.1.    Examples of two-span cable-stayed bridge structures: (*a*) Severin Bridge at Cologne (Germany); (*b*) Kniebrücke (Germany); (*c*) Düsseldorf-Oberkassel (Germany).**

engineers, builders, and contractors to apply new concepts of design and construction. During this period steel was in short supply and a great emphasis was placed on minimum weight design. As a result of this emphasis, orthotropic plate design developed which provided a marriage with cable-stayed design to produce bridges that were in some cases 40% lighter than their prewar counterparts. Efficient use of materials and speed of erection made cable-stayed bridges the most economical type of structure to use for replacements.

The first modern cable-stayed bridge, The Strömsund Bridge, Fig. 1.3, was completed in Sweden in 1955. It is interesting to note that the bridge was built by a German contractor, Demag, in collaboration with a German engineer, Professor F. Dischinger.

In the relatively short period of time from 1955 to 1974 approximately 60 cable-stayed bridges have been built,[3,4] or are being planned, for highway traffic. As of 1974, the distribution of cable-stayed bridges throughout the world was as follows:

| | | | |
|---|---|---|---|
| Germany | 17 | Holland | 2 |
| United States | 8 | USSR | 2 |
| Japan | 7 | Austria | 1 |
| Canada | 4 | Denmark | 1 |
| Great Britain | 3 | India | 1 |
| Italy | 3 | Libya | 1 |
| Argentina | 2 | Sweden | 1 |
| Australia | 2 | Venezuela | 1 |
| France | 2 | Zambia | 1 |

Just less than one-third of the total number have been built in Germany. The remainder are distributed in many countries around the world. The United States' share of the total consists of eight bridges: the first completed highway bridge at Sitka Harbor, Alaska; one in design and one con-

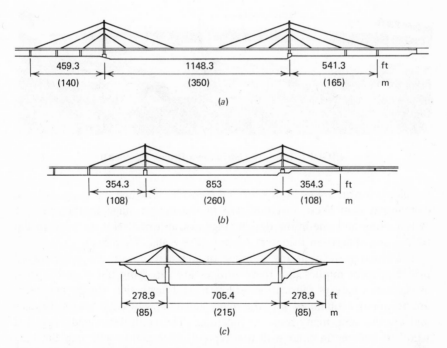

FIGURE 1.2. Examples of three-span cable-stayed bridge structures: (*a*) Duisburg (Germany); (*b*) North Bridge at Düsseldorf (Germany); (*c*) Onomichi (Japan).

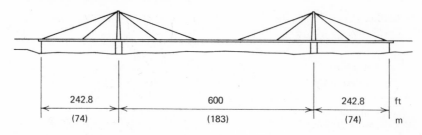

**FIGURE 1.3.    The Strömsund Bridge (Sweden).**

templated for the I-410 bypass around New Orleans; one in design and one contemplated in West Virginia; two in the design stage in the state of Washington; and one in the design stage in California. Not included in the total is the pedestrian bridge at Menomonee Falls, Wisconsin.

The rapid growth in the number of applications of the cable-stayed bridge concept implies that these bridges are satisfying many needs, such as economy, ease of fabrication, erection, and aesthetics. Bridge engineers are becoming acquainted with the many advantages of cable-stayed bridges and are planning many more applications. The contractors and engineers who first become familiar with this type of construction will reap the harvest of future contracts. There appears to be no doubt that the cable-stayed bridge with its many geometrical configurations will be applied in great numbers in the future in the United States.

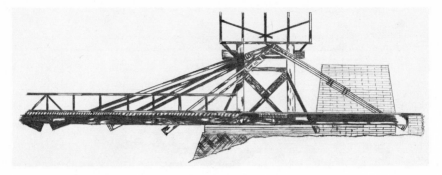

**FIGURE 1.4.  Löscher type timber bridge, 1784. (Courtesy of the British Constructional Steelwork Association, Ltd.)**

## 1.3  PAST EXPERIENCE

Although the concept of a bridge partially suspended by inclined stays dates back to seventeenth-century Venice, the concept of a bridge suspended only by inclined stays is credited to C. J. Löscher, a carpenter of Fribourg who built a completely timber bridge including stays and tower in 1784, Fig. 1.4. Apparently the stayed-bridge concept was not used again until 1817, when two British engineers, Redpath and Brown, built the King's Meadow footbridge, which had an approximate span of 110 ft, using sloping wire cable-stay suspension members attached to a cast iron tower.

Communication among engineers on technical information must have been very good following the English design because in 1821 the French architect, Poyet, suggested a bridge using steel bar stays suspended from high towers, Fig. 1.5.

The stayed bridge might have become a conventional form of construction had it not been for the bad publicity which followed the collapse of two bridges. One was the 259 ft pedestrian bridge crossing the Tweed River near Dryburgh-Abbey, England, which collapsed in 1818 when wind oscillation caused the chain stays to break at the joints.[5] The other bridge credited with delaying the use of cable-stayed bridges collapsed in 1824; it had a 256 ft span and crossed the Saale River near Nienburg.[6]

We can only assume that the technical knowledge of analysis and behavior of materials was insufficient for the successful design and construction of these ill-fated bridges. The inclined members were forged tie bars or chain links made of looped wires. The reason for the failure of the Nienburg bridge was not reported, although the technical literature of the

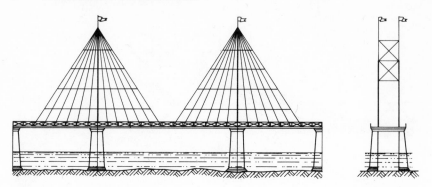

**FIGURE 1.5.   Poyet type bridge, 1821. (Courtesy of the British Constructional Steelwork Association, Ltd.)**

period attributed the collapse to the overloading. Apparently, a crowd of people who gathered on the bridge structure to watch a river festival or boat race caused the collapse but unfortunately the exact reason was not recorded. The famous French engineer, Navier, discussed these failures among his colleagues and his adverse comments are assumed to have condemned the stayed bridge concept to relative obscurity. Whatever the reason, engineers turned to the suspension bridge, which was also emerging, as the preferred type of bridge for river crossings.

**FIGURE 1.6.   Niagara Falls Bridge.**

FIGURE 1.7.    Old St. Clair Bridge, Pittsburgh, Pennsylvania, built in 1860.

The principle of using stays to support a bridge superstructure did not die completely in the minds of engineers. John Roebling incorporated the concept in his suspension bridges, such as the one near Niagara Falls, Fig. 1.6; the Old St. Clair Bridge in Pittsburgh, Fig. 1.7; the Cincinnati Bridge across the Ohio River, Fig. 1.8; and the Brooklyn Bridge in New York, Fig. 1.9. The stays were used in addition to the vertical hangers to support the bridge superstructure. Observations of performance indicated that the stays and hangers were not efficient partners. Consequently, although the stays were comforting safety measures in the early bridges, in the later developments of the suspension bridge the stays were omitted. Evidence of the dual system is also present in the bridge at Wheeling, West Virginia, Fig. 1.10. This bridge was designed and rebuilt by Ellet, without stays. Stays were added later by others, who were influenced by Roebling.

Despite Navier's adverse criticism of the stayed bridge, a few more bridges were built shortly after the collapse of the fatal bridges in England and Germany. The Gischlard-Arnodin cable bridge, Fig. 1.11, used multiple sloping cables hung from two masonry towers. In 1840, Hatley, an Englishman, used chain stays in a parallel configuration, Fig. 1.12, resembling harp strings. He maintained the parallel spacing of the main stays by using a closely spaced subsystem anchored to the deck and perpendicular to the principal load-carrying cables.

Constructed in 1873, the Albert Bridge over the Thames River, Fig. 1.13, with a main span of 400 ft, affords a good example of catenary sus-

**FIGURE 1.8.    The Cincinnati Bridge across the Ohio River.**

pension combined with stays. In this structure the suspension system consisted of stays converging at the tops of the towers. There were three inclined stays on each side of the center span and four stays on each side of the end spans.[6,7]

The virtual banishment of the stayed bridge during the eighteenth and nineteenth centuries can be attributed to the lack of technical knowledge of the theoretical analyses for the internal forces of the total system. The lack of understanding of the behavior of the stayed system and the methods of controlling the equilibrium and compatibility of the various highly indeterminate systems appears to have been the major drawback to the rapid development of the concept. Not only was the theory lacking but the materials of the period were not suitable for stayed bridges. Materials such as

**FIGURE 1.9.  The Brooklyn Bridge in New York.**

timber, round bars, and chains of various types are not the most desirable for the tension forces acting in the stays. These materials exhibit low strengths and cannot be pretensioned to avoid the slack condition resulting from asymmetrical loadings. For the stays to participate in tension at all times without prestressing, it was necessary for the superstructure to have substantial deformations, which endangered the structure as a whole. Perhaps the early collapses were the result of this unsuitable stress condition.

FIGURE 1.10.   Ellet's Wheeling, West Virginia, bridge.

Against this background of lack of theoretical knowledge and less than adequate materials, the German engineer, F. Dischinger,[8] appears to have rediscovered the stayed bridge in 1938. While designing a suspension bridge to cross the Elbe River near Hamburg, Dischinger determined that the vertical deflection of the bridge under railroad loading could be reduced considerably by incorporating cable stays into the suspension system, Fig. 1.14. From these studies and his later design of the Strömsund Bridge in Sweden evolved the modern day cable-stayed bridge.

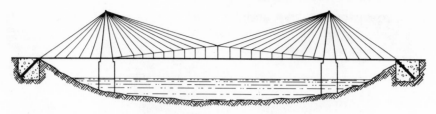

FIGURE 1.11.   Gischlard-Arnodin type sloping cable bridge. (Courtesy of the British Constructional Steelwork Association, Ltd.)

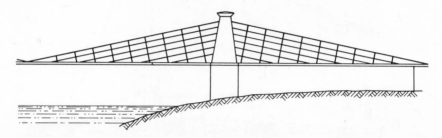

FIGURE 1.12.   Hatley chain bridge, 1840. (Courtesy of the British Constructional Steelwork Association, Ltd.)

Roebling's revived concept of utilizating stays with hangers is reflected in Steinman's bridge across the Tagus River in Portugal, Fig. 1.15. At the top is the present structure, a conventional suspension bridge. At the bottom is the future structure, when cable stays will be added to accommodate additional rail traffic.

## 1.4   UNITED STATES APPLICATIONS

The first modern cable-stayed bridge in the United States was built in 1972. It is a 361 ft, three-span pedestrian bridge located in Menomonee Falls, Wisconsin, Fig. 1.16. The bridge was designed by the Wisconsin Division of Highways Bridge Section.[9] In contrast to the lack of knowledge of prior eras, this bridge was analyzed for internal forces by using a computer program developed at the University of Wisconsin.

The 10 ft wide bridge has a main span of 217 ft with symmetrical end spans of 72 ft. The superstructure consists of two main steel girders of W33 × 130. This particular type of bridge structure was chosen for its natural aesthetic qualities and because of practical and economical conditions at the time of construction. Other systems were considered, such as a

**FIGURE 1.13.    The Albert Bridge. (Courtesy of the British Constructional Steelwork Association, Ltd.)**

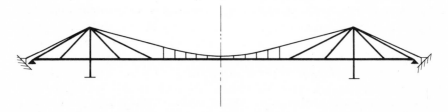

**FIGURE 1.14.    Bridge system proposed by Dischinger.**

6 foot plate girder system, but were unacceptable from cost and aesthetic viewpoints.

This cable-stayed pedestrian bridge is considered to be very attractive because of its general appearance and slim lines. The slimness indicates the manner in which the loads are supported.[9]

The first vehicular cable-stayed bridge in the United States is at Sitka Harbor, Alaska, on Baranof Island in the southeast panhandle, Fig. 1.17.

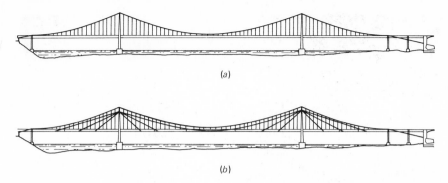

FIGURE 1.15.   The Salazar bridge (*a*) Elevation of present bridge; (*b*) Elevation of future bridge.

The bridge replaces a ferry that went to Japonski Island (the site for a hospital, residential area, a boarding school operated by the Bureau of Indian Affairs, and a jet port). The bridge was designed by the State of Alaska Department of Highways under the direction of William L. Gute.[10]

The cable-stayed bridge was selected for the site because it was most suited to the scenic harbor and most economical for the practical application in that location. Although the cable-stayed bridge concept is relatively new to the United States, the design satisfies all of the American Association of State Highway and Transportation Officials (AASHTO) requirements wherever possible. For those portions of the design not included in the AASHTO specifications, the designer resorted to his best engineering judgment and the past experience of the engineers in foreign countries.

The symmetrical bridge has a center span of 450 ft and two side spans of 150 ft. The roadway is 30 ft wide and the one sidewalk is 5 ft wide. The piers are concrete and the towers are steel. The cable stays are inclined at a steeper angle than has been used in German bridges and are anchored over the approach piers; this technique satisfies the greater stiffness requirements of the United States Bridge Specifications. The stays consist of three 3 inch structural strands, using the conventional galvanized strands and fittings. Although this design calls for three strands per stay, it is theoretically possible for one strand to support the full dead and live loads. Other design and construction details will be discussed in later chapters.

## 1.5   WHY NOT MORE APPLICATIONS IN THE UNITED STATES?

Although the modern cable-stayed bridge has been in existence in Germany since 1955, its adoption in the United States has been long delayed

FIGURE 1.16. Menomonee Falls, Wisconsin, pedestrian bridge. (Courtesy of the Wisconsin Division of Highways.)

and overdue. The Menomonee Falls pedestrian bridge and the Sitka Harbor Bridge for highway vehicles have finally made the concept a sound reality for American engineers. Perhaps the main reason for the long delay in adoption has been the pressure on bridge engineers to meet schedules and construction deadlines for federal, state, and county bridges. However, the Federal Highway Adminstration has adopted a progressive attitude toward this type of structure. Other reasons discussed below are contributary causes.

The cable-stayed bridge is a statically indeterminate system to a very high degree which is somewhat difficult and tedious to analyze with a rea-

**FIGURE 1.17.   Sitka Harbor Bridge.**

sonable degree of accuracy. Therefore, to ensure the greatest precision possible, computer programs must be developed that include all of the parameters. A number of computer programs are now being developed. When they are completed, the drudgery of design may be reduced considerably.

The AASHTO Bridge Specifications do not include provisions for the cable-stayed bridge; consequently, engineers have been reluctant to venture into a design that may not be acceptable to the approving body. Those engineering firms that do decide on a cable-stayed bridge must adapt their design to satisfy the current AASHTO specifications wherever possible, and be prepared to explain their design analysis and any construction details not currently included in the specifications. Some engineering firms feel that they do not want to waste valuable time in a project that may never materialize, and, therefore, do not go beyond the conventional types of bridges. When the AASHTO specifications are broadened to include the various cable-stayed bridge types, more of these bridges will undoubtedly be built in the United States.

One of the most noticeable structural differences between the cable-stayed bridge and the truss and girder types lies in the field of flexibility. Cable-stayed bridges have considerably greater inherent flexibility. The live load deflection of the Severin Bridge at Cologne is 1/225 of the span

—3½ times more flexible than that allowed by the AASHTO specifications. Additional studies may eliminate this disparity in design of the stayed bridge type.

A further design problem results from the fact that U.S. practice requires greater protection for the stay cables against traffic mishaps and salt corrosion than does European practice. Because of these restrictive conditions, it may be necessary to place end anchorages of the cables beyond the edge of the deck.

Added to the design difficulties is the lack of data on the costs of construction of a new type bridge structure. Contractors cannot base their cost estimates on previous experience. As a result, many contractors will overestimate the costs in order to hedge against unforeseen problems that may arise during construction. Learning new techniques and methods is an expensive process, when detailed data is not readily available. However, innovative contractors with construction know-how can produce a realistic cost estimate. This cost estimate may be greatly improved by careful attention of the designer to details and by the care he exercises in illustrating and describing the total design and erection procedure as he sees it. Close cooperation between the contractor and designer will ensure mutual understanding of the details of construction and the structural behavior of the bridge system.

Based on past experience, the cable-stayed bridge appears to have many advantages. An important advantage is the economy of the total design and construction. The main reason for its economy is the fact that conventional construction methods may be used. There are no extreme special erection techniques to be developed. In today's atmosphere of ecological awareness and the desire to be in harmony with nature, the cable-stayed bridge possesses that built-in quality of "form following function." Its inherent beauty emanates from this quality and the bridge blends harmoniously with the natural surroundings.

Other attributes of the cable-stayed bridge are the ease of maintenance, since corrosive action can take place in only a few locations; cables and towers are small and, therefore, easy to fabricate and erect; use of box girders eliminates the need for external bracing; and in general appearance the structure is slim and it is shallow in elevation profile.

As cooperation between bridge design engineers and contractors develops further, the current obstacles of restrictive specifications and preliminary inaccurate cost estimates will be overcome. This type of cooperation will increasingly stimulate more engineers and contractors to design and build the modern cable-stayed bridges. The future applications of the various types of cable-stayed bridges are countless and will no doubt dominate the intermediate and long-span ranges of ravine and river crossings.

Pedestrian bridges and highway overpasses will be developed for reasons of safety and beauty. Center and side piers of multiple span overpasses can be eliminated in favor of one tower or pylon, sufficiently removed from the traveled roadway so that head-on collisions against the piers are no longer a hazardous situation. Thus safety and beauty are combined with economy to reduce the fatalities on the highways.

## 1.6  BRIDGES UNDER CONSIDERATION

During the past five years we have witnessed a rapid increase in the number of cable-stayed bridges under consideration for construction in the United States. Bridge engineers, both private and governmental, have become sufficiently interested to investigate its possibilities as an economical and aesthetic bridge structure. The growing interest stems, no doubt, from the successful completion of the Sitka Harbor Bridge and the Menomonee Falls pedestrian bridge.

Contractors can look forward to many opportunities to build cable-stayed bridges when the current investigations are ready for bids. The following cable-stay bridge projects are either in the bid, design, or con-

FIGURE 1.18.   Pasco-Kennewick Intercity Bridge model. (Courtesy of Arvid Grant, Arvid Grant and Associates, Inc., Olympia, Washington.)

**FIGURE 1.19.** West Seattle freeway bridge. (Courtesy of J. E. Arnberg, City of Seattle Department of Engineering.)

templation stage as of this writing and are listed here to illustrate the ever-growing interest.

1. Intercity bridge between the cities of Pasco and Kennewick, in the state of Washington, crossing the Columbia River, Fig. 1.18.
2. West Seattle Freeway Bridge. A city bridge in Seattle, Washington, Fig. 1.19.
3. Two structures in the state of Louisiana for the interstate 410 bypass around New Orleans.

**FIGURE 1.20.** Luling I-410 bridge. (Courtesy of T. R. Kealey, Modjeski and Masters.)

**FIGURE 1.21.    Chalmette I-410 bridge. (Courtesy of T. R. Kealey, Modjeski and Masters.)**

  a. The Luling Bridge, Fig. 1.20 (this artist's rendering indicates a stiffening truss; however, the bridge is being designed as a twin trapezoidal box girder deck).
  b. The Chalmette Bridge, Fig. 1.21.

**FIGURE 1.22.    East Huntington Bridge. (Courtesy of E. Lionel Pavlo, E. Lionel Pavlo Engineers.)**

**FIGURE 1.23.   Proposed Weirton-Steubenville Bridge. (Courtesy of Michael Baker, Jr., Inc.)**

4. The state of West Virginia is contemplating two bridges.
   a. A bridge structure crossing the Ohio River between East Huntington, and Proctorville, Ohio, Fig. 1.22.
   b. Another bridge spanning the Ohio River connecting Wierton and Steubenville, Ohio, Fig. 1.23.
5. A bridge structure at Meridian, California, crossing the Sacramento River to replace the through truss swing bridge will be a unique cable-stayed swing bridge. Upon completion, it may be the first of its type, Fig. 1.24.

**FIGURE 1.24.   Cable-stay swing span; Meridian, California. (Courtesy Department of Transportation, Division of Highways, Office of Structures, State of California.)**

FIGURE 1.25. Louisiana stayed bridge (Courtesy of Sidney L. Poleynard, State of Louisiana, Department of Highways.)

FIGURE 1.26. South Myrtle Creek Bridge, Washington. (Courtesy of John H. Garren, FHWA Region 10.)

21

FIGURE 1.27. Coos River Bridge, Washington. (Courtesy of John H. Garren, FHWA Region 10.)

FIGURE 1.28. Quinault River Bridge. (Courtesy of Arvid Grant, Arvid Grant and Associates, Inc.)

## 1.7  EARLY APPLICATIONS IN THE UNITED STATES

Although the Menomonee Falls pedestrian bridge and the Sitka Harbor Bridge are considered to be the first contemporary cable-stayed bridges in the United States, bridges employing the general concept have been built previously in the United States.

Figure 1.25 shows a stayed swing span in Louisiana, circa 1929. The stayed girder concept is not unfamiliar to the Pacific Northwest as indicated by Figs. 1.26 and 1.27.

In 1953 the Aloha Lumber Company built a cable-stayed logging bridge across the Quinault River in the State of Washington, Figs. 1.28 through 1.31. This structure was built in 1953 and designed by Frank Milward, Aloha Company logging superintendent. The structure collapsed on September 24, 1964 as a result of a failure of one of the 2¼ in. diameter cables and was rebuilt. The structure collapsed again in August of 1973 and has been revamped. It has a center span of 256 ft.

On July 5, 1957 a stayed structure, Figs. 1.32 and 1.33, crossing the Yakima River at Benton City (10 miles west of Richland, Washington) was opened to traffic. Designed by Homer M. Hadley, the structure has a total length of 400 ft with a center span of 170 ft. A 60 ft central drop-in

**FIGURE 1.29.   Quinault River Bridge. (Courtesy of Arvid Grant, Arvid Grant and Associates, Inc.)**

FIGURE 1.30.　Quinault River Bridge. (Courtesy of Arvid Grant, Arvid Grant and Associates, Inc.)

FIGURE 1.31.　Quinault River Bridge. (Courtesy of Arvid Grant, Arvid Grant and Associates, Inc.)

24

FIGURE 1.32. Benton City Bridge, Washington. (Courtesy of John H. Garren, FHWA Region 10.)

FIGURE 1.33. Benton City Bridge, Washington. (Courtesy of John H. Garren, FHWA Region 10.)

or suspended span of 33 in. wide flange beams is supported by transverse concrete beams which in turn are supported by the 10 in., 112 lb wide flange stays. Continuous longitudinal concrete beams comprise the remainder of the structure and receive support at their extremity in the center span, from the transverse concrete beams and stays. The 10 in. 112 lb wide flange sections were also used for the pylon legs and were encased in concrete. The stays were set with their flanges vertical to match the flanges of the pylon legs. Stays were connected to the pylon by plates and high-tensile bolts in single shear.[11]

## REFERENCES

1. Hopkins, H. J., *A Span of Bridges: An Illustrated History*, Praeger Publishers, 1970, New York and Washington.
2. Kavanagh, T. C., Discussion of "Historical Developments of Cable-Stayed Bridges" by Podolny and Fleming, *Journal of the Structural Division*, ASCE, Vol. 99, No. ST 7, July 1973, Proc. Paper 9826.
3. Podolny, W., Jr. and Fleming, J. F., "Historical Developments of Cable-Stayed Bridges," *Journal of the Structural Division*, ASCE, Vol. 98, No. ST 9, September 1972, Proc. Paper 9201.
4. Podolny, W., Jr., "Cable-Stayed Bridges," *Engineering Journal*, American Institute of Steel Construction, First Quarter 1974.
5. Leonhardt, F. and Zellner, W., "Cable-Stayed Bridges: Report on Latest Developments," Canadian Structural Engineering Conference, 1970, Canadian Steel Industries Construction Council, Toronto, Ontario, Canada.
6. Feige, A., "The Evolution of German Cable-Stayed Bridges—An Overall Survey," *Acier-Stahl-Steel* (English version), No. 12, December 1966, reprinted *AISC Engineering Journal*, July 1967.
7. Thul, H., "Cable-Stayed Bridges in Germany," *Proceedings of the British Constructional Steelwork Association Conference on Structural Steelwork*, held at the Institution of Civil Engineers, September 26–28, 1966, the British Constructional Steelwork Association Ltd., London, England.
8. Dischinger, F., "Hängebrücken für Schwerste Verkehrslasten," *Der Bauingenieur*, March 1949.
9. Woods, S. W., Discussion of "Historical Developments of Cable-Stayed Bridges," by Podolny and Fleming, *Journal of the Structural Division*, ASCE, Vol. 99, No. ST 4, April 1973.
10. Gute, W. L., "First Vehicular Cable-Stayed Bridge in the U.S.," *Civil Engineering*, ASCE, Vol. 43, No. 11, November 1973.
11. Hadley, H. M., "Tied-Cantilever Bridge—Pioneer Structure in U.S.," *Civil Engineering*, ASCE, January 1958.

# 2

# Bridge Component Configurations

## 2.1  GENERAL DESCRIPTION

The cables extending from one or more towers of the cable-stayed bridge support the superstructure at many points along the span. The cable system is ideal for spanning natural barriers of wide rivers, deep valleys, or ravines, and for vehicular and pedestrian bridges crossing wide interstate highways because there are no piers that will form obstructions. For the most part, cable-stayed bridges have been built across navigable rivers where navigation requirements have dictated the dimensions of the spans and clearance above the main water levels.

The most successful span arrangements are of three basic types; they may be categorized as: two spans, symmetrical or asymmetrical, Fig. 1.1; three spans, Fig. 1.2; or multiple spans, Fig. 2.1.

In an economical stayed bridge design the span proportions, tower height, number and inclination of cables, and type of superstructure must be evaluated in conjunction with each other. For the two-span asymmetrical bridge structure, a partial survey of the existing bridges, Table 2.1, indicates that the longer span ranges from 60 to 70% of the total length. Two exceptions are the Batman and Bratislava bridges, Fig. 2.2, whose

TABLE 2.1.   Ratio of Larger Span to Total Length of Two-Span Structures

| Structure | Larger Span | | Total Length | | Ratio |
| --- | --- | --- | --- | --- | --- |
| | ft | m | ft | m | |
| Severin (Germany) | 987.5 | 301 | 1482.9 | 452 | 0.67 |
| Karlsruhe (Germany) | 574.1 | 175 | 958.7 | 292 | 0.60 |
| Kniebrücke (Germany) | 1049.9 | 320 | 1689.6 | 515 | 0.62 |
| Mannheim (Germany) | 941.6 | 287 | 1351.7 | 412 | 0.70 |
| Maya (Japan) | 457.3 | 139.4 | 685 | 208.8 | 0.67 |
| East Huntington (U.S.A.) | 900 | 274.3 | 1350 | 411.5 | 0.67 |
| Batman (Australia) | 689 | 210 | 853 | 260 | 0.81 |
| Bratislava (Czechoslovakia) | 994.1 | 303 | 1240.2 | 378 | 0.80 |

longer spans are 80% of the total length of the bridge structure. The reason for the longer spans is the fact that the back stays are concentrated into a single back stay anchored to the abutment, rather than being distributed along the short span.

A similar survey of numerous three-span cable-stayed bridge structures, Table 2.2, indicates that the center span is approximately 55% of the total length of bridge. The remainder is usually equally divided between the two end spans. An investigation of bridges with multiple spans indicates that the spans are normally of equal length, with the exception of

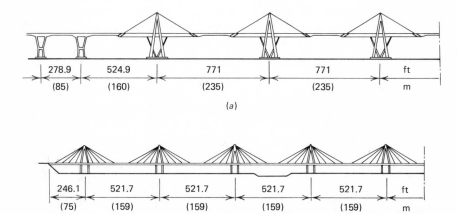

FIGURE 2.1.   Examples of multispan cable-stayed bridge structures: (a) Maracaibo (Venezuela); (b) Ganga Bridge (India).

**TABLE 2.2.    Ratio of Center Span to Total Length of Three-Span Structures**

| Structure | Center Span | | Total Length | | |
|---|---|---|---|---|---|
| | ft | m | ft | m | Ratio |
| Papineau (Canada) | 787.4 | 240 | 1377.9 | 420 | 0.57 |
| Duisburg (Germany) | 1148.3 | 350 | 2148.9 | 655 | 0.53 |
| Rees (Germany) | 836.6 | 255 | 1519 | 463 | 0.55 |
| Bonn (Germany) | 918.6 | 280 | 1706 | 520 | 0.54 |
| Düsseldorf-North (Germany) | 853 | 260 | 1561.7 | 476 | 0.55 |
| Leverkusen (Germany) | 918.6 | 280 | 1614.2 | 492 | 0.57 |
| Norderelbe (Germany) | 564.3 | 172 | 984.2 | 300 | 0.57 |
| Arakawa River (Japan) | 524.9 | 160 | 920.6 | 280.6 | 0.57 |
| Suehiro (Japan) | 820.2 | 250 | 1542 | 470 | 0.53 |
| Onomichi (Japan) | 705.4 | 215 | 1263.1 | 385 | 0.56 |
| Toyosato (Japan) | 708.7 | 216 | 1236.9 | 377 | 0.57 |
| Strömsund (Sweden) | 600.4 | 183 | 1086 | 331 | 0.55 |
| Sitka (U.S.A.) | 450 | 137.2 | 750 | 228.6 | 0.60 |
| Luling (U.S.A.) | 1235 | 376.4 | 2225 | 678.2 | 0.56 |
| Pasco-Kennewick (U.S.A.) | 981 | 299 | 1794 | 546.8 | 0.55 |
| Menomonee Falls (U.S.A.) | 217 | 66.1 | 361 | 110 | 0.60 |

the flanking spans which are adjusted to connect with the approach span or abutments. In this type of bridge the cables are arranged symmetrically on both sides of the towers. For convenience of fabrication and erection, the bridge structure has "drop-in" sections for the center portions of the span. The ratio of drop-in span length to total span length ranges from 20%, when a single stay emanates from each side of the tower, to 8%, where multiple stays emanate from each side of the tower.

The versatile cable-stayed bridge concept lends itself to a large variety of geometrical configurations. The arrangement of the cables, type of superstructure, and style of the towers can be easily adjusted to suit the numerous requirements of site conditions and aesthetics for highway and pedestrian bridges. A detailed description and discussion of the advantages and disadvantages of the many geometrical forms for cable arrangement and type of towers is presented in this chapter.

## 2.2    TRANSVERSE CABLE ARRANGEMENT

In the transverse direction to the longitudinal axis of the bridge, the cables may lie in either a single or a double plane and may be symmetrically or

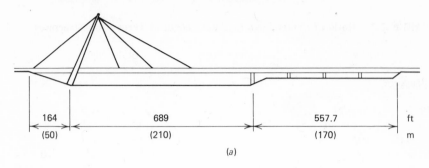

|  164 |  689 |  557.7 | ft |
| (50) | (210) | (170) | m |

(a)

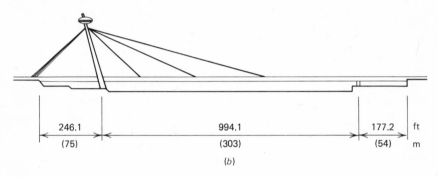

|  246.1 |  994.1 |  177.2 | ft |
| (75) | (303) | (54) | m |

(b)

**FIGURE 2.2.** Examples of two-span cable-stayed bridge structures: (a) Batman (Australia); (b) Bratislava (Czechoslovakia).

asymmetrically placed, and may lie in oblique or vertical planes. These basic arrangements are illustrated in Fig. 2.3.[1,2] A unique arrangement having cables lying in three independent vertical planes was proposed for the Danish Great Belt Bridge Competition, Fig. 2.4.

### 2.2.1  Single-Plane System

The single-plane cable arrangement is generally used with a divided road-way deck with the cables passing through the median strip and anchored below the roadway. This arrangement is not only economical but aesthetically pleasing as well.

With the cables lying in the plane of the median strip, the motorist enjoys an unobstructed view of the natural scenery as he drives across the bridge. For conventional roadways very little additional width is required in the deck to accommodate the cables. However, for narrow width median strips additional deck width may be required in order to allow sufficient space for the towers. A single planar cable system requires single towers

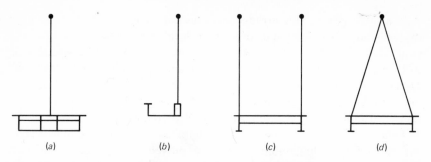

**FIGURE 2.3.**  Transverse cable arrangement. ($a$) Single plane—vertical; ($b$) single plane—vertical/lateral; ($c$) double plane—vertical; ($d$) double plane—sloping.

or pylons at pier supports that are placed in the median strip, thus maintaining a minimum width of the roadway superstructure.

A possible disadvantage of the single-plane cable system is the fact that a relatively large concentrated cable force is transferred to the main superstructure girder, thereby requiring a larger connection and girder to support the cable force. Additional reinforcement and stiffening of the deck, web plates, and bottom flange will normally be required in order to distribute the concentrated load uniformly throughout the cross section of the superstructure members.

**FIGURE 2.4.**  Danish Great Belt Competition. (Courtesy of White Young & Partners.)

In a single-plane cable arrangement, the cables support vertical or gravity loads only. The torsional forces that develop because of the asymmetrical vehicular loading and/or wind forces must be resisted by a torsionally stiff box girder in order to transmit the unbalanced forces to the piers. These additional stiffness requirements for the superstructure may increase the costs but these costs may be counterbalanced by the advantages of simplified fabrication, erection, and added aesthetics.

Although the single-plane cable system has been used symmetrically, with respect to the longitudinal centerline, on vehicular bridges, it has been constructed off-center or asymmetrically for pedestrian bridges. In the asymmetrical applications, the plane of the cables is at the edge of the walkway. Because the walkway loadings are small, the unbalanced system produces only small torsional forces that are easily resisted by the walkway structure.

### 2.2.2    Double-Plane System

The two principal double planar cable systems are: one system consisting of a vertical plane located at each edge of the superstructure and another system in which the cable planes are oblique, sloping toward each other from the edges of the roadway and intersecting at the towers along the longitudinal centerline of the deck. In the double-plane oblique system the term "plane" is to be interpreted loosely. If the stays are connected at more than one level at the tower or if the roadway has any vertical or horizontal curvature each "plane" geometrically forms a warped surface in space. The tower in the oblique double plane arrangement is generally of the A-frame type in order to receive the sloping cables that intersect along the centerline of the roadway.

Using the two-plane cable system, the anchorages may be located either on the outside of the deck structure or within the limits of the deck roadway. With the cable anchorages on the outside of the deck, an advantage is gained, because no portion of the deck roadway is required for the connection fittings. A disadvantage is the fact that additional reinforcement may be required to transmit the eccentric cable loadings of shear and moment into the main girders of the superstructure.

For those applications in which the cable anchorage lies within the limits of the bridge deck, the overall width of the deck must be increased for the full length of the bridge in order to provide room for the anchorage fittings. This additional width of roadway deck usually results in an increased cost for the superstructure.

### 2.2.3   Three-Plane System

A design for a three-plane cable system was submitted in competition for the Danish Great Belt Bridge by the English consulting firm of White, Young and Partners. The design requirements were for three lanes of vehicular traffic and a single rail line in each direction. The solution, illustrated in Fig. 2.4, employed three vertical planes of cables, one in the median strip and the other two on the exterior edges of the bridge deck. The concept is appropriate for use in urban areas, where it may be necessary to include mass transit center lanes or special bus lanes as well as three or four vehicular lanes in each direction.

### 2.3   LONGITUDINAL CABLE ARRANGEMENTS

The arrangement of the cables in the longitudinal direction of the bridge may vary according to the designer's sense of proportion of clear spans and tower heights. For shorter span lengths a single forestay and backstay may be sufficient to satisfy the loading requirements. For longer center and anchor spans a variety of cable arrangements satisfy not only the engineering requirements but result in a pleasing aesthetic geometrical configuration as well.

Basically, there are four cable configurations in general use throughout the world for cable-stayed bridges. The idealized arrangements are indicated in Fig. 2.5, and it may also be assumed that all the configurations or types are applicable to either the single or double planar cable systems. These basic systems are referred to as radiating, harp, fan, and star systems.

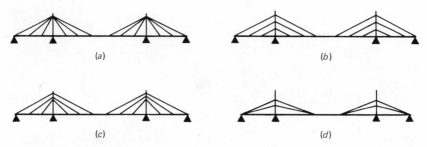

(a)          (b)

(c)          (d)

**FIGURE 2.5.   Longitudinal cable arrangements: (*a*) radiating; (*b*) harp; (*c*) fan; (*d*) star.**

The radiating type, or a converging system, is an arrangement wherein the cables intersect or meet at a common point at the top of the tower, Fig. 1.18.

The harp type, as the name implies, resembles harp strings—the cables are parallel and equidistant from each other. The required number of cables are spaced uniformly along the tower height and, as a result, also connect to the roadway superstructure with equal spacings, Fig. 2.6.

The fan type is a combination of the radiating and harp types. The cables emanate from the top of the tower with equal spacings and connect with equal spacings along the superstructure. Because of the small spacings concentrated near the top of the tower, the cables are not parallel, Fig. 2.7.

In the star arrangement, the cables intersect the tower or pylon at different heights and then converge on each side of the tower to intersect the roadway structure at a common point. The common intersection in the anchor span is usually located over the abutment or end pier of the bridge, Fig. 2.8.

**FIGURE 2.6.   Theodore Heuss-Brücke, Düsseldorf. (Courtesy of Beratungsstelle für Stahlverwendung, H. Odenhousen.)**

FIGURE 2.7.    Friedrich-Ebert Bonn Nord Bridge. (Courtesy of Beratungsstelle für Stahlverwendung, H. Odenhousen.)

The selection of cable configuration and number of cables is dependent on the length of span, type of loadings, number of roadway lanes, height of towers, economy, and the designer's individual sense of proportion and aesthetics. As a result, some bridges have relatively few cable-stays while others may have many stays intersecting the deck such that the cables provide a continuous elastic supporting system. Fig. 2.9 illustrates a few of the radiating and harp type multistay cable bridges that have been built in Germany.

Cost factors have a great influence on the selection of the cable arrangements. Using only a few cable-stays results in large cable forces, which require massive and complicated anchorage systems connecting to the tower and superstructure. These connections become sources of heavy concentrated loads requiring additional reinforcement of webs, flanges, and stiffeners to transfer the loads to the bridge girders and distribute them uniformly throughout the structural system.

When only a few cables support the deck structure, deep girders are required to span the long distance between the cable intersection points. A large number of stays simplifies the cable anchorages to the bridge girders

**FIGURE 2.8.** Norderelbe. (Courtesy of Beratungsstelle für Stahlverwendung, H. Odenhousen.)

and distributes the forces more uniformly throughout the deck structure without major reinforcements to the existing girders and floor beams. Therefore, a large number of cables can provide continuous support, thus permitting the use of a shallow depth girder that also tends to increase the stability of the bridge against dynamic wind forces.[3]

Some engineers prefer the radiating cable arrangement—where all cables converge at the top of the tower—because the cable stays are at a maximum angle of inclination to the bridge girders. In this arrangement,

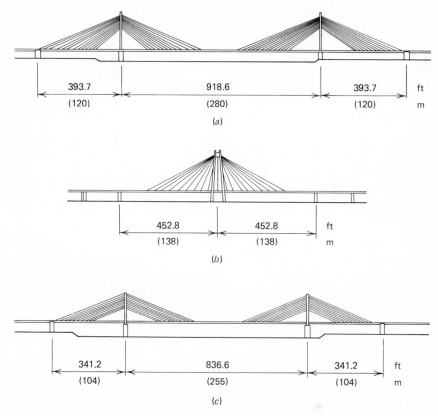

**FIGURE 2.9.** Examples of multistayed cable-stayed bridge structures: (*a*) Bonn (Germany); (*b*) Ludwigshafen (Germany); (*c*) Rees (Germany).

the cables are nearly in an optimum position to support the gravity dead and live loads and simultaneously produce a minimum axial component acting on the girder system.[4]

When using a double plane cable system, the harp configuration may be preferred over the radiating type because it minimizes the visual intersection of cables when viewed from an oblique angle. Thus the motorist may find the harp system more attractive. The harp system, with the cable connections distributed throughout the height of the tower, results in an efficient tower design compared with the radiating system, which has all the cables at the top of the tower. The concentrated load at the top of the tower produces large shears and moments along the entire height of the tower, thus increasing its cost. In addition, the large concentrated cable force at the top of the tower usually presents difficulties in anchoring the

**FIGURE 2.10.    George Street Bridge over the river Usk at Newport, Monmouthshire.**

cables to the tower or over a saddle, thus complicating the transfer of the vertical loads.

The fan arrangement represents a compromise between the extremes of the ·harp and radiating system and is useful when it becomes difficult to accommodate all the cables at the top of the tower.

The star system has only been used on the Norderelbe Bridge in Hamburg, Fig. 2.8. The principal reason for its use is its unique aesthetic appearance. The additional tower height above the cable connection is purely decorative; it serves no structural purpose whatsoever. The cables are not distributed along the bridge deck, instead the cables on each side of the tower converge at the same point in the span. In this arrangement, two small cables function as a single large cable. The two cables can be more efficient to construct and result in a more pleasing appearance than a single cable.

Using the four basic longitudinal cable configurations, a great variety of combinations are possible. The George Street Bridge over the Usk River at Newport, Monmouthshire, England[5] is a hybrid system using a combination cable arrangement on the center span side of the tower and a harp arrangement on the anchor span side, Fig. 2.10.

Another hybrid system won first prize in the Great Belt Bridge competition held in Denmark,[6] Fig. 2.11. This bridge combined the fan and star configurations. The fan arrangement was used for the center span and the star arrangement was used for the end span.

## 2.4    TOWERS

The cable towers are often referred to as pylons, and these terms will be considered interchangeable in this text. Similar to the cable systems, the towers are of many shapes and varieties to accommodate different cable arrangements, bridge site conditions, design requirements, aesthetics, and economics.

In their simplest form, the towers may be a single cantilever to support a single plane arrangement of cables, Fig. 2.7, or two cantilever towers to

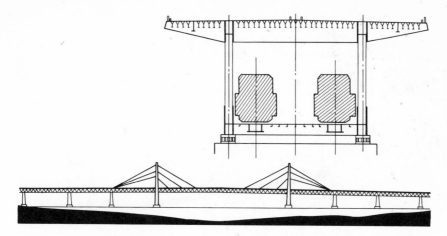

**FIGURE 2.11.    Great Belt Bridge award winner.**

support the double plane cable system, Fig. 2.6. The towers may be fixed
or hinged at the base, depending on the magnitude of the vertical loads
and distribution of cable forces along the tower height.

Other tower forms suitable for cable-stayed bridges are the portal
frame, Fig. 2.12, and A-frame types, Fig. 2.13, either hinged or fixed at
the base. The decision to use a fixed or hinged base for the tower connection either to the pier or the superstructure must be based on knowledge
of the magnitude and relationship of the vertical and horizontal forces acting on the tower. A fixed base induces large bending moments at the base
of the tower, whereas a hinged base does not and may be preferred. However, the increased rigidity of the total structure resulting from the fixed
base of the towers may offset the disadvantage of the large bending
moments. Another consideration is that a fixed base may be more practical to erect and may be less costly than inserting a heavy pinned bearing, which requires the tower to be externally supported until the cables
are connected. The design engineer and contractor should discuss these
considerations early in the design stage of the project in order to arrive at
the most economical solution.

In selecting a specific type of tower, the designer must consider several
factors. For example, when a large clearance is required below the superstructure, the A-frame has a decided disadvantage—a large pier width is
required to accommodate the legs of the frame. In some instances, a modified A-frame with a short top cross member may be the best solution
considering all the factors involved, Fig. 2.14a.

**FIGURE 2.12.    Galipeault Bridge, Montreal, Canada.**

A variation of the A-frame is a narrow diamond-shape tower with the roadway structure at the center of the diamond, Fig. 2.14*b*. This form was selected for the proposed Southern Bay Crossing in San Francisco.[7,8] A delta shape or modified diamond, Fig. 2.14*c*, was used for the bridge at Köhlbrandbrücke, Hamburg, Germany. Imagination combined with engineering economics can produce many variations of tower designs, each type satisfying particular design conditions and requirements better than other conventional types. Familiarity with various construction methods and

FIGURE 2.13.  Toyosato-Ohhashi Bridge, Osaka, Japan.

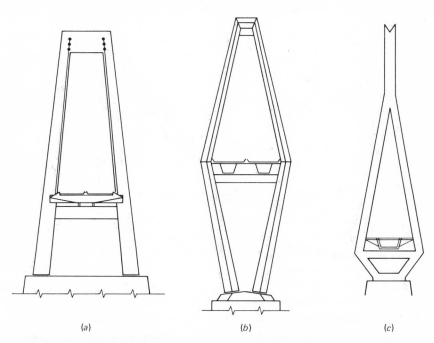

*(a)*

*(b)*

*(c)*

FIGURE 2.14.  Alternate tower types: (*a*) modified A-frame; (*b*) diamond; (*c*) modified diamond or delta.

41

techniques can assist the design engineer and contractor to develop the best tower design.

When several long spans are required, the towers may be designed as two or more portal frames or A-frames or a combination of both. The towers of the Maracaibo and Ganga Bridges, Fig. 2.1, consist of two A-frames that support the cables and superstructure. The prize winning Great Belt Bridge, Fig. 2.4, has three portal frames that support the cables and deck structure. The portal frames are joined at the top by a cross member.

The towers are normally constructed of cellular sections and are fabricated of structural steel or reinforced concrete. The concrete towers are built where steel is in short supply and recourse is made to local natural materials. Details on fabrication and erection are discussed in Chapters 8 and 9.

The height of the tower is determined from several considerations, such as the relation of tower height to span length, the type of cable arrangement, and the general aesthetic proportions of all the spans and towers visualized as an entity. The size and number of cables is determined by the geometrical configuration of the bridge and the type of loading to be imposed on the structure.

## 2.5   CABLE SYSTEM SUMMARY

A tabular summary of the various cable arrangements is presented in Fig. 2.15. The table is a matrix of the four basic cable types and the number

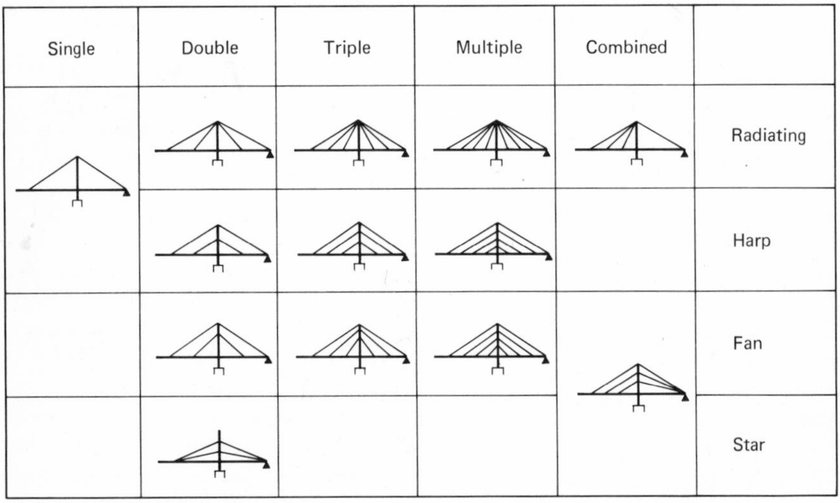

| Single | Double | Triple | Multiple | Combined | |
|--------|--------|--------|----------|----------|--|
| | | | | | Radiating |
| | | | | | Harp |
| | | | | | Fan |
| | | | | | Star |

**FIGURE 2.15.   Matrix of longitudinal configurations. (Adapted from ref. 9.)**

of cables extending from one side of the tower. It can be seen that many other variations and combinations are possible. Therefore, the cable-stayed bridge is not simply one bridge type, but many different types based on an extremely versatile concept of bridge design.

A descriptive summary of some of the typical bridges that have been built is presented in Fig. 2.16. The illustrations indicate the longitudinal cable arrangements and whether the system lies in single, double, or oblique plane configuration. The type of tower is also indicated by the transverse section for each bridge.

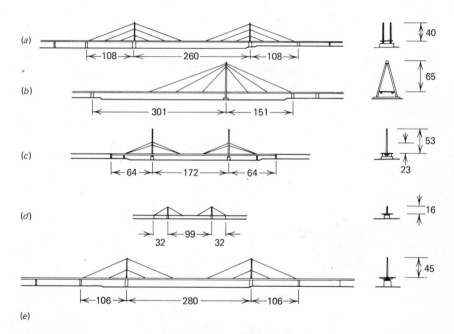

FIGURE 2.16.   Cable-stayed roadbridges in Germany. (From ref. 3, Leonhardt and Zellner, by permission of the Canadian Steel Industries Construction Council.) (*a*) Düsseldorf-North, 1958; (*b*) Cologne, 1960; (*c*) Hamburg, 1962; (*d*) Düsseldorf, 1963; (*e*) Leverkusen, 1964. (*f*) Karlsruhe, 1965; (*g*) Bonn, 1966; (*h*) Rees, 1967; (*i*) Ludwigshafen, 1968; (*j*)Kniebrücke-Düsseldorf, 1969; (*k*) Duisburg, 1970; (*l*) Manheim, 1971; (*m*) Düsseldorf-Oberkassel, 1972, Cable-stayed roadbridges in different countries; (*n*) Strömsund (Sweden); (*o*) Maracaibo (Venezuela); (*p*) Saint-Florent (France); (*q*) Papineau (Canada); (*r*) Hawkshaw (New Brunswick); (*s*) Batman (Australia); (*t*) Ganga-Bridge (India); (*u*) Onomichi (Japan); (*v*) Bratislava (Czechoslovakia).

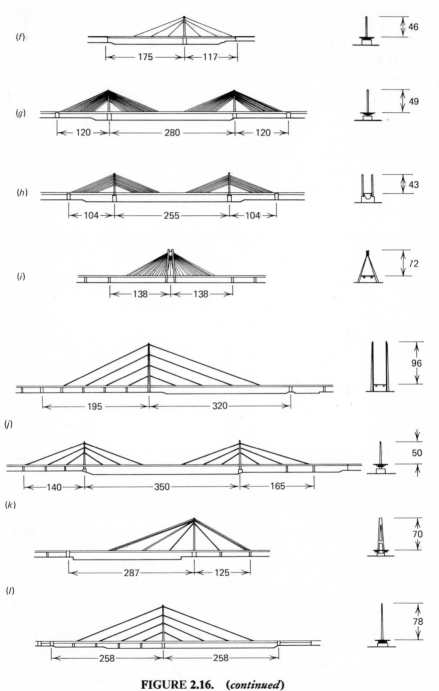

**FIGURE 2.16.** (*continued*)

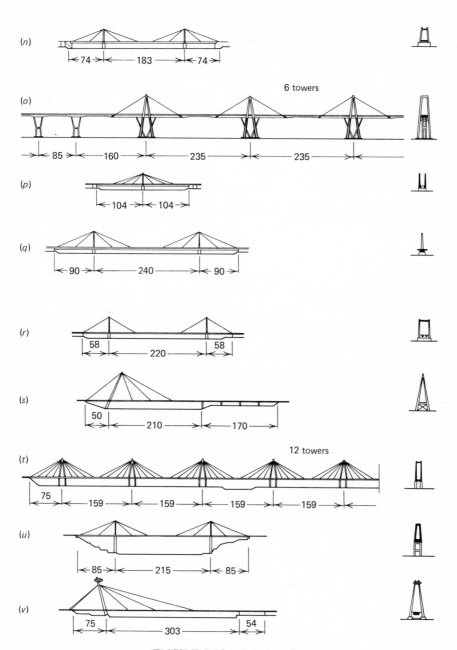

**FIGURE 2.16.** (*continued*)

## 2.6  SUPERSTRUCTURE TYPES

The superstructures for cable-stayed bridges take as many forms as there are structural systems. Basically, however, two types of girders have been used most frequently: the stiffening truss and the solid web types. Past experience with the two systems indicates that the stiffening truss type is seldom used in current designs. The stiffening trusses require more fabrication, are relatively more difficult to maintain, are more susceptible to corrosion, and are somewhat unappealing.

Solid web girders for various types of bridge deck cross sections are illustrated in Fig. 2.17.[4,10] The range of cross sections include the basic two main plate girders or multiple plate girders, Fig. 2.17a and b. These arrangements have the disadvantage of a low value for torsional rigidity.

An increase in torsional rigidity is achieved by using box type cross sections, Fig. 2.17c and d. They may range from the single cell or multicell box with rectangular sides to a similar trapezoidal type with sloping sides. In each of these types the roadway width extends beyond the edges of the single boxed girders.

When the roadways require a large number of traffic lanes, the transverse width requires several box-girder systems to support the deck structure, Fig. 2.17e and f. Twin boxes, either of the rectangular or trapezoidal shapes, have been used to advantage when large deck widths are required.

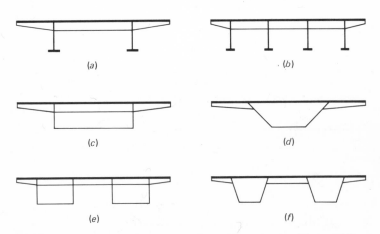

**FIGURE 2.17.**  Girder types: (a) twin I girders; (b) multiple I girders; (c) rectangular box girder; (d) trapezoidal box girder; (e) twin rectangular box girder; (f) twin trapezoidal box girder.

The proportion of the girder depth to the length of span from a survey of 20 cable-stayed bridges, Table 2.3, indicates a ratio which varies from 1:40 to 1:100.

**TABLE 2.3.   Girder Depth to Span Ratio**

| Structure | Transverse Cable Planes | No. Cable Spans | Longitudinal Cable Configuration | Girder Depth/ Span |
|---|---|---|---|---|
| Strömsund (Sweden, 1955) | 2 | 3 | Converging | 1/58 |
| Brucksal (Germany, 1956) | 2 | 3 | Converging | 1/53 |
| Nordbrücke (Germany, 1958) | 2 | 3 | Harp | 1/81 |
| Severin (Germany, 1959) | 2 | 2 | Fan | 1/66 |
| Elbe (Germany, 1962) | 1 | 3 | Star | 1/57 |
| Shinno (Japan, 1963) | 2 | 2 | Converging | 1/40 |
| Jülicherstrasse (Germany, 1964) | 1 | 3 | Converging | 1/60 |
| Leverkusen (Germany, 1965) | 1 | 3 | Harp | 1/64 |
| Maxau (Germany, 1966) | 1 | 2 | Fan | 1/62 |
| Maya (Japan, 1966) | 1 | 2 | Fan | 1/50 |
| Ludwigshafen (Germany, 1967) | 2 | 2 | Converging | 1/55 |
| Rees (Germany, 1967) | 2 | 3 | Harp | 1/73 |
| Bonn (Germany, 1967) | 1 | 3 | Fan | 1/61 |
| Onomichi (Japan, 1968) | 2 | 3 | Converging | 1/67 |
| Kniebrücke (Germany, 1969) | 2 | 2 | Harp | 1/100 |
| Papineau-Leblanc (Canada, 1969) | 1 | 3 | Converging | 1/67 |
| Toyosata (Japan, 1970) | 1 | 3 | Fan | 1/72 |
| Arakawa (Japan, 1970) | 1 | 3 | Harp | 1/67 |
| Ishikari (Japan, 1972) | 2 | 3 | Fan | 1/80 |
| Sitka Harbor (U.S.A., 1973) | 2 | 3 | Single-Stay | 1/75 |

In contrast to the conventional ratio of girder depth to span length, the Kniebrücke Bridge, Fig. 1.1, is a two-span asymmetrical structure with a girder depth to span ratio of 1:100. This extremely high ratio is achieved by anchoring the back stays to the girder directly over the piers. This method of anchoring the back stays produces a greater stiffness in the main span, thus permitting the shallow girder.

Experience in the design of the bridge decks has led to increased use of the orthotropic box-girder arrangement with the trapezoidal configuration. This box type cross-sectional shape with cantilever extensions to include the full roadway width also has superior aerodynamic stability. For single plane cable systems, the box girder is preferred because of its increased torsional stiffness.

## REFERENCES

1. Podolny, W., Jr. and Fleming, J. F., "Historical Development of Cable-Stayed Bridges," *Journal of the Structural Division*, ASCE, Vol. 98, No. ST 9, September 1972, Proc. Paper 9201.
2. Podolny, W., Jr., "Cable-Stayed Bridges," *Engineering Journal*, American Institute of Steel Construction, First Quarter, 1974.
3. Leonhardt, F. and Zellner, W., "Cable-Stayed Bridges: Report on Latest Developments," *Proceedings of the Canadian Structural Engineering Conference, 1970*, Canadian Steel Industries Construction Council, Toronto, Ontario, Canada.
4. Simpson, C. V. J., "Modern Long Span Steel Bridge Construction in Western Europe," *Proceedings of the Institution of Civil Engineers*, 1970 supplement (ii).
5. Brown, C. D., "Design and Construction of the George Street Bridge over the River Usk, at Newport, Monmouthshire," *Proceedings of the Institution of Civil Engineers*, September 1965.
6. Anon., "Great Belt Bridge Award Winner," *Consulting Engineer*, England, March 1967.
7. Seim, C. F., Larson, S., and Dang, W., "Design of the Southern Crossing Cable-Stayed Girder," Preprint 1352, ASCE National Water Resources Engineering Meeting, Phoenix, Arizona, January 11–15, 1971.
8. Seim, C. F., Larson, S., and Dang, W., "Analysis of Southern Crossing Cable-Stayed Girder," Preprint 1402, ASCE National Structural Engineering Meeting, Baltimore, Maryland, April 19–23, 1971.
9. Feige, A., "The Evolution of German Cable-Stayed Bridges—An Overall Survey," *Acier-Stahl-Steel* (English version), No. 12, December 1966, reprinted *AISC Journal*, July 1967.
10. Feige, A., "Steel Motorway Bridge Construction in Germany," *Acier-Stahl-Steel* (English version), No. 3, March 1964.

# 3

# Economic Evaluation

## 3.1 INTRODUCTION

The selection of a specific type of bridge to cross a river, ravine, or high-way is not an automatic determination. Many factors must be considered before a final decision is made. In some instances, the factors affecting the design are similar to, if not the same as, those previously considered at another location or site, so several bridges of the same type are chosen.

The principal factors to be considered are the relationship of span lengths of various segments of the bridge, the number of piers and place-ment for safety, the aesthetic considerations for the site, and, finally, the relative costs of bridges of comparable acceptable proportions and type.

Many types of bridges share similar aesthetic and safety considerations, but relative costs of bridges depend on the number and length of spans and number of piers that affect the method of construction. Studies of comparative costs of cable-stayed bridges and other types of bridges are few; consequently, a design engineer must perform a detailed investigation of the economics of the total structure until sufficient data is available to make general decisions quickly.

The contractor also lacks specific cost data upon which to base his esti-mate and must rely on the detailed design drawings and written specifica-tions for his basic information. Therefore, to arrive at a realistic cost

**49**

estimate, it is advisable for the designer and contractor to communicate ideas at an early stage in the design process. The method of fabrication and erection can affect both the design and the costs and may decide whether the cable-stayed bridge is the most economical one or not. Contractors must be willing to study and evaluate various methods of erection in order to arrive at a meaningful cost estimate.

Although two cable-stayed bridges have been built in the United States, Sitka Harbor, Alaska,[1] and Menomonee Falls, Wisconsin,[2] they do not supply sufficient data on which to base general criteria for the economical considerations for this type of bridge. Therefore, general studies performed by European engineers and feasibility studies of several American bridges are presented here to illustrate the potential range of economical application of cable-stayed bridges.

## 3.2   INTERMEDIATE SPANS

The open competitive design system that exists in Germany has produced numerous feasibility studies which have resulted in actual construction of many cable-stayed bridges. The bridges have main spans ranging in length from 500 to 1200 ft. This span range was determined to be economical in the postwar period when many damaged bridges were replaced.

Often a survey and study of existing bridges can reveal meaningful data with respect to the general application of a particular type of bridge and the geometrical proportions best suited to that application. In his survey of the bridges in Germany, Thul[3] compared the center span length to the total length of the bridge for three-span continuous girder bridges, cable-stayed bridges, and suspension bridges, Fig. 3.1. This investigation may be considered a general study on the economical range of applications for the various types of bridges surveyed.

Limits of economical application appear to be 700 ft for the center span of a three-span continuous girder bridge, with ratios of center span to total length ranging from 30 to 50%. The suspension bridge begins to be economical for a center span of 1000 ft, with a ratio of center span to total length ranging from 60 to 70%. The cable-stayed bridge fills the void left by the continuous girder and suspension bridges in the range of center span from 700 to 1000 ft, with a corresponding center span to total length range of 50 to 60%.

In his comparative study, Thul has shown that the cable-stayed concept can be economical for bridges with intermediate spans. However, with greater experience in design and construction, the application of longer main spans of cable-stayed bridges has increased. Because other studies

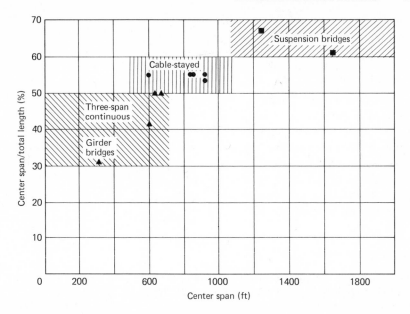

**FIGURE 3.1. Bridge type span comparison. (Courtesy of the British Constructional Steelwork Association, Ltd. From ref. 3.)**

have indicated that longer center spans for cable-stayed bridges are possible, the supremacy of the conventional suspension bridge may well be challenged.

In another study of the economics of cable-stayed bridges with respect to other bridge types, a comparison was made of the weight of structural steel in pounds per square foot of roadway deck versus center span length. The study was made for girder bridges, suspension bridges, and cable-stayed bridges using an orthotropic steel superstructure. The data are presented graphically in Fig. 3.2 and are the result of a study by P. R. Taylor, a Canadian engineer.[4]

A comparison of steel deck weights indicates that the cable-stayed bridge again fills the void between the continuous girder and suspension bridges. The data for the girder bridges fall into two distinct paths and may be the result of the different methods of design and different arrangements of the cross-sectional girders. Taylor recognized the difference in the ratio of material to labor costs in Europe and North America and concluded that for Canada highways cable-stayed bridges with center spans ranging from 700 to 800 ft were 5 to 10% more economical than other types of comparable bridges.

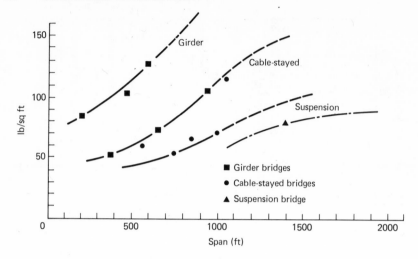

**FIGURE 3.2.** Weight of structural steel in lb/sq ft. of deck for orthotropic steel bridges. (Courtesy of Engineering Journal (Canada). From ref. 4.)

Limited experience to date has indicated that cable-stayed bridges with center spans less than 500 ft are most suitable for pedestrian bridges. The total economical range of the various types of cable-stayed highway and pedestrian bridges has not been fully examined. Therefore, it is incumbent upon designers and contractors to develop the necessary data by careful study and evaluation of each new application as it presents itself. The general economy appears to be present but, for the moment, it must be evaluated separately for each individual application.

### 3.3   LONG SPANS

The economic survey by Taylor, Fig. 3.2, has a reference point for a cable-stayed bridge that is higher than one would expect for that magnitude of center span. It appears that this singular point is apparently based on the data taken from the Kniebrücke Bridge at Düsseldorf, Fig. 3.3, which is an asymmetrical bridge with one tower. The data in Fig. 3.2 is for a center span of 1050 ft with a corresponding weight of deck structural steel of 115 pounds per square foot.

If the Kniebrücke Bridge were considered to be one-half of a symmetrical two-tower arrangement, with a center span of approximately 2000 ft, and the data replotted against previous data, Fig. 3.4, a different conclusion may be drawn.[5] The cable-stayed bridge is then seen to com-

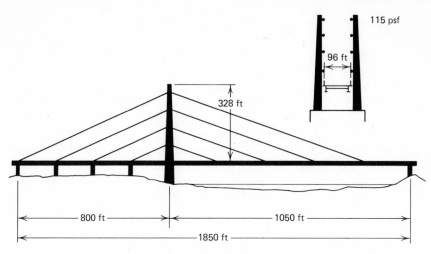

**FIGURE 3.3.   Kniebrücke, Düsseldorf.**

pete favorably with the suspension bridge of comparable center span. From this limited study it appears reasonable to assume optimistically that cable-stayed bridges may penetrate the complete range of spans now dominated by suspension bridges. In fact, the feasibility of a cable-stayed bridge with a center span of approximately 2000 ft is being considered in some preliminary bridge designs. Improved and imaginative methods of

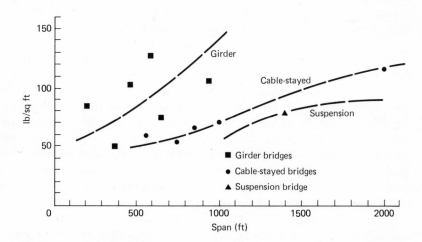

**FIGURE 3.4.   Weight of structural steel in lb/sq ft of deck for orthotropic steel bridges. (From ref. 5.)**

construction may tip the economic scale in favor of the cable-stayed bridge.

When Thul[6] wrote: "It is considered highly unlikely or unrealistic to build bridges with very long spans using cable-stayed construction. Such span lengths will be reserved for suspension bridges because there are considerable difficulties in construction of cable-stayed bridges," he apparently did not foresee the effects of improved technology and modern techniques of erection and construction, as perceived by Leonhardt.[7,8,9] Leonhardt concluded that cable-stayed bridges are particularly suited for spans in excess of 2000 ft and may even be constructed with spans of more than 5000 ft. The arguments Leonhardt presents for his conclusions are discussed below.

### 3.4   STIFFNESS COMPARISON

The studies conducted by Leonhardt and Zellner are based on a comparison of a suspension and cable-stayed span of similar proportions and assumptions for hinged joints as indicated in Fig. 3.5. The span is considered to be a center span of a bridge system; its cables are fixed at the terminals with respect to vertical and horizontal movements as though they were attached to a fixed abutment. Other common factors are: cable stress limited to 102,000 psi, modulus of elasticity of the cable is 29,200 ksi, hinges located at all connections of the stayed cables and hangers to the stiffening girders, normal forces acting on the girder are neglected in the deflection calculations, termination of the cables is at a height of $\frac{1}{6}$ the span above the girders, and live load is assumed to be 60% of the dead load.

The deflection diagram for the effect of live load on the entire span is indicated in Fig. 3.6. For this arrangement of cables and assumptions, as noted above, the deflection for the suspension system is 77% of that for the cable-stayed system. However, when only one-half the span is loaded, the deflection of the suspension system is 4.6 times that of the stayed system, Fig. 3.7. The reason for the large variation is the relatively large asymmetrical deformation of the main cable in the suspension system as it adjusts itself to a new position of equilibrium. In the stayed system, the vertical loads are supported by the cables, which transfer the loads directly to the tower or saddles without the large deformations taking place.

Therefore, to reduce the deflections of the suspension system it may be concluded that a larger bending stiffness is required than for the cable-stayed system. This effect helps to explain why a suspension bridge must use a high bending stiffness in the stiffening truss, resulting in a propor-

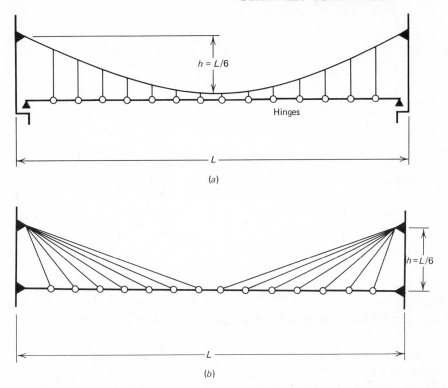

**FIGURE 3.5.**  Suspension versus cable-stay: (*a*) suspension; (*b*) cable-stay. (From ref. 9.)

tionately higher dead weight requirement. In a corollary study, the height of the cable terminus was reduced to 1/10 of the span, and although the suspension system was stiffer, the deflection was approximately three times that of the cable-stayed system.

Leonhardt also investigated the effect of a concentrated live load on both cable systems and concluded that the stayed system has less deflection. The relative deflection curves are indicated in Fig. 3.8 for a uniform live load of 1.2 times the unit dead load placed over a length equal to 0.067 of the span. The deflection of the suspension system is 5.5 times that of the stayed system. These results suggest that a cable-stayed system is stiffer than a suspension system of comparable span.

The first studies by Leonhardt and Zellner assumed fixed terminals for the cables in both systems. However, in actual structures the cables are connected to towers that deflect under load and, consequently, the girder deflection patterns will vary with the deflection of the towers.

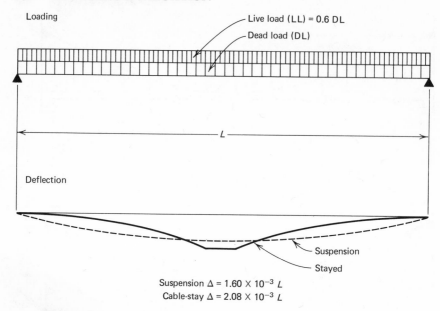

FIGURE 3.6. Deflection—Suspension versus cable-stay. (From ref. 9.)

A study of the total suspension bridge system with a comparable cable-stayed system, Fig. 3.9, was also performed by Leonhardt and Zellner. The assumptions for this study are the same as those for the previous study with the added consideration that the towers are hinged at the girder or truss level. The hinged tower assumption permits the tower to move laterally unrestrained as required by the loads and geometry of the structures. For consistency in the comparison, the end span dimension is assumed to be 0.4 of the center span.

The deflection curves indicated in Fig. 3.9, are based on a live load 0.6 times the dead load distributed over the center span. The suspension span has greater deflections in the center and end spans compared to the stayed system. The maximum center span deflection of the suspension system is 1.55 times that of the stayed system when the tower rotation is considered. With no tower displacement the suspension system had a deflection less than the stayed system which now appears to be inaccurate when the tower displacement is included.

The increase in deflection for the center span of suspension systems is also indicative of the increase in deflections in the end spans. Calculations indicate that the deflection of the end spans is 3.85 times greater in the suspension system than in the stayed system. The tops of the towers of the suspension system deflect horizontally 2.6 times more than those of the

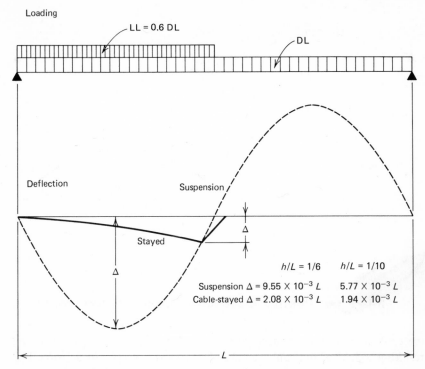

FIGURE 3.7.  Deflection—Suspension versus cable-stay. (From ref. 9.)

stayed system. The deflections of the girder and tower can be decreased proportionately by increasing the stiffness of the back stays in the end spans.

## 3.5  CABLE WEIGHT

The weight of cables in a cable-stayed bridge becomes one of the more important parameters in the cost feasibility study for a particular bridge. The weight and cost of the foundations, superstructure, and towers generally can be estimated closely by previous experience with similar types of construction. The weight and cost of the cables is a new factor for engineers and contractors to consider, and, unfortunately, estimating data is scarce.

Leonhardt[9] has developed a relationship in which the amount of cable steel required for a given cable force is considered as a function of the ratio of the tower height to the center span. The effect of the weight of

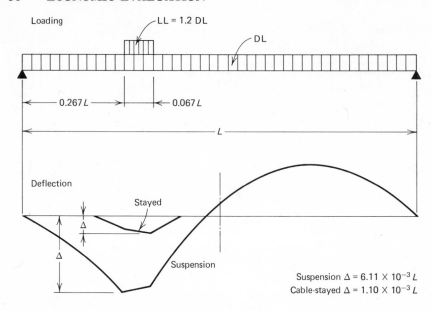

**FIGURE 3.8.   Deflection—Suspension versus cable-stay. (From ref. 9.)**

the cable and any load concentrations are neglected. The equation for the resulting weight of the cable required to support a given tensile force is

$$W = \frac{q\gamma L^2}{\sigma} C \tag{3.1}$$

where $W$ = weight of steel in cables in pounds
 $q$ = total load (dead load plus live load)
 $\gamma$ = specific weight of cable steel
 $\sigma$ = allowable cable stress in psi
 $L$ = length of main span in feet
 $C$ = dimensionless coefficient depending on bridge type

The cable weight equation is applicable to the classical suspension bridge and the cable-stayed harp and radiating types. The constant $C$ varies for each type and takes the following forms:

For the suspension bridge:

$$C_S = \frac{2L_1 + L}{2L} \left[ \sqrt{16 + \frac{1}{u^2} \left( \frac{1}{4} + \frac{2u^2}{3} \right)} + \frac{2u}{3} \right] \tag{3.2}$$

where $L$ = center span length
 $L_1$ = side span length

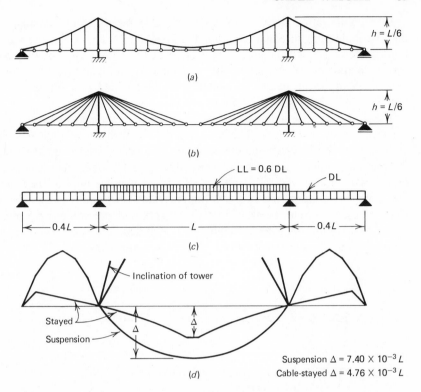

**FIGURE 3.9.** Deflection—Suspension versus cable-stay: (*a*) suspension; (*b*) cable-stay; (*c*) loading; (*d*) deflection. (From ref. 9.)

For the cable-stayed harp:

$$C_H = u + \frac{1}{4u} \qquad (3.3)$$

For the cable-stayed radiating:

$$C_R = 2u + \frac{1}{6u} \qquad (3.4)$$

Where $u$ is the ratio of tower height above the deck to the length of the center span expressed as $h/L$.

To provide comparative cable weights, it was necessary to assume the hanger weights were included in the total weight of cable for the suspension system, but to exclude the quantity of cable steel from the ends of the side spans to the anchorages.

In a comparative study of the cable weights for different systems, the only variable term in equation 3.1 is the constant $C$. Therefore, the variation of the constant will also be indicative of the comparison of the steel weight in the cables. A plot of the coefficients $C$ for varying values of $u$ (the ratio of tower height to center span length $h/L$) is illustrated in Fig. 3.10, for the three types of structures to be discussed.

The end spans are assumed to be 0.4 times the center span for each of the three types of bridges. The lowest point of each curve indicates the optimum minimum value for the coefficient $C$, which is also indicative of the minimum cable steel weight. The value of $u$ for both the suspension bridge and cable-stayed radiating type is approximately equal to 0.28, and the cable-stayed harp type has a minimum value of $u$ equal to 0.5. These values do not include the weight of the towers and the stiffening girders. When these additional steel weights are included, the most economical ratio of $h/L$ for the cable-stayed bridges is approximately 0.2 or 1:5,

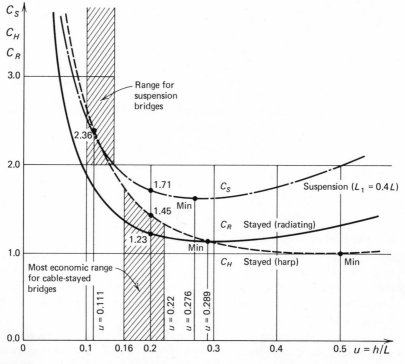

**FIGURE 3.10.   Economic Comparison—Suspension versus cable-stay. (From ref. 9.)**

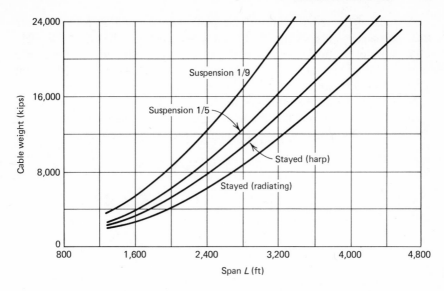

**FIGURE 3.11. Cable Weight Comparison. (From ref. 9.)**

while that for the suspension type is 0.125 or 1:8. However, to obtain greater stiffness from the cables for suspension bridges, a value of $u$ of 0.111 or 1:9 is preferred.

The corresponding values of $C$ are

suspension system, $u = 1:9$  $C_S = 2.36$
suspension system, $u = 1:5$  $C_S = 1.71$
cable-stayed radiating type,
$\qquad\qquad u = 1:5$  $C_R = 1.23$
cable-stayed harp type,
$\qquad\qquad u = 1:5$  $C_H = 1.45$

To illustrate the comparison of cable weights for the three types of bridges, Leonhardt plotted the cable weights for various span lengths, Fig. 3.11, using the coefficients $C$ as determined from Fig. 3.10. The values for cable weight plotted in Fig. 3.11 were determined for the following data:

1. Side span = 0.4 times the center span
2. Dead load plus live load = 27 kip per ft
3. Bridge width = 124 ft
4. Cable unit weight = 499 lb per cu ft
5. Allowable cable stress = 102 ksi
6. Neglect cable dead weight

A study of the relationships of the plotted curves in Fig. 3.11 indicates that the suspension system with a value of $u$ equal to $\frac{1}{9}$ has approximately 60% more cable steel weight than the cable-stayed radiating type with a $u$ ratio of 1:5. This value is for the complete range of span lengths from 1200 to 4000 ft, which can be interpreted from the plot.

In order to refine the estimates for the steel weight of the several cable systems, Leonhardt modified the basic equation for cable weight to include the dead weight of the cable. The equation then takes the form,

$$W = \frac{q\gamma L^2}{\sigma} CK \tag{3.5}$$

where $K$ is the modification factor and is expressed as follows for the three systems:

$$K_S = 1 + \frac{\gamma L^2}{\sigma(2L_1 + L)} C_S + \left[\frac{\gamma L^2}{\sigma(2L_1 + L)} C_S\right]^2 + \cdots \tag{3.6}$$

$$K_H = 1 + \frac{\gamma L}{3\sigma} C_H + \frac{\gamma^2 L^2}{8\sigma^2} C_H^2 + \cdots \tag{3.7}$$

$$K_R = 1 + \frac{\gamma L^2 k}{\sigma C_R} + \frac{\gamma L}{6\sigma C_R} + \frac{\gamma L}{80\sigma k^2 C_R} + \frac{3\gamma^2 L^2}{160\sigma^2 k} + \cdots \tag{3.8}$$

where $k = u$.

A plot of the adjusted values for steel weight versus span length including the modification factor is indicated in Fig. 3.12.

Values for the modification factor are tabulated in Table 3.1. The values of $K$ in Table 3.1 are calculated for classical suspension bridges with ratios of tower height to span length of 1:9 and for the cable-stayed types with ratios of 1:5 for the two spans of 3280 ft (1000 m) and 9842 ft (3000 m). The addition of the cable dead weight increases the cable steel requirements for the suspension bridge by 17% but by only 5% for the

**TABLE 3.1**    **Modification Factors**

| | | Cable-Stay | |
|---|---|---|---|
| $L$ (ft) | Suspension $K_S$ | Harp $K_H$ | Radiating $K_R$ |
| 3280 | 1.170 | 1.057 | 1.048 |
| 9842 | 1.750 | 1.190 | 1.151 |

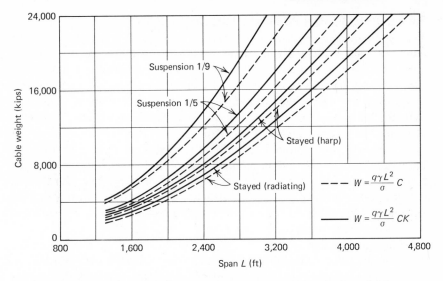

**FIGURE 3.12.** Modified Cable Weight Comparison. (From ref. 9.)

cable-stayed bridges with a center span of 3280 ft. For the longer span of 9840 ft the requirements for the suspension system increase by 75% as opposed to a 15 to 20% increase for the stayed bridges.

One may think that a 9840 ft span borders on the ridiculous, but a proposal to bridge the Messina Straits between Sicily and Italy with a span of 3000 m (9840 ft) has been considered, Fig. 3.13. If built, the bridge will have a center span of approximately 1.8 miles.

**FIGURE 3.13.** Proposed Messina Straits Bridge. (From ref. 9.)

## 3.6    COMPARISON WITH OTHER BRIDGE TYPES

Dubrova[10] has presented some interesting data on the economics of nine types of bridge construction in the Soviet Union. Dubrova evaluated five concrete and four steel bridges. The concrete bridges are cable-stayed, arch-cantilever, arch, rigid frame suspension, and continuous. The continuous type consists of box-girder construction erected by the cantilever method. The arch-cantilever, Fig. 3.14, is constructed as a cantilever for dead load and pin-connected at midspan for live load shear transfer without moment resistance. When moment capability is built into the midspan connection, the structure reacts as an arch for live loads. The rigird frame suspension bridge, Fig. 3.15, is constructed as a cantilever with a drop-in suspended center section. The steel bridges are cable-stayed, conventional suspension, arch, and a continuous type.

Although the relative costs of construction in the USSR differ from the costs in the United States, the economic study by Dubrova is useful in developing a comparative relationship of the relative costs of the various types of bridges.

The Dubrova's economic evaluation included the costs of the piers and the erection procedures combined with the cost of the superstructure. The study of the costs of the different erection methods, illustrated in Fig. 3.16, indicates a variation of 300% between the cantilever and pontoon as-

**FIGURE 3.14.    Arch-cantilever. (From ref. 10.)**

**FIGURE 3.15. Rigid frame suspension bridge. (From ref. 10.)**

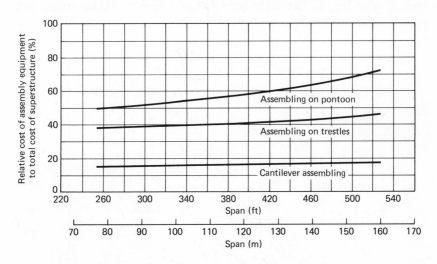

**FIGURE 3.16. Relation of assembly equipment and total cost to span of bridge. (From ref. 10.)**

semblies. The plot indicates a decided advantage for the cantilever method of construction.

Another study was concerned with the amount of concrete used in the superstructure as a function of the span length. A graphical representation, Fig. 3.17, indicates the volume of concrete per square foot of bridge deck plotted against the span length of the bridge.

For the span lengths investigated, ranging from 200 to 1000 ft, the cable-stayed bridge required the least volume of concrete. The other types of bridges in order of least concrete usage are arch-cantilever, arch, rigid frame suspension, and continuous. The variation in concrete volume required for the various bridge types indicates a wide spread between the lightest cable-stayed system and the continuous system, especially as the span lengths increase beyond 800 ft.

An investigation of similar bridge types using a structural steel superstructure is illustrated in Fig. 3.18. The plot indicates the amount of steel in pounds per square foot of bridge surface versus span lengths ranging from 200 to 1800 ft. As in the previous study of concrete usage, the cable-stayed system is the most economical in the span range of 600 to 1000 ft, and the conventional system becomes the most economical beyond the 1000 ft span length. The other types follow a similar ranking order, with respect to concrete usage, Fig. 3.17.

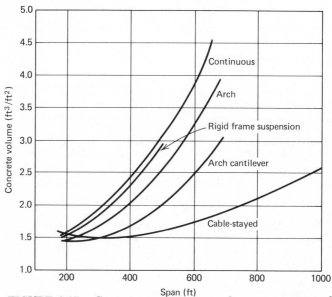

**FIGURE 3.17.  Concrete superstructure volume versus span of bridge. (From ref. 10.)**

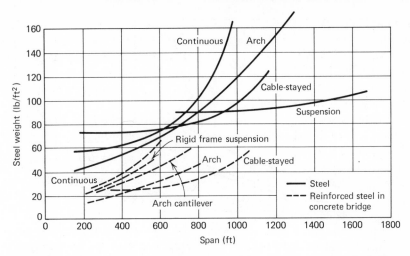

**FIGURE 3.18. Steel superstructure weight versus span of bridge. (From ref. 10.)**

The data based on the current costs of design and construction appear to reinforce the concept that suspension bridges are the most economical for the longer spans. However, Leonhardt indicated that cable-stayed bridges can be directly competitive with the classical suspension bridge when innovative methods of design and construction are considered. Figure 3.18 also indicates the amount of reinforcing steel used in the five bridges plotted in Fig. 3.17. The steel weight relationships follow very closely the concrete relationships with the exception of the interchange of the arch and arch-cantilever bridges.

A separate study of the amount of concrete required for the piers of various types of bridges is illustrated in Fig. 3.19. The plot indicates the volume of pier concrete in cubic feet per square foot of bridge deck versus span lengths ranging from 150 to 1000 ft. The figure is a composite, it includes the bridges with both steel and concrete superstructures. The solid lines are steel structures, and the dashed lines represent the concrete superstructure.

As evidenced in the previous studies, the cable-stayed and suspension systems are the most economical for both types of superstructures. Furthermore, this study indicates that these systems require less pier volume for the complete range of span lengths. Evidently, this fact is the result of less total weight for the superstructures.

When Dubrova combined the cost data relationships for the individual components he determined the total cost of the bridge in terms of the unit area of the bridge deck, Fig. 3.20. As one would expect, the cable-stayed

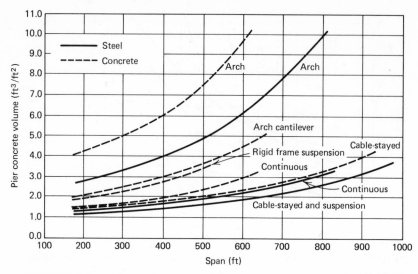

**FIGURE 3.19.    Pier concrete volume versus span of bridge. (From ref. 10.)**

and suspension systems show up as the most economical types. The cable-stayed system falls in the range of span lengths from 400 to 1000 ft and the suspension system takes over beyond the 1000 ft span.

Dubrova's investigation in terms of current bridge construction practices in the USSR indicates that continuous box girders erected by the cantilever method are most economical for the range of spans from 150 to 500 ft. The cable-stayed system with a concrete superstructure is most economical to 800 ft, while the cable-stayed system with a steel superstructure is economical to a span of 1000 ft. Beyond the 1000 ft span length the classical suspension bridge becomes the most economical type.

However, it is important to bear in mind that these are idealized studies assuming current costs and methods, which are ever-changing and influencing the designs chosen. Designers and contractors should be alert to constant innovations in bridge design and construction methods. What may appear to be standard practice one day may become obsolete the next day as a result of imaginative and innovative contractors and designers.

### 3.7    SITKA HARBOR BRIDGE

The economic feasibility study for the Sitka Harbor Bridge considered six different types of bridges before a final decision was made to adopt the cable-stayed system. The various bridge types evaluated are indicated in

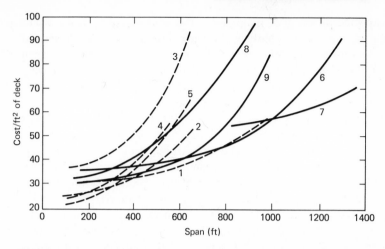

FIGURE 3.20.   Cost per sq ft deck versus span. (From ref. 10.) Concrete:
(*1*) cable stay; (*2*) arch cantilever; (*3*) arch; (*4*) rigid frame suspension; (*5*)
continuous. Steel: (*6*) cable-stay; (*7*) suspension; (*8*) arch; (*9*) continuous.

Table 3.2, which ranks the types on a relative basis using the cable-stayed
system as a base of 1.00.

TABLE 3.2   Sitka Harbor Bridge—Cost Study

| Type | Description | Cost Ratio (Cable-Stayed Girder = 1.00) |
|------|-------------|-----------------------------------------|
| I | Plate girder with fenders | 1.15 |
| II | Plate girder continuous | 1.13 |
| III | Orthotropic box girder | 1.04 |
| IV | Through tied arch | 1.04 |
| V | Half through tied arch | 1.06 |
| VI | Cable-stayed box girder | 1.00 |

Gute[1] discusses the advantages and disadvantages of each type studied in
order to arrive at a comparable value for each type.

The plate girder system classified as Type I required a main span of 250
ft (76 m) which had to be skewed in order to accommodate the fender
system along the sides of the navigation channel. Because the piers and
fenders would be in 52 ft (15.8 m) of water this was the most expensive
system. Other considerations that detracted from the Type I system was

the expected difficulty of maintaining the fender system which would ultimately become an unsightly wall, especially at low tide. A second consideration for this bridge was the idea of increasing the main span to 450 ft (137 m), which would move the main piers out of the deep water and beyond the limits for navigation thus eliminating the fenders. However, the cost would still be high as a result of the long main span.

Bridge Types II and III had spans of 300, 450, and 300 ft (91.4, 137.0, and 91.4 m). This turned out to be too long and, therefore, too costly for the continuous plate girder. The orthotropic deck box girder, Type III, was calculated to be 4% higher than the cable-stayed system. However, its shortcoming is the large depth of girder required at midspan. Design considerations indicated a midspan depth for the superstructure to be 14 ft (4.3 m) compared to a 6 ft (1.8 m) depth required for the tied arch and cable-stayed system. The difference of 8 ft (2.4 m) would be reflected in a lower cost for the approaches because of the reduction in the grade line. This added cost must be included as a part of the cost of the total bridge structure and approaches.

The possible use of a through truss or cantilever truss was discounted on the basis of expected maintenance difficulties in a sea atmosphere and because it would be less aesthetically appealing at that particular site.

Types IV, V, and VI make use of small short piers and reduced side spans of 150 ft (46 m). The two tied arch systems would require high superstructures in the center of the channel, which also serves as the approach path for seaplane operations. The possible hazards eliminated these types from further considerations. In view of all the considerations and cost factors the cable-stayed bridge system was finally selected. The actual cost of the cable-stayed bridge was $1,900,000, which is $52 per square foot of deck surface. Representative prices include $145 per cubic yard for Class "A" concrete; 22¢ per pound for reinforcing steel; 46¢ per pound for structural steel; and $1.63 per pound for cables and fittings.

### 3.8   NEW ORLEANS BYPASS, I-410

A proposal to construct a new segment of the highway system around the southern edge of the New Orleans metropolitan area is under consideration by the Louisiana Department of Highways. The new segment is referred to as the New Orleans Bypass for interstate route I-410. This alignment necessitates two new crossings over the Mississippi River, one near Chalmette and the other near Luling.

The feasibility design by Modjeski and Masters[11,12] considered several bridge types: conventional suspension, steel cantilever, and cable-stayed. The study indicated that the cable-stayed bridge, with center spans of 1600

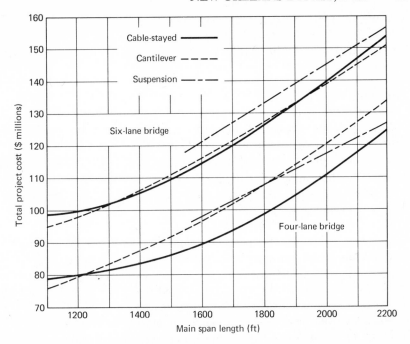

**FIGURE 3.21.** Comparative construction cost estimate, New Orleans bypass—I-410. (Courtesy of T. Robert Kealey. From refs. **11 and 12.**)

to 2100 ft, is the most favorable type for these locations. The results of the economic feasibility study is indicated in Fig. 3.21 as the total project cost in millions of dollars versus the main span length.

The plot permits a convenient comparison of the three types of bridges for a six-lane roadway and a four-lane roadway. It is to be noted that the total project cost for the six-lane suspension bridge is significantly higher than the cantilever or cable-stayed systems. The cost differential between the cantilever and cable-stayed systems is negligible and both are acceptable. For the four-lane roadway, the cable-stayed bridge has the decided cost advantage compared with either the suspension or cantilever types.

The high cost of the conventional suspension bridge compared with the cantilever and cable-stayed types is primarily attributable to the cost of the substructure. These high costs are for the cable anchorages and foundations required and are greatly affected by the relatively poor soil conditions present in the Mississippi River delta. As a result of the study and the close economical comparison of the cantilever and cable-stayed bridges the final selection was based on the importance of other related factors in each type of bridge and the particular location.

## 3.9  PASCO-KENNEWICK INTERCITY BRIDGE

The economic evaluation of this structure considered five alternate structural designs:

1. Constant depth steel plate girders with a precast composite deck.
2. Cable-stayed girder with a deck constructed of precast concrete.
3. Continuous constant depth posttensioned concrete box girder, constructed on shore and pushed into position.
4. Variable depth posttensioned concrete box girder constructed segmentally by the cantilever method.
5. Asymmetrical steel box girder cable-stayed main span with concrete box-girder approach spans.

Studies were based on an approximate overall length of structure of 2480 ft, four traffic lanes and two sidewalks, a minimum vertical navigation clearance of 50 ft above the 50-year flood level over a horizontal channel distance of 350 ft. It is to be noted that after a final choice was made some minor changes were made in the design.

The above alternates were considered the most feasible and were studied in detail. Other alternates were studied but were then discarded as unfeasible. These included steel orthotropic plate deck girders, cable-stayed steel orthotropic plate girders, and various span configurations of steel plate girders combined with steel or concrete girder approach spans.

A brief description of the five principal alternates as presented by the consultants, Arvid Grant and Associates, Inc. in professional collaboration with Leonhardt and Andra, in their preliminary design report[13] are summarized below.

The steel plate girder design, alternate 1, Fig. 3.22, consisted of eight continous spans with expansion joints only at the abutments. Span arrange-

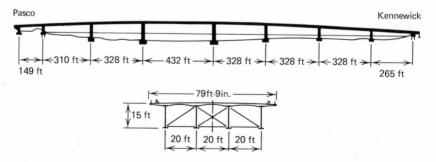

**FIGURE 3.22.   Pasco-Kennewick Intercity Bridge, Alternate 1. (From ref. 13.)**

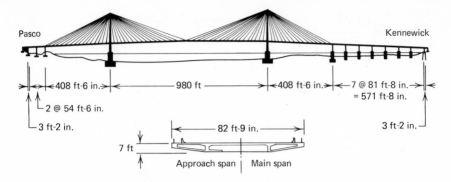

**FIGURE 3.23.    Pasco-Kennewick Intercity Bridge, Alternate 2. (From ref. 13.)**

ment starting at the Pasco abutment was 149-310-328-423-3 at 328-265 ft for a total length of 2468 ft. The superstructure consisted of four lines of girders 20 ft on centers with a constant depth of 15 ft. Fixed bearings were located at the piers adjacent to the center 432 ft span with all other bearings being expansion bearings. The deck was envisioned as precast units posttensioned longitudinally before being made composite with the deck.

The preliminary design for alternate 2, Fig. 3.23, contemplated an overall length of structure of 2484 ft. The 1797 ft main unit was to be supported by a cable-stay configuration radiating from the pylons in two vertical planes. The deck structure was continuous from abutment to abutment, with expansion joints only at the abutments. Parallel wire strands supported the precast deck every 27 ft. The deck was only 7 ft deep and consisted of continuous triangular box beams at the edges connected by cross beams at 9 ft spacing. The unit was open in the cable supported portion, (i.e., it had no enclosing bottom flange) and the approach spans were fully closed to partially closed approaching the cable-stay portion. The cross beams were prestressed except in the approach spans where only longitudinal prestressing was required. Main longitudinal webs were prestressed at the middle of the main span and at the ends of the end spans (cable-stayed) where the axial force of the cables was small. Mild steel reinforcement was utilized in both directions.

Alternate 3, Fig. 3.24, proposed a single cell concrete box with five interior spans of 374 and end spans of 299 ft. The superstructure box had a constant depth of 15 ft, an overall width of 79 ft 9 in., and bottom flange width of 38 ft, with the transverse prestressing in the top flange. Longitudinal prestressing was to be in two stages, first stage for construction and launching was a concentric force of 4400 tons positioned in the webs and flanges, second stage prestressing required 11,000 tons installed

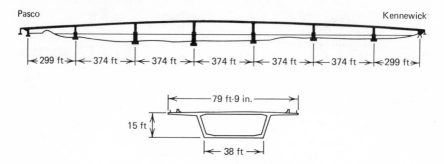

**FIGURE 3.24.    Pasco-Kennewick Intercity Bridge, Alternate 3. (From ref. 13.)**

externally in the box. The superstructure would be constructed in successive 75 ft length units at one embankment and progressively pushed out until the opposite abutment was reached. Temporary falsework bents at the center of each span and a launching nose attached to the forward end of the superstructure would be required to reduce cantilever stresses during erection.

Posttensioned balanced cantilever segmental construction, alternate 4, Fig. 3.25, consisted of a main span of 500 ft, three 250 ft side spans on each side, and 233 ft end spans. The main span and two flanking spans were haunched and all other spans were constant depth. Free cantilever construction without falsework was contemplated. The main span and two flanking spans had a depth varying from 10 ft 6 in. to 23 ft, all other spans had a constant depth of 10 ft 6 in. Posttensioning was provided in the top of the box for cantilever erection stresses and, after closure, continuity tendons would be provided in the bottom of the box. The deck was posttensioned transversely, and diagonal and vertical tendons were required in the web for shear stresses.

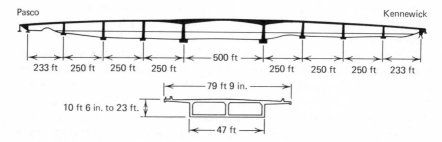

**FIGURE 3.25.    Pasco-Kennewick Intercity Bridge, Alternate 4. (From ref. 13.)**

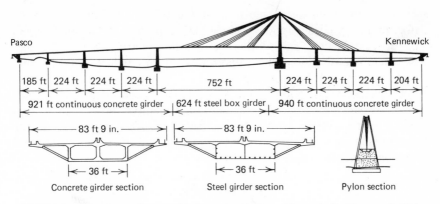

Pasco                                                                    Kennewick

| 185 ft | 224 ft | 224 ft | 224 ft | 752 ft | 224 ft | 224 ft | 224 ft | 204 ft |

921 ft continuous concrete girder | 624 ft steel box girder | 940 ft continuous concrete girder

|← 83 ft 9 in. →|    |← 83 ft 9 in. →|

|← 36 ft →|    |← 36 ft →|

Concrete girder section        Steel girder section        Pylon section

**FIGURE 3.26.   Pasco-Kennewick Intercity Bridge, Alternate 5. (From ref. 13.)**

Design alternate 5, Fig. 3.26, consisted of four concrete girder approach spans on the Pasco side, one at 185 ft and three at 224 ft; a cable-stayed steel box-girder composite concrete deck main span of 752 ft; and four approach spans on the Kennewick side, three at 224 ft and one at 204 ft. This alternate was an asymmetric structure with a single A-frame pylon, a radiating stay arrangement in elevation, and a single transverse vertical plane located in the median.

The economic comparison of these five alternates using alternate 2 (the final design choice) as a base is shown in Table 3.3.

**TABLE 3.3   Pasco-Kennewick Bridge—Economic Comparison**

| Alternate | Description | Cost Ratio |
|-----------|-------------|------------|
| 1 | Steel plate girder | 1.005 |
| 2 | Cable-stayed concrete box girder | 1.000 |
| 3 | Concrete box girder—push-out method | 0.952 |
| 4 | Concrete box girder—cantilever method | 0.981 |
| 5 | Cable-stayed steel box girder | 1.019 |

As seen from the above estimated construction cost comparison (including substructure) there is no conclusive economic argument for the approval of any one design. Therefore, satisfactory functional requirements, anticipated long-term performance, construction and design requirements, as well as the estimated initial costs must also be evaluated.

Functional requirements should consider channel clearance, approach grades, aesthetics, and overload capacity. Long-term performance considerations are maintenance and structure durability. Construction and design requirement considerations include familiarity of construction method, ease of construction, risk during construction, local labor and materials, overall construction time, opportunity for cost reduction in the final design process, and design complexity.

Obviously, consideration of the above items is dependent on the particular site conditions, local environmental conditions relative to natural hazards, along with the local and national economic environment at the time the estimate is made, as well as any short- and long-term economical conditions that may affect the final cost.

## REFERENCES

1. Gute, W. L., "Design and Construction of the Sitka Harbor Bridge," Meeting Preprint 1957, ASCE National Structural Engineering Meeting, April 9–13, 1973, San Francisco, California.
2. Woods, S. W., Discussion of "Historical Development of Cable-Stayed Bridges," by Podolny and Fleming, *Journal of the Structural Division*, ASCE, Vol. 99, No. ST 4, April 1973.
3. Thul, H., "Cable-Stayed Bridges in Germany," *Proceedings of the Conference on Structural Steelwork*, held at the Institution of Civil Engineers, September 26–28, 1966, The British Constructional Steelwork Association Ltd., London, England.
4. Taylor, P. R., "Cable-Stayed Bridges and Their Potential in Canada," *Engineering Journal* (Canada), Vol. 52/11, November 1969.
5. Podolny, W., Jr., "Economic Comparisons of Stayed Girder Bridges," *Highway Focus*, Vol. 5, No. 2, August 1973, U.S. Department of Transportation, Federal Highway Administration, Washington, D. C.
6. Thul, H., "Schrägseilbrücken," Preliminary Report, Ninth IABSE Congress, Amsterdam, May 1972.
7. Leonhardt, F., "Seilkonstrucktonen und seil verspannte Konstruktionen," Introductory Report, Ninth IABSE Congress, Amsterdam, May 1972.
8. Leonhardt, F. and Zellner, W., "Cable-Stayed Bridge: Report on Latest Developments," *Proceedings of the Canadian Structural Engineering Conference, 1970*, Canadian Steel Industries Construction Council, Toronto, Ontario, Canada.
9. Leonhardt, F. and Zellner, W., "Vergleiche zwichen Hängebrücken und Schrägkabelbrücken für Spannweiten über 600 m," *International Association for Bridge and Structural Engineering*, Vol. 32, 1972.
10. Dubrova, E., "On Economic Effectiveness of Application of Precast Reinforced Concrete and Steel for Large Bridges (USSR)," *IABSE Bulletin*, 28, 1972.
11. "Feasibility Study of Mississippi River Crossings Interstate Route 410," Report to Louisiana Department of Highways in Cooperation with Federal Highway Administration, Modjeski and Masters, Consulting Engineers, Harrisburg, Pennsylvania, July 1971.
12. Kealey, T. R., "Feasibility Study of Mississippi River Crossings Interstate 410," Meeting Preprint 2003, ASCE National Structural Engineering Meeting, April 9–13, 1973, San Francisco, California.
13. "Pasco-Kennewick Intercity Bridge Preliminary Design Report," December 1972, Arvid Grant and Associates, Inc., Engineers, Olympia, Washington.

# 4

# Concrete Superstructures

## 4.1 INTRODUCTION

Using cable stays to support a bridge structure is a versatile concept because there are no restrictions on the type of material that may be used for the superstructure. As a consequence, bridge superstructures have been built of reinforced concrete, prestressed concrete, and steel. Each material has its advantages and disadvantages for a particular site and engineers and contractors must evaluate all related factors before making a final choice.

Feasibility studies for a given site should include not only the type of material but also the type of superstructure form such as single or multiple box girders, or multiple girder systems. The location of the bridge site in relation to readily available materials, fabrication facilities, and erection convenience may influence the decision in favor of one particular style of bridge and, therefore, the material to be used.

The first modern prestressed concrete cable-stayed bridge was built over Lake Maracaibo in Venezuela in 1962 and was designed by Professor Riccardo Morandi of Rome University. There is some evidence that a bridge designed by Torroja and completed in 1925 at Tempul Aqueduct

77

in Spain had a concrete superstructure. Dr. Morandi, however, has carried the concept further and has designed several bridges with concrete superstructures which have been built in various countries of the world.

The following sections discuss several bridges that have been built in the world and some which are currently under construction. The bridges have superstructures with reinforced concrete or prestressed concrete components.


## 4.2   LAKE MARACAIBO BRIDGE, VENEZUELA

The Lake Maracaibo Bridge, one of the longest cable-stayed bridges in the world, has a superstructure of reinforced and prestressed concrete construction. The structural analysis and detailed plans were performed by the Maracaibo Bridge Joint Venture, consisting of Precomprimido-C.A., of Caracas, Venezuela, and Julius Berger, A.G., of Wiesbaden, Germany.

The Morandi concrete design was selected over 12 other competitors which were designed with steel superstructures. The reasons stated by the Government Commission for the selection of the Maracaibo Bridge Joint Venture design are:

1. Reduction of maintenance costs—the concrete design could withstand local climatic conditions.

2. The aesthetics of the design proportions.

3. Greater use of local materials, therefore, less foreign exchange for imported materials.

4. Greater use of local engineering talent and labor.

The total length of the Maracaibo Bridge is 5.4 miles (8.7 km), which includes five main navigation openings. The superstructure over the navigation openings consists of prestressed concrete cantilevered cable-stayed girders with suspended spans having a total length of 771 ft (235 m), Fig. 4.1.

The navigational requirements stipulated a horizontal clearance of 656.2 ft (200 m) and a vertical clearance of 147.6 ft (45 m). The central spans over the navigation channels were intentionally designed as statically determinate systems in order to preclude possible damage from earthquakes or unequal foundation movements. Therefore, the five main spans are divided into cantilever sections supporting suspended center portions. Each suspended portion consists of four prestressed T-girder sections, Fig. 4.2.

FIGURE 4.1.   General view, Lake Maracaibo Bridge. (Courtesy of Julius Berger-Bauboag Aktiengesellschaft. Ref. 1.)

The pier foundation width of 113.5 ft (34.6 m), Fig. 4.3, was determined to satisfy the asymmetrical loading conditions caused by the placement of a center section during erection and the unbalanced traffic loads during vehicular movements. The cantilever arm of the superstructure extends a distance of 252.6 ft (77 m) beyond the pier foundation and, instead of deep girders associated with conventional cantilevers, cable-stays are used as supporting members. These stays are located near the end of the cantilever arm where the suspended span is connected. The towers that anchor the stays are 303.5 ft (92.5 m) high in order to provide the necessary slope of the cables. By using stays to support the cantilever arm which in turn supports a 150.9 ft (46 m) suspended center section, the designers were able to satisfy the navigational requirements of a clear span of 656.2 ft (200 m). Other bridge concepts and types would have required deeper girders, thus raising the roadway elevation and increasing the cost of the approaches.

The cantilever span is supported on X-type frames, and the cable-stays are supported by two A-type rigid frames with a portal member connecting them at the top. The X- and A-frames are completely independent of each

**FIGURE 4.2.    Erection of Suspended T girder section Lake Maracaibo Bridge. (Courtesy of Julius Berger-Bauboag Aktiengesellschaft. Ref. 1.)**

other, Fig. 4.3, because the X-frames are directly below the roadway and the A-frames bypass the exterior of the roadway structure. The roadway is supported by the continuous cantilever which is a three cell prestressed concrete box girder, 16.4 ft (5 m) deep and 46.7 ft (14.22 m) wide. The primary prestressing force is provided by the horizontal component of the cable-stays, so that conventional reinforcement only is required to resist the additional loads on the structure. Supplemental prestressing tendons were required to develop resistance to the negative moments occurring at the X-frame supports and the transverse cable-stay anchorage beams.[1]

The pier cap consists of the three cell box girder with the X-frames projecting into the boxes to become transverse diaphragms, Figs. 4.4 and 4.5.

When the pier is completed, service girders are raised into position for construction of the remainder of the cantilever arm. During this phase of construction bending moments are produced in the box girder by the service trusses and weight of the cantilever arm, thus requiring additional concentric prestressing tendons in the box girder, Fig. 4.5. Therefore, to avoid overstressing the X-frames during this operation, temporary horizontal ties are installed and tensioned by hydraulic jacks, Figs. 4.5 and 4.6 (see Chapter 9). The steel service trusses used for form work are supported at one end by the completed pier cap and at the channel end by rigid bents

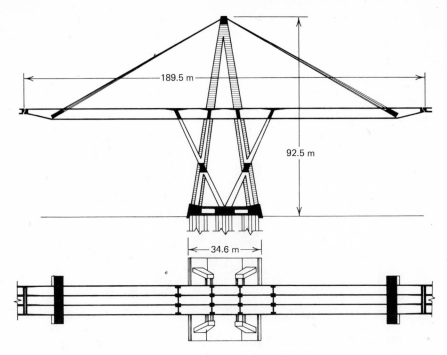

**FIGURE 4.3. Main span tower and X-frames Lake Maracaibo Bridge. (Courtesy of Julius Berger-Bauboag Atkiengesellschaft. Ref. 1.)**

placed in the channel and moved along as construction progressed, Fig. 4.7.

The anchorage for the cable stays are placed in a prestressed inclined transverse girder, which is 73.8 ft (22.5 m) long, Fig. 4.8. The reinforcing cages for these members were fabricated on shore in a position corresponding to the inclination of the cables in the completed structure.

The transverse anchor girders weigh 60 tons and contain 70 prestressing tendons. The cable-stays are housed in thick-walled steel pipes, Fig. 4.9, welded to steel plates at the ends. A special steel spreader beam was used to erect the fabricated cage in its proper position in the roadway structure, Fig. 4.10.

## 4.3 POLCEVERA VIADUCT, ITALY

The Polcevera Viaduct in Genoa, Italy, was designed by Dr. Morandi and is similar in appearance and construction to his Lake Maracaibo Bridge, Fig. 4.11. The bridge is a high-level viaduct 3600 ft (1100 m) long with the roadway above the terrain at an elevation of 181 ft (55 m). The three

**FIGURE 4.4.    Pier cap with X-frames, Lake Maracaibo Bridge. (Courtesy of Julius Berger-Bauboag Aktiengesellschaft. Ref. 1.)**

cable-stayed spans have lengths of 664, 689, and 460 ft (200, 210, and 140 m)[2] with center suspended spans of 118 ft (36 m).

The bridge supports the Genoa-Savona highway over an area containing railway yards, roads, industrial plants, and the Polcevera Creek. The top of the A-frame that supports the cables is 139.5 ft (42.5 m) above the roadway elevation. The tower is similar to the A-frame of the Lake Maracaibo Bridge; it has a longitudinal girder at the roadway level and a transverse connecting girder at the top. The roadway deck structure is a five cell box girder of reinforced concrete. The transverse anchorages for the cables are also box girders that require the cables to be subdivided into smaller units above the roadway and then passed through the webs of the girders to be anchored. The cables are composed of pretensioned steel strands encased in a protective concrete shell cover.

## 4.4   WADI KUF BRIDGE, LIBYA

Dr. Morandi designed the Wadi Kuf Bridge in Libya, Fig. 4.12, to be very similar to the Lake Maracaibo and Polcevera Creek Bridges by using the

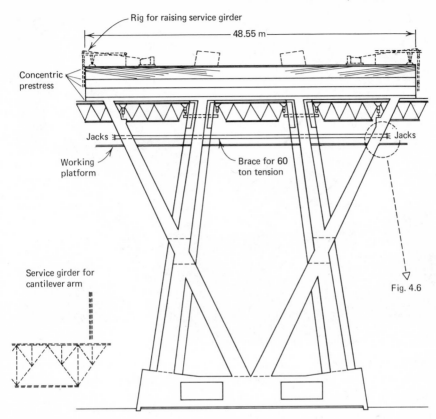

Rig for raising service girder

48.55 m

Concentric prestress

Jacks

Jacks

Working platform

Brace for 60 ton tension

Service girder for cantilever arm

Fig. 4.6

**FIGURE 4.5.   Pier cap of a main span and service girder, Lake Maracaibo Bridge. (Courtesy of Julius Berger-Bauboag Aktiengesellschaft. Ref. 1.)**

same type of A-frame and superstructure supports for all three structures. The bridges have similar proportions of tower height to span and use the same arrangement for the single cable stays on each side of the roadway.

The Wadi Kuf Bridge consists of only three spans, the center span is 925 ft (280 m) long and the two end spans are each 320 ft (98 m), for a total length of 1565 ft (475 m). The simply supported drop-in center portion of the main span consists of three double-T beams 180 ft (55 m) in length; each beam weighs approximately 220 tons.[2]

The familiar A-frame towers are 459 and 400 ft (140 and 122 m) high and the roadway deck is 597 ft (182 m) above the lowest point of the valley beneath the structure. The superstructure is a single cell box girder of variable depth with cantilever flanges forming the 42.7 ft (13 m) deck.[3]

**FIGURE 4.6.** Brace members bear against X-frames after being tensioned by hydraulic jacks, Lake Maracaibo Bridge. (Courtesy of Julius Berger-Bauboag Akteingesellschaft. Ref. 1.)

The contractor made good use of traveling forms to construct the box girder and deck, using the balanced cantilever technique to build on both sides of the towers at the same time. Traveling forms were used because extreme height and difficult terrain made other conventional construction methods impossible or too costly. The adopted procedure required temporary cable stays to support the cantilever arms during the construction sequence as the superstructure progressed in both directions from the tower. When the superstructure extended sufficiently, the permanent stays were installed and the structure was subsequently completed in the same manner.

## 4.5  MAGLIANA VIADUCT, ITALY

The Magliana Viaduct is on the road between Rome and Fiumicino Airport. Although Dr. Morandi designed this bridge, it is not similar to the

FIGURE 4.7. Placing service girder for forming cantilever girders, Lake Maracaibo Bridge. (Courtesy of Julius Berger-Bauboag Aktiengesellschaft. Ref. 1 )

FIGURE 4.8. Fabrication of anchorage beam, Lake Maracaibo Bridge. (Courtesy of Julius Berger-Bauboag Aktiengesellschaft. Ref. 1.)

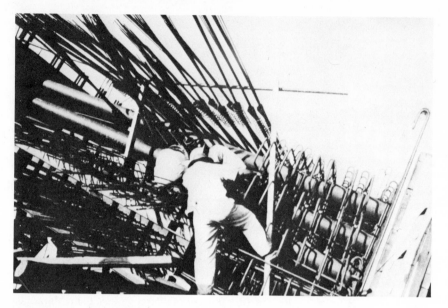

**FIGURE 4.9.**   Housing for cable-stays, Lake Maracaibo Bridge. (Courtesy of Julius Berger-Bauboag Aktiengesellschaft. Ref. 1.)

Morandi bridges discussed in previous sections. Extreme local conditions dictated a slightly different type of bridge structure, as noted by the single portal tower and single cable stays on each side of the roadway, Fig. 4.13. It was necessary to build the bridge over a swamp formed by a bend in the Tiber River.

The bridge structure has a total length of 652 ft (198.6 m) with the two asymmetrical side spans of 476 and 176 ft (145 and 53.6 m). The longer span of 476 ft (145 m) consists of a 206 ft (63 m) suspended span extending from a pier support to the 42.7 ft (13 m) cantilever overhang of the 269 ft (82 m) cable supported section, Fig. 4.13. An additional complication to the design is the 1558 ft (475 m) radius required for the horizontal curvature for the roadway at this site.[4]

In contrast to the fixity of the X-shaped piers and A-frame towers characteristic of Morandi's other designs, the Magliana Bridge is a fully articulated structure. The portal tower is hinged at the base and is offset 7.8 ft (2.37 m) at the top as it inclines toward the anchor abutment. The full articulation is achieved by installing hinges on both ends of the cantilever span, one at the base of the tower and the other at the cantilever support for the anchor span. The hinges are large-radius, steel-lined surfaces on concrete that extend for the full width of the roadway deck, Fig. 4.14.

**FIGURE 4.10.**  Erection of reinforcing cage for prestressed transverse anchoring girder, Lake Maracaibo Bridge. (Courtesy of
Julius Berger-Bauboag Aktiengesell-schaft. Ref. 1.)

The transverse anchorage beam for the stays is a box section 26 ft
(8 m) deep and 8.8 ft (2.7 m) wide, which is similar to the corresponding
Polcevera Viaduct beam. The long-span stay cable is divided into two
parts to accommodate the anchorage into each web of the transverse box
section, Fig. 4.15. The anchorage beam is prestressed with 76 cables of
sixteen 0.2 in. (5 mm) diameter wires, Fig. 4.16.[4]

FIGURE 4.11.  General view, Polcevera Creek Bridge.

FIGURE 4.12.  General construction view, Wadi Kuff Bridge. (Courtesy of Dr. Morandi.)

88

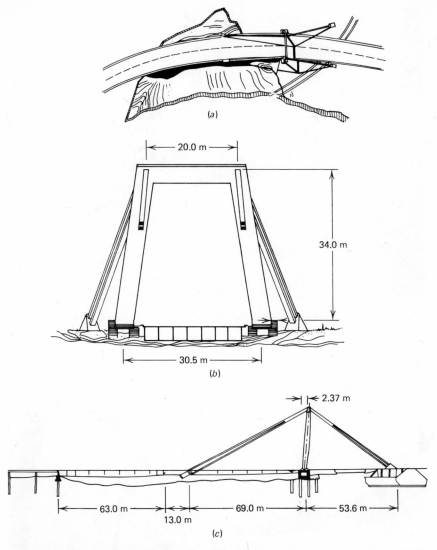

**FIGURE 4.13.** Magliana Viaduct: (*a*) plan; (*b*) transverse section, (*c*) longitudinal section. (Courtesy of L'Industria Italiana del Cemento. Ref. 4.)

FIGURE 4.14. Steel-lined hinge, Magliana Viaduct. (Courtesy of L'Industria Italiana del Cemento. Ref. 4.)

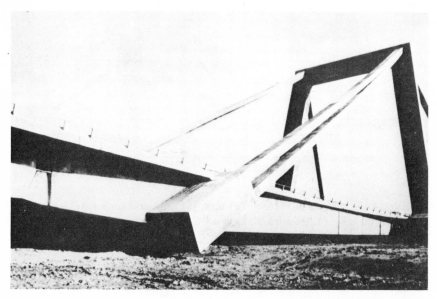

FIGURE 4.15. Fore-stay, Magliana Viaduct. (Courtesy of L'Industria Italiana del Cemento. Ref. 4.)

**FIGURE 4.16.    Prestress tendons in anchorage beam, Magliana Viaduct. (Courtesy of L'Industria Italiana del Cemento. Ref. 4.)**

The cable stays consist of parallel prestressing wires, which were considered to be an innovation at the time of construction of this bridge. Today, parallel wire cables are used even for the large conventional suspension bridges. The intersection of the cable stays and the transverse prestressing tendons at the transverse anchorage beam requires careful attention to details and dimensions in order to accomplish the anchorages adequately, Fig. 4.17.

## 4.6    DANISH GREAT BELT BRIDGE, DENMARK

The competition for a suitable bridge design in Denmark produced many new concepts and architectural styles. The design requirements specified three lanes for vehicular traffic in each direction and a single railway line in each direction. The rail traffic was based on speeds of 100 mph (161 km/hr).[5] Navigational requirements stipulated that the bridge deck be 220 ft (67 m) above water level and the clear width of the channel was to be 1130 ft (345 m).

The firm of White, Young and Partners submitted a design, Fig. 4.18, embodying the principles of a cable-stayed bridge combined with conven-

**FIGURE 4.17.   Juncture of stay with anchorage beam, Magliana Viaduct. (Courtesy of L'Industria Italiana del Cemento. Ref. 4.)**

tional approaches of girders and piers with normal spans. Although the bridge design was not selected as the competition prize winner, it did receive a third place award.

The principal feature of the bridge design is the three-plane longitudinal alignment of the cable stays. The deck consists of two parallel single cell prestressed concrete box girders, Fig. 4.19. The rail traffic is supported within the box on the bottom flange and the roadway is of conventional

FIGURE 4.18.   Artist's rendering, Danish Great Belt Bridge. (Courtesy of White, Young & Partners.)

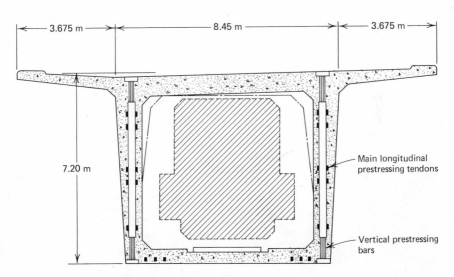

FIGURE 4.19.   Section through deck beam at expansion and construction joint, Danish Great Belt Bridge. (Ref. 5.)

design on the top of the 27.75 ft (8.45 m) box and its cantilever flanges of 12 ft (3.675 m). The total depth of the box is 23.5 ft (7.20 m).[5]

The piers and towers are designed to be cast in place to support the deck units which are to be precast at various locations on shore and floated to the bridge site to be erected into position. The maximum weight of a single box unit is estimated to be 2200 tons. All units of the superstructure are of reinforced and prestressed concrete.

## 4.7   RIVER FOYLE BRIDGE, IRELAND

The proposed bridge, Fig. 4.20, was planned to cross the River Foyle at Madam's Bank near Londonderry, Northern Ireland. The concept for the bridge consists of dual three-lane roadways with a centrally located walkway.

The cable system is a single-plane arrangement located along the centerline of the superstructure which uses an inverted Y pylon and the main central deck girder for the cable-stay anchorage. The main span has a vertical clearance of 105 ft (32 m) to satisfy navigational requirements. Although the cable-stayed design was not selected for construction it does suggest a unique concept for the site. The final approved design is a self-anchored suspension bridge with a concrete deck spanning 826.7 ft (252 m). The designers, Mott, Hay and Anderson, Consulting Civil Engineers, felt that a detailed design should involve a model analysis of the cable-

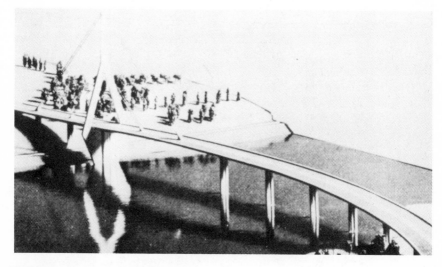

FIGURE 4.20.    Artist's rendering, River Foyle Bridge. (Courtesy of Civil Engineering and Public Works Review. (Ref. 6.)

stayed proposal but other factors recommended the adoption of the conventional suspension bridge.

A description of the proposed cable-stayed bridge and its general dimensions are discussed for conceptual ideas and suggestions for other designers and contractors to review before proceeding with their own plans.

The approach spans from the west abutment to the tower were to be one at 164 ft (50 m), and two at 229.7 ft (70 m). The center span on the east side of the tower is 689 ft (210 m) followed by six approach spans of 229.7 ft (70 m) and one at 164 ft (50 m).

The superstructure was designed as a single trapezoidal prestressed box girder with side cantilevers of constant depth and continuous over the total length of the bridge. The box girders were to be precast units, bonded together in place at the joints with an epoxy resin.

The tower was a unique design, an inverted Y-shaped pylon that was to be cast in place up to the deck level for the placement of the deck and then completed by casting the remaining A-frame portion.

The 115 ft (35 m) approach spans were to be constructed by the balanced cantilever method to meet the main span. A lower back-stay from the A-frame tower was to be anchored to an approach pier and a temporary stay cable positioned 115 ft (35 m) from the tower into the main span. The cantilever procedure was then to be continued for another 115 ft (35 m). The first permanent cable stay was to be located 229.7 ft (70 m) from the tower and the deck subsequently cantilevered another 115 ft (35 m). This erection operation was to be continued until the entire center span was cantilevered 574 ft (175 m) from the tower and all permanent cables were positioned and installed. When the box girder was completely erected, the transverse ribs and deck slabs were to be placed and tied down.

The permanent stays were to consist of parallel wire cables using conventional prestressing anchorages. The dead end anchorages were to be located at the deck attachments with the jacking anchorages at the tower. The two back-stays were to be anchored to the piers, thus increasing the rigidity of the entire bridge structure.[6]

## 4.8    INTERCONTINENTAL PEACE BRIDGE, ALASKA–SIBERIA

An independent group has proposed that a cable-stayed bridge, Fig. 4.21, be built across the Bering Straits linking Alaska and Siberia, and thus providing a transportation route that would connect five continents. The bridge would be a symbol of international peace and cooperation, because geographically the Bering straits are approximately equidistant from Hono-

**FIGURE 4.21.**    Artist's rendering, Inter-Continental Peace Bridge. (Courtesy of T. Y. Lin.)

lulu, San Francisco, Tokyo, Moscow, New York, London, Paris, and Berlin.[7] At present, the bridge is in the conceptual stage and a final design has not yet been developed.

A proposed concept is a combined highway and railway bridge 50 miles (80.5 km) in length requiring approximately 260 spans of 1000 ft (304.8 m) each. When built this bridge would be of record-breaking length. The total length is not as formidable as one would first imagine it to be. It is not an unrealistic goal to be shelved and set aside easily. The Diomede Islands divide the length approximately in half and the feasibility of using a 25 mile (40.25 km) long bridge has been demonstrated by the long bridge of Lake Pontchartrain, Louisiana. Other bridges such as the Wadi Kuf and the Danish Great Belt have also demonstrated that the technology and innovative imagination of designers and contractors can achieve spans of 1000 ft (304.8 m) or more.

It is interesting to note that the maximum water depth across the Bering Straits is 180 ft (55 m) and that the average depth is 150 ft (45.7 m) which is not an unreasonable depth for pier construction.

There have been several bridges built with piers at a depth of 160 ft (48.8 m) such as the Naragansett Bay Bridge at Newport, Rhode Island, and the Mackinac Bridge in Michigan. Other experience in deep waters has been the numerous offshore drilling platforms which have been built to a depth of 340 ft (103.6 m) with some designs now being planned for depths of 1000 ft (304.8 m).[7]

A proposed conceptual single-plane cable-stayed design envisions the piers to be prefabricated in a unit along the shore, floated into position and sunk into their permanent locations. Borings have indicated that there is a 20 ft (6.1 m) sediment cover over bedrock and as a consequence the piers would have to be anchored to the rock by prestressing cables. Because the piers will be subject to ice pressures from all directions, they are designed to be circular in cross section to resist compressive forces, and their curving slope at the water line will deflect and break the ice floes as they push forward and upward along the curved surface of the pier.

The proposed superstructure will be built as a prefabricated assembly of 2000 ft (610 m) units which will comprise two spans of the bridge. The unit will be transported by barges which will support the 30,000 ton (27,215 metric tons) assemblage at the quarter points.[8]

## 4.9   PASCO-KENNEWICK INTERCITY BRIDGE, U.S.A.

The first cable-stayed bridge with a concrete superstructure to be built in the United States may be the proposed Pasco-Kennewick Intercity Bridge crossing the Columbia River in Washington, Fig. 4.22. The bridge is de-

**FIGURE 4.22.**   Artist's rendering, Pasco-Kennewick Intercity Bridge. (Courtesy of Arvid Grant.)

signed by the engineering firm of Arvid Grant and Associates of Olympia, Washington, in professional collaboration with Leonhardt and Andra, Stuttgart, Germany.

The final design of this structure evolved from the configuration indicated in alternate 2, Fig. 3.23, to that indicated in Fig. 9.80 whereby the overall length of the superstructure increased by 4 ft (1,22 m) to 2488 ft (758.3 m), the center span increased by 1 ft (0.3 m) to 981 ft (299 m), the cable-stayed supported end spans decreased by 2 ft (0.6 m) to 406 ft 6 in. (123.9 m). The Pasco approach was changed to one span of 126 ft (38.4 m) while the Kennewick approach was changed to three spans at 148 ft (45.1 m) and one span at 124 ft (37.8 m). The total deck width was decreased to 79 ft 10 in. (24.3 m).

The deck is continuous without joints from abutment to abutment, being fixed at the Pasco abutment and with an expansion joint at the Kennewick abutment. The deck structure receives no vertical support at the pylons other than that from the cable-stays. The deck is supported at the pylon for lateral loads.

The piers at the ends of the 1794 ft (546.8 m) cable-supported section are connected to the deck structure in a manner whereby the piers take the vertical component of back-stay cable force but do not accept the horizontal component transmitted directly to the deck structure.

The approach spans are posttensioned, four cell box girders that are cast in place from the abutments to a point cantilevering 52 ft 6 in. (16 m) into the cable-supported end spans. A 60 ft (18.3 m) length, 30 ft (9.14 m) on each side of the pylons, is cast in place on falsework at each pylon and temporarily fixed to the pylons. The balance of the deck structure is composed of precast segmental deck units. The precast deck unit is cast as a roadway width and 27 ft (8.2 m) in length and consists of two triangular boxes at each cable plane connected by posttensioned transverse ribs at 9 ft (2.74 m) centers. Joints between precast elements are match cast.

The stays consist of a varying number of parallel $\frac{1}{4}$ in. (0.635 cm) diameter, ASTM A421 high-tensile wire, button-headed at each end. Stays are anchored at the pylon top and at the girder. The stress range in the cables is small as the dead load is approximately 90% of the total load, producing a maximum stress range of approximately 10 ksi (703 kg/cm$^2$). For corrosion protection the stays are encased in a polyethylene pipe, which is then filled with concrete grout similar to a posttensioned tendon. The grout is installed after final adjustment and installation of the roadway wearing surface.

The pylon is a concrete portal frame of cellular construction. The legs are rectangular, tapering on three sides with the inside face (adjacent to the roadway) being vertical.

## 4.10   CHACO/CORRIENTES BRIDGE, ARGENTINA
### (General Manuel Belgrano Bridge)

The Chaco/Corrientes Bridge (also referred to as the General Manuel Belgrano Bridge) crosses the Paraná River between the provinces of Chaco and Corrientes in northeast Argentina and represents an important link in one of the highways between Brazil and Argentina, Fig. 4.23. It has a center navigation span of 803 ft 10 in. (245 m), side spans of 537 ft (163.7 m), and a number of 271 ft (82.6 m) approach spans on both the Chaco and Corrientes sides of the river. The vertical clearance in the main span above flood level is 115 ft (35 m).[9,10]

In appearance the structure very much resembles the Lake Maracaibo Bridge. However, the similarity begins and ends with its appearance. Although the structure has the same dominating portal A-frame pylon, it does not have the X-frame supporting the deck structure in the vicinity of the pylon. Instead, inclined struts are used from the base of the pylon legs to the underside of the deck, Fig. 4.24). Although the pier cap section of the deck (between inclined struts) is cast in place, the cantilever girder is precast segmentally and posttensioned (the Maracaibo Bridge was cast in

FIGURE 4.23.   General view, Chaco/Corrientes Bridge. (Courtesy of Normer Gray. Ref. 9.)

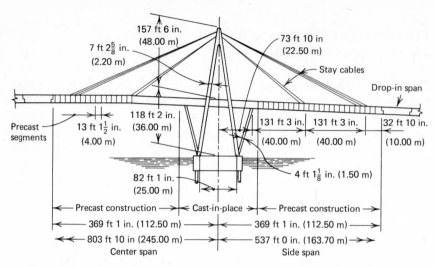

**FIGURE 4.24.** Longitudinal geometry, Chaco/Corrientes Bridge. (Courtesy of Civil Engineering, ASCE. Ref. 10.)

place). The drop-in spans are cast in place as opposed to precast sections in the Maracaibo Bridge. Further, in elevation, this structure has two stays (in each plane) radiating from each side of the pylon as opposed to one in the Maracaibo Bridge.

The deck structure consists of two longitudinal hollow boxes 8 ft 2½ in. (2.5 m) in width and with a constant depth of 16 ft 6 in. (3.5 m), which support precast roadway deck elements, Fig. 4.25. The precast girder ele-

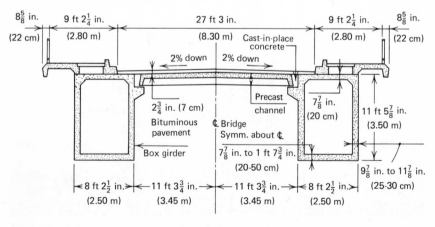

**FIGURE 4.25.** Deck cross section, Chaco/Corrientes Bridge. (Courtesy of Civil Engineering, ASCE. Ref. 10.)

ments were match cast on the river bank in lengths of 13 ft. $1\frac{1}{2}$ in.
(4.0 m), with the exception of shorter units at the point of stay attachment, which contain an inclined transverse anchorage beam. Units were cast
by the long-line method on a concrete foundation with the proper camber
built in. Each unit was cast with three alignment keys, one in each web
and one in the top flange. The units were erected as balanced cantilevers
with respect to the pylon to minimize erection stresses. After a unit was
hoisted, an epoxy joint material was placed over all of the butting area
and then the unit was placed against the already erected unit and tensioned.[9]

Because of the length, weight, and acute angle of inclination of the
stays, a saddle at the top of the pylon was impractical; the stays were
therefore anchored at the top of the pylon, Fig. 9.78.[10]

## 4.11 MAINBRÜCKE BRIDGE, GERMANY

The Mainbrücke Bridge is a prestressed concrete cable-stayed bridge that
connects the Fabwerke Hoechst's chemical industrial complex on both
sides of the River Main in West Germany, Fig. 4.26. The total length of
structure is 984.3 ft (300 m) with a river span of 485.6 ft (148 m). It
carries two three-lane roads separated by a railway track and pipe lines.

The railroad track and pipe lines are in the median between the two

FIGURE 4.26. General view, Mainbrücke Bridge. (Courtesy of Richard Heinen.)

cantilever towers and are supported on an 8.5 ft (2.6 m) deep, torsionally stiff box girder. The centerline of the longitudinal webs of the box girder coincide with the centerline of the individual cantilever pylons and are 26.25 ft (8 m) apart. Transverse cross beams at 9.8 ft (3 m) centers form diaphragms for the box and cantilevers, which extend 39 ft (11.95 m) on one side and 36 ft (11 m) on the other side of the central box to support the two roadways, Fig. 4.27. The end of the transverse cross beams are connected by secondary longitudinal beams, which improve the load distribution of concentrated loads, Fig. 4.27.

Figure 4.28 shows a view of the partially completed structure and Fig. 4.29 shows a view looking down the stays at the deck. Each stay is composed of 25-$\frac{5}{8}$ in. (16 mm) diameter Dywidag bars encased in a metal sheath, which is grouted for corrosion protection similar to posttensioned concrete construction.

**FIGURE 4.27. View of deck at pylon, Mainbrücke Bridge. (Courtesy of Richard Heinen.)**

This structure is believed to be the world's longest prestressed concrete bridge carrying railroad traffic. The completed structure is shown in Fig. 4.30.

**FIGURE 4.28. Partially completed structure, Mainbrücke Bridge. (Courtesy of Richard Heinen.)**

FIGURE 4.29. View of deck from pylon top, Mainbrücke
Bridge. (Courtesy of Richard Heinen.)

**FIGURE 4.30.** Completed structure, Mainbrücke Bridge. (Courtesy of Richard Heinen.)

## REFERENCES

1. Anon., "The Bridge Spanning Lake Maracaibo in Venezuela," Bauverlag GmbH., Wiesbaden, Berlin, 1963.
2. Morandi, R., "Some Types of Tied Bridges in Prestressed Concrete," First International Symposium, Concrete Bridge Design, ACI Publication SP-23, Paper SP 23-25, American Concrete Institute, Detroit 1969.
3. Anon., "Longest Concrete Cable-Stayed Span Cantilevered over Tough Terrain," *Engineering News-Record*, July 15, 1971.
4. Morandi, R., "Il Viadotto dell Ansa della Magliana per la Autostrada Roma—Aeroporto di Fiumicino," *L'Industria Italiana del Cemento* (Rome), No. 38, March 1968.
5. Anon., "Morandi—Style Design Allows Constant Suspended Spans," *Consulting Engineer*, March 1967.
6. Anon., "River Foyle Bridge," *Civil Engineering and Public Works Review* (London), Vol. 66, No. 780, July 1971.
7. Anon., "Intercontinental Peace Bridge," ICPB, Inc., San Francisco, 1970.
8. Lin, T. Y., Kulka, F. and Yang, Y. C., "Basic Design Concepts for Long Span Structures," Meeting Preprint 1727, ASCE National Structural Engineering Meeting, Cleveland, Ohio, April 24–28, 1972.
9. Gray, N., "Chaco/Corrientes Bridge in Argentina," *Municipal Engineers Journal*, Paper No. 380, Vol. 59, Fourth Quarter, 1973.
10. Rothman, H. B. and Chang, F. K., "Longest Precast-Concrete Box-Girder Bridge in Western Hemisphere," *Civil Engineering*, ASCE, March 1974.

# 5

# Steel Superstructures

## 5.1 INTRODUCTION

The previous chapter discussed cable-stayed bridges constructed of cast-in-place or precast concrete with prestressed elements in some cases. For the most part, however, cable-stayed bridges have been of steel construction.

The first modern cable-stayed bridge of steel construction was the Strömsund Bridge in Sweden, constructed in 1955. This structure had two principal plate girders in the longitudinal direction, a double-plane vertical cable arrangement transversely, and a radiating configuration in elevation.

The pylon was of the portal frame type. From this beginning a multitude of concepts have evolved that have utilized orthotropic decks; twin box girders; single, torsionally rigid spine box girders; single-plane cable arrangement transversely with various configurations in elevation. The single cantilever and A-frame pylon also evolved.

The following sections contain a selection of steel cable-stayed bridges presented in chronological order to indicate the constant evolution of geometric concepts.

## 5.2   STRÖMSUND BRIDGE, SWEDEN

As mentioned in Chapter 1, the Strömsund Bridge, Fig. 5.1, is the first modern implementation of the concept of supporting a bridge deck with inclined cable stays. For its time (1955), this structure is a monumental achievement in the ingenuity of its design. Transversely the cable stays are in two vertical planes while in elevation the stays are of the radiating or converging configuration. Each stay consists of four locked coil strands (see Chapter 6) that anchor into the pylon head and into transverse anchorage box beams between the main girders, Figs. 5.2 and 5.3 (see Chapter 7). Jacking of the stays and adjustment is accomplished at the deck level.

The pylons are portal frames with inclined legs, Fig. 9.27. The portal frame is independent of the two main longitudinal plate girders and is supported at its base by rocker bearings. These bearings are oriented such that they provide a rotation or hinge action in the longitudinal direction of the structure but provide a rigidity or fixity in a transverse direction to the bridge. The cable stays provide a restraint, in the longitudinal direc-

FIGURE 5.1.   General view, Strömsund Bridge. (Courtesy of Der Stahlbau. Ref. 2.)

**FIGURE 5.2.    Pylon head, Strömsund Bridge. (Courtesy of Der Stahlbau. Ref. 2.)**

tion, at the top of the pylon. In this manner a pendulum movement of the pylon is permitted in the longitudinal direction.[1,2]

## 5.3    THEODOR HEUSS BRIDGE, GERMANY

Theodor Heuss Bridge, also known as the North Bridge over the Rhine at Düsseldorf, was completed in 1958 and was the first long-span, cable-stayed bridge built in Germany, Fig. 2.6. It has a main span of 853 ft and side spans of 354.3 ft, Fig. 5.4. The stays are arranged in two vertical planes transversely and are of the harp configuration in elevation. The three parallel harp stays attach at the third points of the four vertical

**FIGURE 5.3.    Anchorage box beam, Strömsund Bridge. (Courtesy of Der Stahlbau. Ref. 2.)**

pylons. Stays are supported on saddles and thus continuous through the pylons. Saddles for the center stays are fixed, while the upper and lower saddles are supported on movable bearings.

Pylons are of the single cantilever type and rise about 131 ft above the roadway. These towers are fixed to the stiffening girder at the base, forming an inverted portal frame.

In cross section, the width of the deck is 88.9 ft and consists of two box girders with an orthotropic deck spanning them, Fig. 5.4. Dimensions of the box girder are approximately 10.5 ft in depth and 5.25 ft in width, which produces a ratio of depth to center span of 1:81. The walkways are of reinforced concrete and cantilever out from the main box girders.

The top plate of the orthotropic deck has a minimum thickness of 0.55 in. Welded to the top surface of the plate are flat bars (1.1 in. by 0.24 in.) set on edge and spaced 6 in. on center in a herringbone pattern. The purpose of these bars is to hold the 2 in. asphalt wearing surface in position. Tests using alternating loads indicated that there was no hazard from fatigue, since the bars were welded to the deck plate.

One of the considerations in the planning of the structure was a requirement that the center span would be erected by the cantilever method without falsework. The harp type cable-stay configuration permitted the cables

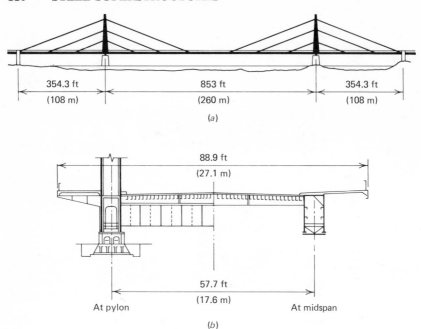

**FIGURE 5.4. Theodor Heuss Brücke, Düsseldorf, Germany. (*a*) elevation; (*b*) cross section.**

to act as guy-ropes during erection. A design requirement was that the cable thrust be proportionately distributed over the entire cross section. If the box girders had been cantilevered out and attached to the stays first, and the deck plates added later, the girders would have had to absorb individually part of the axial load, and the stress distribution over the entire cross section would not have been proportional. The deck units, box girders, and deck plate were preassembled in 118 ft long sections, floated into position, and attached to the stays. Thus the requirement for proportional distribution of stress across the cross section was realized.[3]

## 5.4   SEVERIN BRIDGE, GERMANY

The Severin Bridge in Cologne, Figs. 1.1 and 5.5, represents the first of the asymmetric cable-stayed structures; it has spans of 987.5 and 495.4 ft. This structure was a second prize winner in a design competition. This particular structural design was adopted over a first prize girder bridge because it did not require a river pier in the vicinity of the left bank as

FIGURE 5.5.  Overall view, Severin Bridge.

FIGURE 5.6.  A-frame pylon, Severin Bridge.

111

did the girder concept. Navigation requirements in the Rhine River dictated the elimination of such a pier. The two most striking features of the structure are its asymmetry and the unusual (at this time) triangular pylon.

The choice of a triangular A-shaped pylon was based on aesthetic as well as structural considerations. From the structural point of view, the A-frame is more stable than the portal frame or individual cantilever type of pylons. Stiffness in the face of horizontal forces such as wind is considerably improved by the inclination of the legs. With the A-frame, the horizontal portal member can be eliminated, and a weight savings in the pylons can be effected. The cable stays converge to a single point on the apex of the tower, Fig. 5.6, which is advantageous to the three-dimensional space frame structural concept. Further, there are no visual intersections of the stays, which might be distracting. The legs of the tower are splayed such that they clear the deck structure. Finally, the A-frame pylon harmonizes with the tall spires of the adjacent cathedral. Aesthetic considerations precluded a tall tower in the vicinity of the spires. Thus, the adoption of the asymmetric concept of one pylon on the right bank acting as a counterpoint to the cathedral and the old city on the left bank. The giant A-frame pylon rises to 252.6 ft above its fixed base on the pier and 200.8 ft above the deck, and serves as a monument to the medieval bishop after whom this structure is named. The geometrical concept of this structure fulfills its utilitarian purpose and at the same time achieves a harmony with its surroundings.[4]

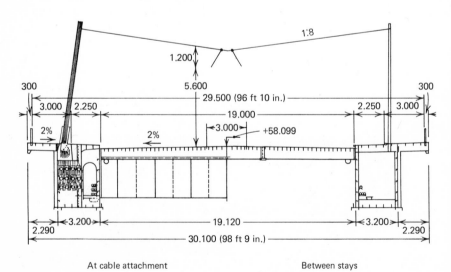

At cable attachment                    Between stays

**FIGURE 5.7.    Cross section, Severin Bridge. (Courtesy of Acier-Stahl-Steel. Ref. 4.)**

The deck consists of twin box girders with an orthotropic deck and carries four lanes of traffic, a tramway, bicycle paths, and walkways. The cable stays attach to the box girders almost at the dividing line between the bicycle path and the walkway, which cantilevers out from the box girder, Fig. 5.7. In cross section all components of the deck are structurally active with the exception that an effective width for analysis had to be determined for the orthotropic deck with respect to the box girder. The box girders are 10.5 ft wide and vary in depth from 9 ft $10\frac{1}{2}$ in. at the ends of the bridge to 15 ft in the largest span, producing a depth to span ratio of 1:66. The maximum design deflection under maximum live load conditions is approximately 4 ft 5 in. in the river span, which is a ratio of deflection to span length of 1:223.

## 5.5  NORDERELBE BRIDGE, GERMANY

The development of the torsionally rigid box girder led to the concept of a cable-stay bridge using a single central box girder, a single vertical cable plane located in the median, and individual cantilever pylons. The first such structure was the Norderelbe highway bridge over the Elbe at Hamburg, Fig. 2.8. It has spans of 210, 564, and 210 ft, as shown in Fig. 5.8.

The cross section of the deck consists of four main longitudinal girders spaced 25 ft 7 in. on center and stiffened by transverse diaphragms. An orthotropic deck acts as a top flange and connects all four girders. In the center span the two inside girders are connected by a bottom plate which produces a torsionally rigid box section. In the side spans the bottom plate is replaced by a horizontal diagonal bracing system. The girder depth is approximately 9 ft 10 in., producing a depth to span ratio of 1:57.

The pylons rise about 174 ft above the deck and about 98 ft above the top cable-stay support. As a result of the poor soil conditions and the size of the piers, the pylons were rigidly connected to the box girder in the longitudinal direction and to cross girders in the transverse direction.

Stays are attached to the pylons at about 56 and 76 ft and converge downward to the deck. The upper saddles are fixed, while the lower ones are on pendulum type rocker bearings. Structure weight is about 66 pounds per square foot.[5,6,7,8,9]

## 5.6  RHINE RIVER BRIDGE AT MAXAU, GERMANY

The first asymmetrical, single vertical plane, central torsional box girder cable-stayed structure is the Rhine Bridge at Maxau, Figs. 5.9 through 5.11. There are two spans of 575 and 383 ft, with the single pylon rising

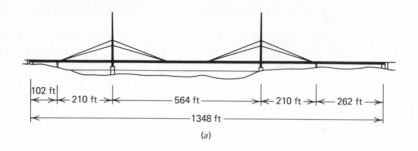

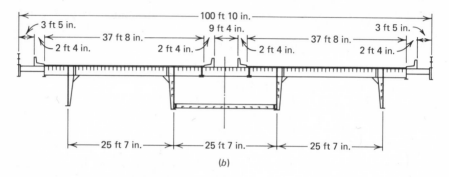

**FIGURE 5.8.** Norderelbe, Hamburg, Germany: (*a*) elevation; (*b*) cross section.

**FIGURE 5.9.** Overall view, Rhine River Bridge at Maxau. (Courtesy of Der Stahlbau. Ref. 10.)

**FIGURE 5.10.** View from top of pylon, Rhine River Bridge at Maxau. (Courtesy of Der Stahlbau. Ref. 10.)

141 ft above roadway elevation. The three stays are arranged in the fan configuration.

Stays are continuous through the pylon and are clamped to saddles within the pylon. The upper saddle is allowed to move in the longitudinal direction, while the middle and lower saddles are fixed. The pylon is restrained within the girder and has a rocker bearing on the pier.

In cross section the total deck width is 115.8 ft with the torsionally rigid spine box girder having a width of 39 ft 4 in. and an average depth of about 9 ft 9 in. (resulting from superelevation). The orthotropic deck cantilevers out from the central box 38 ft 3 in. on each side. Because of the long cantilever, longitudinal edge girders are utilized for load distribution.

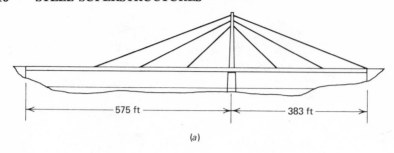

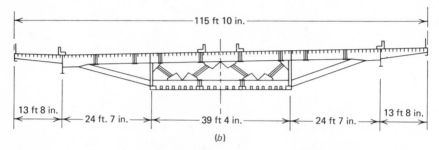

**FIGURE 5.11.  Rhine River Bridge at Maxau: (*a*) elevation; (*b*) cross section. (Courtesy of The British Constructional Steelwork Association, Ltd. Ref. 11.)**

Erection of the superstructure was from the right bank by the cantilever method (see Chapter 9) and utilized temporary piers. The pylon was erected next with the stays following, starting from the top stay. Temporary catwalks were used in the stay erection. The stays were tensioned by lowering the superstructure at the temporary piers and abutments and by jacking the saddle supports.[10,11]

## 5.7   WYE RIVER BRIDGE, GREAT BRITAIN

The cable-stayed structure crossing the Wye River at Chepstow, Fig. 5.12, has received little attention in the literature, perhaps because it is dominated by the nearby Severn Suspension Bridge. The structure has a main span of 770 ft, side spans of 285 ft, and approach or viaduct spans on either side ranging from 182 to 210 ft, Fig. 5.13. In a transverse direction it utilizes a single vertical plane in the median and consists of a single stay emanating from each side of the pylon. Each stay consists of 20 spiral strands built up into a triangular form.

Pylons are steel box columns that rise 96 ft above the roadway and are located in the median. The pylon weight and vertical component of stay

**FIGURE 5.12.    Wye Bridge with Severn Bridge in background. (Courtesy of The British Constructional Steelwork Association, Ltd.)**

force is taken by a hinged bearing below the deck. The load is then transmitted to a heavy steel diaphragm which in turn transmits it to the portal pier below. The pylons are hinged at their base in the longitudinal direction of the bridge. Cable-stays are rigidly attached to the pylon top.

The design utilizes a single trapezoidal box girder with projecting cantilevers on each side for an overall width of 100 ft 6 in., Figs. 5.13 and 5.14. The box is 55 ft wide and 10 ft 6 in. deep. The bottom flange width of the box is 32 ft and the flange is supported on portal frame piers. Erection was by the cantilever method. The deck section, consisting of the steel box and two side cantilevers, was assembled at the site from fabricated plate and was then transported over the completed portion of the deck to the cantilever end. Erection was from both sides to the middle. In general, the assembled units were 56 ft in length and weighed 120 tons.

The average weight of the structure is 2 tons per foot, or about 40 pounds per square foot. With respect to the center span the depth-to-span ratio is 1:73.

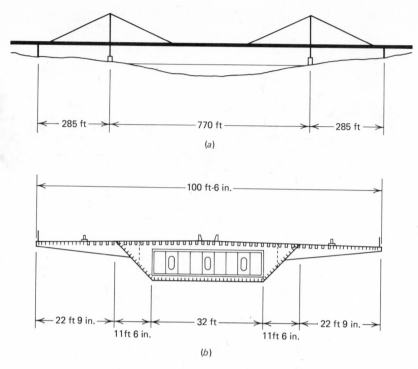

**FIGURE 5.13.** Wye Bridge: (*a*) elevation; (*b*) cross section. (Courtesy of The British Constructional Steelwork Association, Ltd.)

## 5.8  RHINE RIVER BRIDGE AT REES, GERMANY

The cable-stay structures illustrated thus far have had relatively few cable-stays and the girder might be considered as continuous, since it is elastically supported at the distinct points of cable anchorages, which are spaced relatively far apart. The Rees Bridge, Figs. 2.9 and 5.15 (and the following two examples) are what might be termed multicable systems. In contrast to the previous structures, these systems have numerous closely spaced cables, thus providing a continuous elastic support to the girder which approximates a girder supported on a continuous elastic foundation. This system evolved from a desire for a simpler transmission of forces from the cables to the girder. The structural advantage is that the relatively smaller cable force component at the attachment is distributed over a greater length of bridge girder.

The Rees Bridge employs two vertical cable planes in a harp configuration with four individual cantilever pylons. The center span is 837 ft with

**FIGURE 5.14.**    View of underside of girder, Wye Bridge. (Courtesy of The British Constructional Steelwork Association, Ltd.)

equal side spans of 341 ft. The pylons rise 136 ft above the roadway deck and are rigidly fixed to the deck structure.

The deck superstructure consists of two main plate girders 11 ft 6 in. in depth, producing a depth to span ratio of 1:73 of the main span, Fig. 5.16. The main girders support I-section transverse ribs at 10 ft 6 in. spacing and the orthotropic deck plate.

Each vertical plane has 10 cable stays on each side of the pylon, Fig. 5.17. The stays are spaced 8.2 ft apart at the pylon and 21 ft apart at the girder. The cable plane is approximately 2 ft 6 in. outside the web of the main girder and the stays are attached to cross girders.[12]

FIGURE 5.15.   General view, Rees Bridge. (Courtesy of Beratungsstelle für Stahl-verwendung, H. Odenhousen.)

FIGURE 5.16.   Erection view showing plate girders, Rees Bridge. (Courtesy of Wolfgang Borelly.)

FIGURE 5.17. View of cable-stays, deck, and pylon, Rees Bridge. (Courtesy of Wolfgang Borelly.)

## 5.9 FRIEDRICH-EBERT BRIDGE, GERMANY

This structure, Figs. 2.7 and 2.9, crosses the Rhine River at Bonn, Germany, and is sometimes referred to as the Bonn-Nord Bridge. As mentioned in the previous example, the concept of suspending the girder at a large number of points has been extended in this structure. A single central vertical stay plane is used in this structure with 20 cables on each side of the pylon. The cables are spaced at 3 ft $3\frac{1}{2}$ in. apart along the pylon and 14 ft 9 in. apart along the girder, Fig. 5.18. The stays form a fan configuration in elevation. Because of the small size of the stays, varying in diameter from approximately 3 to $4\frac{3}{4}$ in., their appearance is not obtrusive and an overall appearance of lightness is achieved.

The structure has a center span of 918.6 ft and side spans of 393.7 ft. Pylons are 161 ft high above the bridge deck; they penetrate the girder and are independent of the girder. Fixity in both directions at the pier top is achieved with prestressed anchor rods.

The girder is a single cell box girder, Fig. 5.19, similar to that of the Maxau Bridge, Fig. 5.11. The cross-sectional dimensions of the torsionally stiff box girder are 41 ft 4 in. wide by 13 ft 10 in. deep. Depth of girder

**FIGURE 5.18.** View of cable-stay plane, Friedrich-Ebert Bridge. (Courtesy of Beratungsstelle für Stahlverwendung, H. Odenhousen.)

**FIGURE 5.19.** Erection view of box girder, Friedrich-Ebert Bridge. (Courtesy of Wolfgang Borelly.)

to main span ratio is, therefore, 1:67. The orthotropic deck has transverse ribs spaced at approximately 7 ft 6 in. on center. Total width of the deck is 118 ft, with the 38 ft cantilever overhangs supported by inclined struts from the bottom of the box.

Side spans of the structure were erected on temporary piers and the center span was erected by the free cantilever method. Cable stays were attached and tensioned as the erection proceeded.[12]

## 5.10 ELEVATED HIGHWAY BRIDGE AT LUDWIGSHAFEN, GERMANY

In 1968 at Ludwigshafen, Germany, an old dead-end railway station was converted into a modern through station. Road and rail are now routed at five different levels, the highest of which is formed by a cable-stayed girder bridge with a converging cable arrangement, Fig. 5.20. This structure extends the multistay concept to two inclined planes.

The single pylon is unusual in that it is composed of four legs that form an A-frame on all four sides, Fig. 5.21. The pylon rises 246 ft above the railway lines crossed by the bridge and supports two equal spans of

**FIGURE 5.20.   Aerial view, Ludwigshafen Bridge. (Courtesy of *Der Stahlbau.* Ref. 13.)**

approximately 453 ft, Fig. 2.9. The stays converge to the apex of the pylon and are attached to a rhomboid box anchorage, Fig. 5.22.

The orthotropic deck is 80 ft wide and is supported on two longitudinal web girders spaced 54 ft 9 in. apart, Fig. 5.23. Longitudinal plate girders are 8 ft 2 in. in depth corresponding to a depth to span ratio of 1:55.[11,12,13]

## 5.11   ONOMICHI BRIDGE, JAPAN

Although the Onomichi Bridge is not the first cable-stay bridge in Japan, it is the first long-span cable-stayed bridge and also the first in the world to be constructed almost entirely (82%) of corrosion-resistant weathering steel. This structure, Fig. 5.24, has an overall length of 1263 ft and connects the island of Mukai-Jima to the island of Honshu. The center span is 705.4 ft and side spans are 278.8 ft.

Geometrically it consists of two transverse vertical planes which are of the radiating configuration in elevation. The pylons are of the portal frame type. The superstructure consists of an orthotropic plate deck spanning two longitudinal plate girders that are spaced 33 ft 6 in. on center. The plate girders are 10 ft 6 in. deep producing a depth to span ratio of 1:67.

**FIGURE 5.21. Pylon configuration, Ludwigshafen Bridge. (Courtesy of *Der Stahlbau*. Ref. 13.)**

Pylons are mounted on high piers to provide adequate navigation clearance. The top stays are fixed to the pylons, while the lower stays allow movement in the longitudinal direction of the structure. Stay anchorage cross girders span the two longitudinal plate girders. Erection of the center span was by the free cantilever method.[14]

## 5.12  DUISBURG-NEUENKAMP BRIDGE, GERMANY

Duisburg-Neuenkamp is the first major bridge structure in Europe to use all-welded construction with various combinations of U.S. developed high-

**FIGURE 5.22.  Stay anchorage at pylon, Ludwigshafen Bridge. (Courtesy of *Der Stahlbau*. Ref. 13.)**

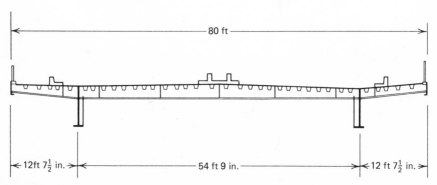

**FIGURE 5.23.  Cross section, Ludwigshafen Bridge. (Courtesy of The British Constructional Steelwork Association, Ltd. Ref. 11.)**

FIGURE 5.24.    Aerial view, Onomichi Bridge.

FIGURE 5.25.    General view, Duisburg-Neuenkamp Bridge. (Courtesy of Gutehoff-nungshüte Sterkrade AG, Friedrich Weisskopf.)

strength steels in the pylons, Figs. 1.2 and 5.25. The bridge has a center span of 1148 ft and an overall length of 2550 ft. Intermediate piers are used in the side spans between the pylon and the abutment to anchor all but one of the side span back-stays. In this manner increased stiffness is provided to the structure.

Cable stays are arranged in a fan configuration and are positioned in a single vertical plane in the median. Each stay is composed of nine locked coil strands in a cross-section grid of 3 by 3. The individual locked coil strands vary in diameter (depending on the force in a particular stay) from $2\frac{1}{4}$ to $3\frac{1}{4}$ in. At intervals of about 56 ft, the stays are clamped by square clamps fabricated from approximately $\frac{3}{8}$ in. plate. The stays may be adjusted at the girder anchorage and are continuous through the pylon, where they are supported on saddles. At each pylon the locked coil strands forming a stay are supported in layers of three in cast steel saddles that have a lead-lined bottom and sides. A lead-lined cover was clamped down on the strand by high-strength bolts, Fig. 5.26. All saddles are rigidly fixed to the pylons.

Individual full-length strands arrived at the construction site fitted with their end sockets and coiled on a drum. They were then fitted to their respective saddles and erected on the pylon. With the use of electric

**FIGURE 5.26.    Cable strands in saddle, Duisburg-Neuenkamp Bridge. (Courtesy of Wolfgang Borelly.)**

winches, the socketed ends were pulled into the box girders and installed in the cable anchorages. One strand in each stay was tensioned to the correct tension and then served as a indicator for the tensioning of the remaining eight strands in the stay. Final tensioning was accomplished by raising or lowering the saddles in the pylons by hydraulic jacks.

The two pylons are 157 ft high and are fixed to the girder. They are a constant dimension of 6 ft 3 in. in the transverse direction of the bridge. In a longitudinal direction they are of a variable dimension. At the top they have a dimension of 10 ft and then taper down to a 6 ft 5 in. dimension at 33 ft above the deck and then increase to an 8 ft dimension at the base. The slender pylons have an aesthetically pleasing appearance. The slenderness was achieved by the use of U.S. licensed high-strength, water-quenched and tempered, fine-grained structural steel. Plate thicknesses are less than $1\frac{3}{4}$ in. The slenderness of the pylons allows a narrow median which in turn reduces the dead weight of the structure. The tower design in this example allows a degree of flexibility such that the bending moment in the towers is reduced considerably.

Dual three-lane roadways, a footpath, and a bicycle path are carried by a 119 ft wide deck, Fig. 5.27. The center two cell box girder is 41 ft 8 in. wide with constant depth webs of 12 ft 4 in., producing a depth to span ratio of 1:93. The deck cantilevers from the sides of the box a distance of 38 ft 8 in. and is supported by box struts spaced at 16 ft 5 in. Cross girder spacing at the top of the box is 8 ft $2\frac{1}{2}$ in. In the bottom, the cross girders

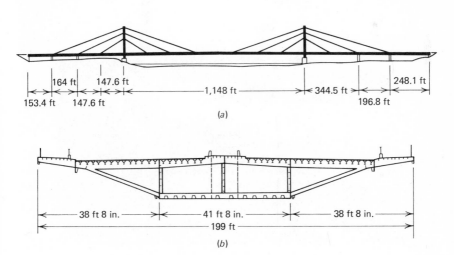

FIGURE 5.27. Duisburg-Neuenkamp Bridge: (*a*) elevation; (*b*) cross section. (Courtesy of Gutehoffnungshütte Sterkrade AG, Friedrich Weisskopf.)

are spaced at 16 ft 5 in., except in those regions subject to high compressive forces additional intermediate cross girders are added.

The orthotropic deck is stiffened by "wine glass" or Y-shaped stringers spaced 1 ft $11\frac{1}{2}$ in. on center. The top of the Y is $11\frac{3}{4}$ in. wide and the space between stringers is $11\frac{3}{4}$ in. The rolled T-section, which forms the stem of the "Y," penetrates the cross girder, and the plates forming the "V" at the top are terminated at the cross girder where they are fillet welded. The bottom flange of the box is stiffened by cold-formed trapezoidal stiffeners.

The torsional rigidity of the two cell box girder is enhanced by solid diaphragms (with the exception of two inspection walkway openings)

**FIGURE 5.28.   Cantilever erection, Duisburg-Neuenkamp Bridge. (Courtesy of Gutehoffnungshütte Sterkrade AG, Friedrich Weisskopf.)**

spaced at intervals varying from 82 to 98 ft. At the two ends of the bridge the diaphragms include the triangular area formed by the inclined struts.

Erection was from both abutments toward the center and employed the cantilever erection method, Fig. 5.28. Auxiliary supports were used in the side spans.[15]

## 5.13   KNIEBRÜCKE BRIDGE, GERMANY

The Kniebrücke at Düsseldorf has an asymmetric harp type cable-stay configuration with a major span of 1050 ft and a minor span of 800 ft, Figs. 1.1, 3.3, and 5.29. The fore-stay harp configuration is in two vertical planes that, along with the pylons, are outside the width of the deck.

The stays consist of 13 locked coil strands that in cross section are arranged in three layers of 4-5-4 strands which form a horizontally elongated hexagon. The stays are anchored to cantilever diaphragms at the girder. In the short span at the points of attachment of the stays, the girder is attached by a linkage to the piers such that the vertical component of stay force is taken by the pier in tension while the horizontal component is transferred to the girder, Fig. 9.55. In this manner the stiffness

FIGURE 5.29.   General view, Kniebrücke. (Courtesy of Beratungsstelle für Stahl-verwendung, H. Odenhousen.)

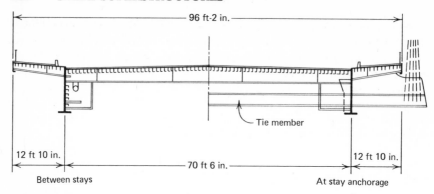

**FIGURE 5.30. Cross section, Kniebrücke. (Courtesy of Beton-Verlag GmbH, Düsseldorf. Ref. 16.)**

of the structure is not affected by the flexural stiffness of the side span and the stiffness of the longer span is enhanced. The stays pass through the pylon and are supported on saddles which are allowed a limited movement during erection, but are fixed in the final system.

The pylons are of the individual cantilever type fixed to the pier, i.e., there is no portal at the top. The pylon height above the pier is 374 ft 4 in. and is 328 ft above the roadway. In cross section the pylon is formed of box sections in a T-shaped arrangement, see Fig. 9.52.

**FIGURE 5.31. Cantilever erection, Kniebrücke. (Courtesy of Beratungsstelle für Stahlverwendung, H. Odenhousen.)**

An orthotropic deck plate spans the two longitudinal plate girders and provides three lanes of traffic in each direction (with no median). Overall deck width is 96 ft 2 in., with the webs of the plate girders spaced 70 ft 5 in. apart, Fig. 5.30. The depth of the plate girders is 9 ft 11 in. and transverse girders are spaced at 7 ft 7 in. Footpaths cantilever out from the web of the plate girder a distance of 12 ft 10 in.

The bridge was erected by the cantilever method. The river span of 1050 ft was erected without auxiliary support, Fig. 5.31.[16]

## 5.14  PAPINEAU-LEBLANC BRIDGE, CANADA

Papineau-Leblanc Bridge spans the Riviere des Prairies north of Montreal and is the longest cable-stayed bridge on the North American continent. It is also the first to use a single vertical plane with a torsionally stiff center box girder, Figs. 5.32 and 5.33. It has a center span of 790 ft and equal side spans of 295 ft. In elevation the two stays emanate from each side of the pylon in a radiating configuration.

Economic comparisons were made for the following five bridge types: (1) prestressed box girders of 275 ft span, (2) three steel arches of 446 ft span with an orthotropic deck, (3) steel plate girders with 335 ft main spans, (4) steel box girders with 260 ft main spans, and (5) the three-span cable-stayed box-girder bridge selected. Cost estimates varied from the low of $5.5 million for the cable-stayed to a high of $6.4 million for the arch. Cost for the completed structure was $5.4 million, $4.6 million for the superstructure and $0.8 million for the substructure.[17]

FIGURE 5.32.   General view, Papineau-Leblanc Bridge. (Courtesy of *Civil Engineering,* ASCE. Ref. 17.)

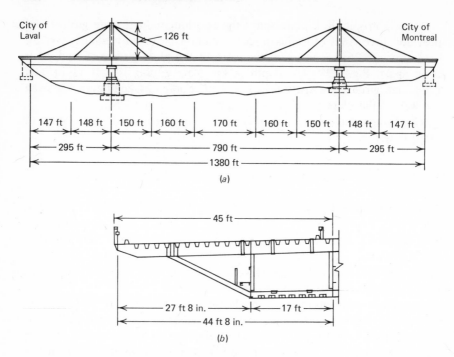

**FIGURE 5.33.   Papineau-Leblanc Bridge: (a) elevation; (b) cross section. (Courtesy of *Civil Engineering*, ASCE. Ref. 17.)**

Upper stays consist of twenty-four $2\frac{5}{8}$ in. diameter bridge strands arranged in two bundles of 12 each. Lower stays are twenty-four $1\frac{5}{8}$ in. diameter strands also arranged in two bundles of 12 each. All stays are continuous through the tower, since they are clamped to saddles. The lower saddles are fixed to the tower, and the upper saddles are allowed to slide in the longitudinal direction to minimize the longitudinal displacement of the top of the tower. Each wire in an individual strand is galvanized and, in addition, each strand has a 0.20 in. thick polyethylene covering extruded directly onto the strand. Connection of the stays to the box girder superstructure is discussed in Chapter 7.

Towers rise 126 ft above the deck and are of a single box section that tapers from 6 by 6 ft at the base to 5 by 5 ft at the top. The towers are fabricated from 2 in. thick ASTM A441 high-strength steel plate. The box section is formed from four unstiffened plates, butt welded at the edges. Each tower is rigidly fixed to the girder, which in turn is supported by sliding rotaflon bearings. Each tower weighs 140 tons and supports a load of 4500 tons.

The deck structure consists of a two cell rectangular box girder 34 ft wide and 11 ft 8 in. deep, producing a depth to span ratio of 1:68. Transverse floor beams at 15 ft spacing are 2 ft 6 in. deep and cantilever out approximately 28 ft on each side of the box. The orthotropic deck and transverse floor beams are supported by diagonal struts. At 45 ft intervals the box girder is stiffened by transverse diaphragms. The main span was erected by the free cantilever method, see Chapter 9.[17,18]

## 5.15  TOYOSATO-OHHASHI BRIDGE, JAPAN

The Toyosato-Ohhashi Bridge was built at the intersection of the Shinjo Yamatogawa Highway with the Yodo River as part of the public works for Expo 70. It is a single vertical plane fan configuration with A-frame pylons and a trapezoidal box girder, Figs. 2.13 and 5.34.

The stays in this structure represent the first use of prefabricated parallel wire strands in Japan. Upper stays consist of 16 strands of 154 wires each. Lower stays consist of 12 strands of 127 wires each. Wire diameter is approximately 0.2 in. Each strand is fabricated of parallel wires bunched in a hexagonal shape. The strands in the stays are compacted into a circular shape such that the upper and lower stays have diameters of approx-

FIGURE 5.34.   General view, Toyosato-Ohhashi Bridge.

imately 11 in. and $8\frac{5}{8}$ in., respectively. In addition to zinc coating, the stays are covered by a synthetic resin wrapping, Fig. 9.9. Stays are continuous over saddles in the pylon, Figs. 9.6 and 9.7. At the girder anchorage, the stress condition was evaluated by a three-dimensional finite element analysis and checked by model testing. Upper saddles in the pylon are fixed, while the lower ones are allowed to move.

A-frame pylons, Fig. 2.13, rise 114 ft 6 in. above the pier. They are hinged in the longitudinal direction and fixed in the transverse direction of the bridge and are designed to withstand earthquake forces.

The three-span continuous girder has spans of 264 ft, 708 ft 8 in., and 264 ft. The spans are trapezoidal box sections, 34 ft 5 in. wide at the top flange and 23 ft wide at the bottom flange. Depth is 9 ft 10 in., about 1/72 of the span. The orthotropic deck is supported by transverse cross beams at 5.9 ft on centers which cantilever out from the box 15 ft 6 in. to produce a total deck width of 65 ft 5 in.[19]

Erection of the bridge and stressing of the stays are discussed in Chapter 9.

## 5.16  ERSKINE BRIDGE, SCOTLAND

The total length of this multispan, all-welded steel box girder bridge is 4334 ft. The main cable-stay span of Erskine Bridge is 1000 ft, with two

**FIGURE 5.35.    General view, Erskine Bridge. (Courtesy of Andrew Lally, American Institute of Steel Construction.)**

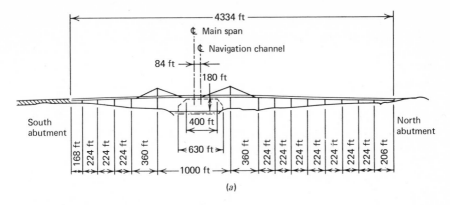

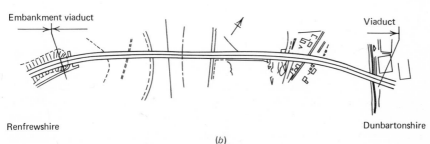

**FIGURE 5.36.** Erskine Bridge: (*a*) elevation; (*b*) plan. (Courtesy of The British Constructional Steelwork Association, Ltd. Ref. 20.)

anchor spans of 360 ft. Approach spans on the south side starting at the abutment are 168 ft and three at 224 ft. On the north side from the abutment approach spans consist of one span at 206 ft and seven at 224 ft, Figs. 5.35 and 5.36.

In cross section, the steel deck is aerodynamically similar to the Severn Suspension Bridge and the Wye Bridge discussed previously, Fig. 5.37. The depth of the cross section is 10 ft $7\frac{1}{2}$ in., producing a depth to span ratio of 1:94. The overall width of the deck for the cable-stayed portion is 102 ft 6 in. because of the narrower median requirements. The deck cantilevers approximately 20 ft 9 in. from each side of the trapezoidal box. The box is of all-welded construction and consists of a $\frac{1}{2}$ in. deck plate throughout. The $\frac{3}{8}$ in. inclined web plates are increased to $\frac{7}{16}$ in. and $\frac{1}{2}$ in. in portions of the main and anchorage spans, and a $\frac{3}{8}$ in. bottom plate increases to $\frac{1}{2}$ in. thickness at the piers and $\frac{3}{4}$ in. at the pylons. V-shaped longitudinal stiffeners for the deck are spaced 2 ft on centers, 8 in. bulb flats on 1 ft 4 in. centers stiffen the bottom flange and the lower web sections, and 5 in. bulb flats on 2 ft centers are used for the upper and central web

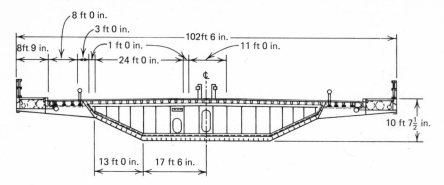

**FIGURE 5.37. Typical cross section, Erskine Bridge. (Courtesy of The British Constructional Steelwork Association, Ltd. Ref. 20.)**

sections. Transverse diaphragms on 14 ft centers are of $\frac{1}{4}$ in. plate (except at the piers, where they are increased to 1 in).

The stay system comprises a single vertical plane in the median with one stay on each side of the pylons. Stays are anchored at the girder, 330 ft on each side of the pylon and are continuous through the pylons being supported on saddles. The stay consists of twenty-four 3 in. diameter strands laid up in four layers of six strands each. Each spiral strand consists of 178 wires with diameters of 0.198 in.

The pylons are single cell boxes that taper from 5 ft 6 in. by 4 ft 6 in. at the base to 4 ft by 3 ft 6 in. at the top. The pylon cell is made up of $1\frac{1}{2}$ in. high-strength plate that is internally stiffened and filled with high-strength concrete (7500 psi). Longitudinal movement is provided at deck level by a rocker bearing. Pylon load is carried through a massive diaphragm in the box girder to bearings on the piers.

Erection was by the cantilever method and proceeded from both ends to meet in the middle. Each succeeding fabricated box is erected on trolleys on the completed portion of the deck and transported to the cantilever end, where a specially designed launching girder, Fig. 5.38, lowers it into position and holds it while it is being welded to the preceding section.[20,21]

## 5.17   BATMAN BRIDGE, AUSTRALIA

Another asymmetric structure with an inclined A-frame tower is the Batman Bridge in Tasmania, Australia, Figs. 5.39, 5.40, and 9.45. Cable stays are in  two sloping planes in a radiating configuration.

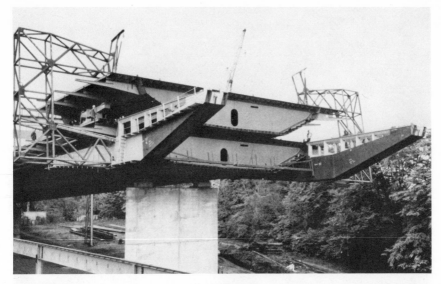

**FIGURE 5.38.    Launching girder, Erskine Bridge. (Courtesy of The British Contructional Steelwork Association, Ltd. Ref. 20.)**

The deck structure is of the conventional stiffening truss type used in suspension bridge structures, Fig. 5.41. The trusses are 12 ft 9 in. deep, i.e., a ratio of depth to span of 1:53, and are 29 ft 9 in. on centers. The chords of the trusses and principal bracing members are box sections with the gussets fabricated integrally with the chord members. Truss diagonals are fabricated H-sections and the verticals are made up of two T-sections that are battened between the stems. Main members are of welded construction. The deck is all steel, fully welded and orthotropic. It consists of $\frac{1}{2}$ in. thick plate stiffened by longitudinal V stiffeners formed from $\frac{5}{16}$ in. plates that span between cross beams at 11 ft 3 in. on centers. These cross beams bear on rubber pads placed on top of the top chord of the trusses.

The A-frame is of box construction with the plates varying from $\frac{1}{2}$ in. to 2 in. in thickness. Legs of the A-frame taper from 13 ft by 7 ft 6 in. to 6 by 7 ft. Plates of the box are stiffened on the inside both vertically and horizontally. The pylon is inclined 20 degrees toward the center, Fig. 9.50.

Stays are made up of $2\frac{3}{8}$ in. diameter locked coil strand. Each strand is cut to length and socketed in the shop. The back-stay in each plane is made up of 16 strands in four layers of four each, the inside fore-stay is 2 strands in one layer, and the center and outside fore-stays consist of 4 strands in two layers of two each.[22,23]

**FIGURE 5.39.** View looking east, Batman Bridge. (Courtesy of Department of Public Works, Tasmania. Ref. 22.)

**FIGURE 5.40.** View looking west, Batman Bridge. (Courtesy of Department of Public Works, Tasmania. Ref. 22.)

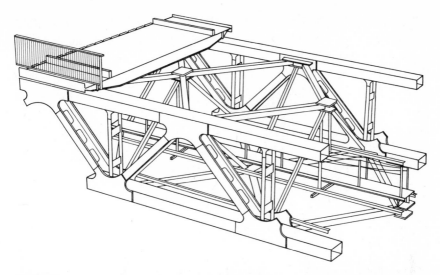

FIGURE 5.41.   Truss arrangement, Batman Bridge. (Courtesy of Department of Public Works, Tasmania. Ref. 22.)


## 5.18   BRIDGE OVER THE DANUBE AT BRATISLAVA, CZECHOSLOVAKIA

This highway bridge crossing the Danube River was opened to traffic in August 1972. It is an asymmetric structure with a major span of 994 ft and an anchor span of 246 ft, Fig. 2.2. The single A-frame pylon is unusual in that it is inclined backward and supports a coffeehouse on top, Fig. 5.42. Perhaps this concept of a restaurant at the top of the pylon may be an additional means of financing bridge structures.

The supporting stay system consists of six pairs of stays, three pairs of fore-stays and three pairs of back-stays. In elevation, the fore-stays are of a radiating configuration while the back-stays are all in the same position anchoring to the abutment, Fig. 2.2. However, in plan the stays are distributed along the portal member at the top of the A-frame, and then converge to the median or abutment anchorages, Figs. 5.43 and 5.44. Stays are continuous from abutment anchorage to girder anchorage and are supported on rocker type saddles at the pylon portal member. The stays are composed of locked coil strands, each strand approximately $2\frac{3}{4}$ in. in diameter. The number of strands and configuration of each strand is indicated in Fig. 5.44. The excess strands indicated in stays D are individually anchored at the pylon by bearing sockets.

**FIGURE 5.42. Inclined tower, Bratislava Bridge. (Courtesy of John H. Garren, FHWA Region 10.)**

The A-frame tower is approximately 278 ft high with tapered rectagular box section legs. One leg contains a high-speed elevator, the other contains a stairwell. The tower legs are fixed at their base, Fig. 5.45.

The deck structure consists of a two cell box girder 41 ft 4 in. wide by 15 ft in depth. Depth to span ratio is 1:66. The orthotropic deck cantilevers 13 ft 9 in. out from the box on each side. Total width of the deck is 68 ft 10 in. The footpaths are below the roadway and cantilever out from the box 11 ft 6 in. on each side, Figs. 5.46 and 5.47.

FIGURE 5.43. Tower showing stay arrangement, Bratislava Bridge. (Courtesy of John H. Garren, FHWA Region 10.)

This structure indicates the versatility of the cable-stay concept for bridge design.[24,25]

## 5.19  SITKA HARBOR BRIDGE, U.S.A.

Sitka Harbor Bridge, Fig. 1.17, is the first vehicular cable-stayed bridge built in the United States. It is located in Sitka, Alaska and connects

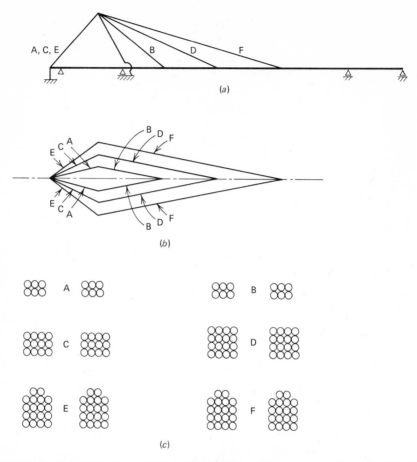

**FIGURE 5.44.** Bratislava Bridge: (*a*) elevation; (*b*) plan; (*c*) stay arrangement. (Courtesy of *Der Bauingenieur*. Ref. 24.)

Baranof Island to Japonski Island. Bridge type selection was discussed in Section 3.7. The bridge is a symmetric three-span structure; it has a 450 ft center span and 150 ft side spans. There are three 125 ft approach spans on the Sitka side and one approach span on the Japonski side of 125 ft. The total length of the structure is 1250 ft. Stay geometry in a direction transverse to the longitudinal axis of the bridge is in two vertical planes. In elevation, only one stay emanates from each side of the pylon, Fig. 5.48. Fore-stays are attached to the girder at the third points of the center span. Back-stays anchor over approach piers. Each stay consists of three 3 in. diameter galvanized bridge strands oriented in a vertical plane. The stays

FIGURE 5.45. Tower base, Bratislava Bridge. (Courtesy of John H. Garren, FHWA Region 10.)

FIGURE 5.46. General view of superstructure, Bratislava Bridge. (Courtesy of B. G. Johnston, University of Arizona.)

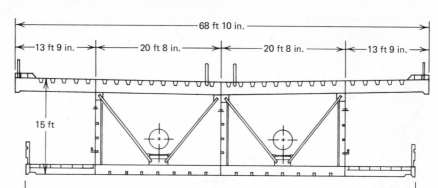

**FIGURE 5.47.   Cross section, Bratislava Bridge. (Courtesy of *Acier-Stahl-Steel*. Ref. 25.)**

are anchored at the girder to 47 ft long, 5 ft diameter tubes that cantilever out from the longitudinal girders, Fig. 5.49. Connection of the cables to the girder and pylon is discussed in Chapter 8.

The cross section of the superstructure, Fig. 5.50, consists of two longitudinal box girders spaced 32 ft 6 in. center to center. Box girders are 2 ft 6 in. in width by 6 ft in depth, providing a depth to main span ratio of 1:75. The girders are of constant depth throughout the 1250 ft length. Box girders are framed by 3 ft deep plate girder floor beams at 25 ft on centers. Stringers spanning between floor beams are 18 in. deep, wide flange sections, Fig. 5.51. Girders, floor beams, and stringers are all composite with the $6\frac{1}{2}$ in. deck slab.

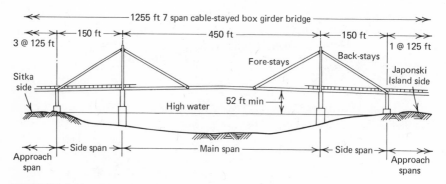

**FIGURE 5.48.   Elevation, Sitka Harbor Bridge. (Courtesy of *Civil Engineering, ASCE*. Ref. 26.)**

FIGURE 5.49.    Cable anchorage at girder, Sitka Harbor Bridge.
(Courtesy of John H. Garren, FHWA Region 10.)

Pylons are of the free standing cantilever type that rise 100 ft above the
pier, Figs. 1.17 and 5.50, and are fixed to the pier. They are single cell
rectangular boxes 3 by 4 ft, with the minor dimension parallel to the
longitudinal axis of the bridge. Plate stiffeners, 6 by $\frac{3}{8}$ in., reinforce the
plates of the box in the vertical and horizontal directions of the pylon.

Deck superstructure steel was erected by conventional methods, without
falsework, from both sides until it reached the pylon piers. Pylons were
then erected. Temporary guys or stays were utilized to support the deck as
it cantilevered out into the main span. After the transverse anchorage
beam was installed 150 ft from the pylons, and the permanent stays installed, the center girders were erected.[26]

Design plans and specifications indicated the tension and camber required at the completion of steel superstructure erection and prior to the
installation of concrete roadway, handrail, and other final attachments.
Tensioning of each strand was accomplished with center hole jacks using
calibrated pressure gauges for determination of the force applied by the
jacks. Upon completion of jacking, the steel box girders at midspan were
approximately 27 in. above final design grade. After installation of the
roadway deck and other miscellaneous items, the final roadway grade was
within 0.03 ft of design grade.

The special provisions required balancing and adjustment of the design
tension in the strands after application of full dead load by use of a calibrated jack. An allowable variation of 5% of mean stress in the strands
was called for with strand stress to be adjusted if necessary. The antici-

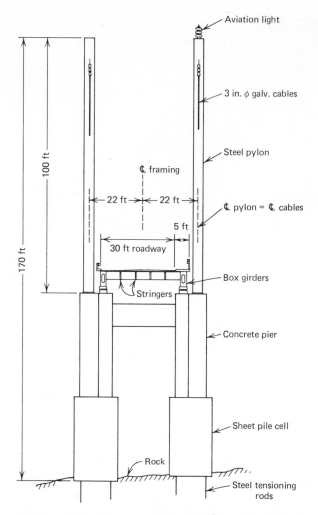

Aviation light

3 in. φ galv. cables

Steel pylon

℄ framing

22 ft — 22 ft

℄ pylon = ℄ cables

5 ft

100 ft

30 ft roadway

170 ft

Box girders

Stringers

Concrete pier

Sheet pile cell

Rock

Steel tensioning rods

**FIGURE 5.50. Cross section of bridge deck and supporting piers, Sitka Harbor Bridge. (Courtesy of *Civil Engineering*, ASCE. Ref. 26.)**

pated method of tension check would have required a reattachment of the center hole tensioning jacks at 24 locations. This method was complicated by the fact that should it become necessary to adjust the tension in one strand the tension in the remaining strands would be affected.

A method of checking the tension by vibrating the strand and measuring the frequency of vibration was proposed by Mr. Albert W. O'Shea, Project Engineer for Associated Engineers and Contractors, Inc. This methodology had been previously used in Germany.[2]

**FIGURE 5.51.** Underside of deck showing framing, Sitka Harbor Bridge. (Courtesy of John H. Garren, FHWA Region 10.)

Before measuring cable stresses by harmonics, a correlation test with a calibrated jack was required in order to confidently use the harmonic method. The strands can actually resist some bending and they are virtually fixed at the socketed ends. This results in a shortening in the effective length of the strands for use in the harmonics formula.

Upon assembly of the jacking apparatus to the strand selected by the Alaska Department of Highways for the correlation test, pressure was applied until the spanner nut came free from the cable anchor assembly (Fig. 8.26) and the entire load of the cable was being carried by the jack. At this point the gauge pressure was read and the load on the cable computed. While pressure was maintained, frequency of vibration was determined by physically inducing an oscillation within the cable and visually counting these oscillations for 1 minute. Timing was done with a stop watch.

After the vibration frequency was determined the load was computed from the following formula:

$$P = \frac{4WL^2F^2}{g} \tag{5.1}$$

where $P$ = Cable tension
$L$ = Cable length
$g$ = 32.2 ft/sec²
$F$ = Vibration frequency
$W$ = Unit weight of the cable

Frequency of vibration was determined at seven different gauge pressures. The load determined by the seven readings averaged 7.14 tons higher than the load determined by hydraulic jacking. This corresponds to an effective cable shortening of about 6 ft. On this basis loading of the other bridge strands was determined by obtaining a frequency of vibration, calculating the loading from the formula, and reducing this figure by 7 tons. All the strands on this structure were checked for actual tension in this manner and found to be balanced within the 5% tolerance. The total check was accomplished in less than a day.

The harmonic method of determining cable stress can be employed at any time with no more equipment than a stop watch. It therefore becomes an advantageous method that can be easily implemented; for example,

1. If for some reason a strand or strands have to be replaced, tensions can be easily checked.

2. Changes in strand tension can be easily determined if the bridge deck is resurfaced or other modifications are made.

3. It is easily implemented by bridge inspectors on routine inspection.

4. In the event of damage to a strand or strands, the stresses in the remaining strands can be readily determined.

## 5.20   NORDBRÜCKE MANNHEIM-LUDWIGSHAFEN BRIDGE, GERMANY

The Nordbrücke Rhine River bridge, also known as the Kurt-Schumacher-Brücke, connnects the cities of Mannheim and Ludwigshafen, Figs. 5.52 and 5.53. In elevation it is an asymmetric structure with a major span of 941.4 ft and a minor span of 480.3 ft. The stays are in the radiating configuration, Fig. 5.54. Transversely the stays are in two sloping planes.

The major span superstructure consists of a 121 ft wide orthotropic deck supported on two rectangular box girders. The center portion of the deck between box griders and through the A-frame pylon is for trams. Two lanes for auto traffic are supported over the box girders on each side of the pylon, and bicycle paths are at the extreme edges of the deck. At the Ludwigshafen side the superstructure widens to 170.3 ft to accommodate ramps on both sides and the center tramway portion depresses below the roadway elevation, Figs. 5.54 and 5.55. The minor span, including the portion over the pylon pier, is a box girder of prestressed concrete construction. A rigid connection is provided between the steel and concrete superstructure construction. The superstructure has a constant depth of 14 ft 9 in. producing a depth to main span ratio of 1:64.

FIGURE 5.52. Aerial view, Nordbrücke Mannheim-Ludwigshafen. (Courtesy of Wolfgang Borelly.)

**FIGURE 5.53.  Night view, Nordbrücke Mannheim-Ludwigshafen. (Courtesy of Wolfgang Borelly.)**

The pylon rises 234 ft 6 in. above the roadway elevation and is of steel construction. Legs of the pylon pierce the prestressed concrete box girder superstructure to be supported by the pylon pier.

Each parallel wire strand in the stays consists of 295 wires approximately 9/32 in. in diameter compacted to a strand of approximately 5 in. in diameter. In each sloping plane the stays take the following strand pattern: top fore-stay, six strands in three layers of two strands each; center fore-stay, four strands in two layers of two strands each; lower fore-stay and back-stay, two strands in one layer of two; top back-stay, 10 strands in five layers of two strands each. Each strand is individually anchored at the pylon. Fig. 5.56 is a view looking up along the top back-stay to the top of the pylon.[27,28,29]

## 5.21  KÖHLBRAND HIGH-LEVEL BRIDGE, GERMANY

In our opinion the Köhlbrand Bridge is one of the most attractive cable-stayed bridges conceived thus far, Figs. 5.57 and 5.58. Construction of this six-lane, high-level bridge over the Köhlbrand, an arm of the River Elbe

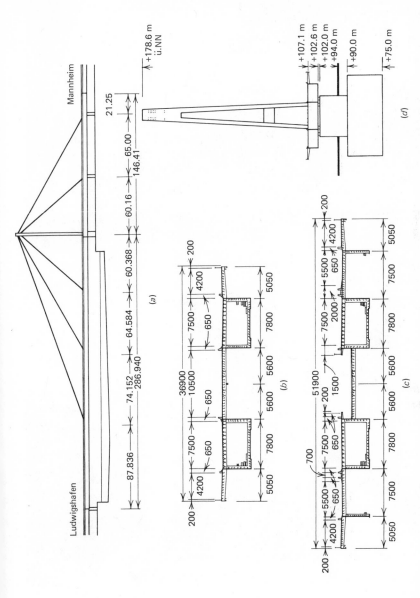

**FIGURE 5.54.** Nordbrücke Mannheim-Ludwigshafen: (*a*) elevation; (*b*) typical cross section; (*c*) cross section in the area of the sloping streetcar ramp; (*d*) cross section at pylon (dimensions are given in meters).

153

FIGURE 5.55.  Erection view, Nordbrücke Mannheim-Ludwigshafen. (Courtesy of Wolfgang Borelly.)

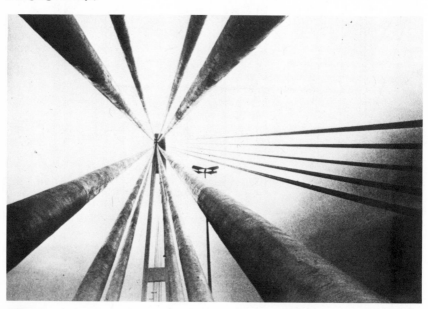

FIGURE 5.56.  Worm's eye view of stays, Nordbrücke Mannheim-Ludwigshafen. (Courtesy of Wolfgang Borelly.)

**FIGURE 5.57.** Aerial view, Köhlbrand high-level bridge. (Courtesy of Wolfgang Borelly.)

at Hamburg, Germany, Fig. 5.59, began in the autumn of 1970. This structure connects the Hamburg Free Port area with a western motorway bypass. Its main span of 1066 ft makes it one of the largest cable-stayed bridges in Germany, Fig. 5.60. The distinctive delta-shaped towers provide for narrow pylon piers, Fig. 5.60. It is anticipated that at a future date a twin structure will be built along side of the completed structure (dotted line, Fig. 5.59).

Geometrically the stays are in two sloping planes and are of the radiating configuration. The 1706 ft long cable-stay portion of this structure is of steel construction and has a center span of 1066 ft and equal anchor spans of 320 ft. The cable-stayed superstructure is a trapezoidal box with side

**FIGURE 5.58.  View of roadway camber, Köhlbrand high-level bridge.  (Courtesy of Wolfgang Borelly.)**

cantilevers and accommodates six lanes of traffic, Fig. 5.60. Eighty-eight locked coil strands, ranging in diameter from $2\frac{1}{8}$ in. to $4\frac{3}{32}$ in., support the superstructure. Other than for lateral support the 1706 ft length of steel box girder is supported by the stays and is independent of the pylons.

The west approach spans total 3438 ft in length, with spans varying from 111 ft 6 in. to 229 ft 8 in. of a single trapezoidal concrete box-girder

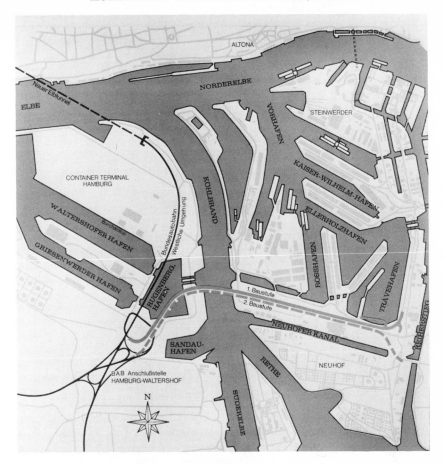

FIGURE 5.59.    Location map, Köhlbrand high-level bridge. (Courtesy of Wolfgang Borelly.)

construction, Fig. 5.60. The east approach spans total 6222 ft. in length, with spans varying from 114 ft 10 in. to 213 ft 3 in. of a two cell concrete box-girder construction, Fig. 5.60.

The steel pylon rises 321 ft 6 in. above the pier, Fig. 5.61. Erection of the pylon required a truck-mounted crane of 1000 ton capacity—one of the largest in Europe, Fig. 5.62. The distinctive shape of the pylon results from the height of the superstructure above water and an especially narrow pylon pier. A much wider pier would have resulted if the legs were extended as in the conventional A-frame. Erection of the steel box girder

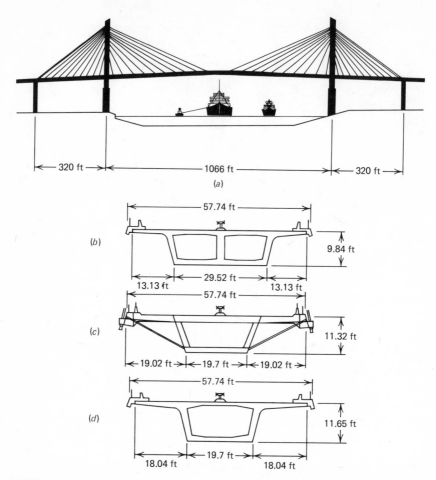

**FIGURE 5.60.** Köhlbrand, High-Level Bridge: (*a*) elevation; (*b*) concrete cross section, east approach spans; (*c*) steel cross section, cable-stay spans; (*d*) concrete cross section, west approach spans. (Courtesy of Wolfgang Borelly.)

units are shown in Figs. 5.63 and 5.64. Figure 5.65 is a composite of views of the completed structure, which is, in our opinion, an excellent example of form following function.

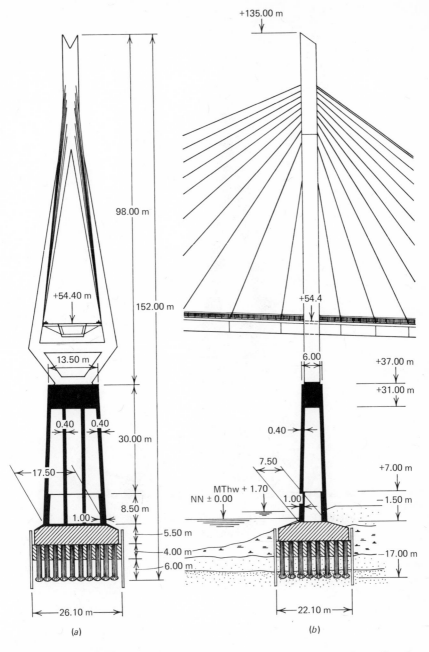

FIGURE 5.61. Köhlbrand High-Level Bridge: (*a*) pylon cross section; (*b*) pylon elevation. (Courtesy of Wolfgang Borelly.)

FIGURE 5.62. Pylon erection, Köhlbrand High-Level Bridge. (Courtesy of Wolfgang Borelly.)

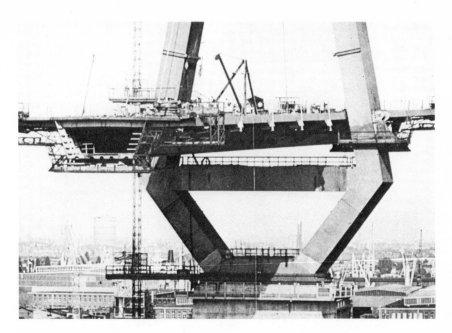

FIGURE 5.63. Cantilever erection, Köhlbrand High-Level Bridge. (Courtesy of Wolfgang Borelly.)

FIGURE 5.64. Box girder erection, Köhlbrand High-Level Bridge. (Courtesy of Wolfgang Borelly.)

FIGURE 5.65. Köhlbrand High-Level Bridge. (Courtesy of Wolfgang Borelly.)

**FIGURE 5.65.** (*continued*)

163

**FIGURE 5.65.**  (*continued*)

## REFERENCES

1. Wenk, H., "Die Strömsundbrücke," *Der Stahlbau*, No. 4, April 1954.
2. Ernst, H. J., "Montage eines seilverspannten Balkens im Gross-Brückenbau," *Der Stahlbau*, No. 5, May 1956.
3. Wintergerst, L., "Nordbrücke Düsseldorf. III. Teil: Statik und Konstruktion der Strombrücke," *Der Stahlbau*, No. 6, June 1958.
4. Fischer, G., "The Severin Bridge at Cologne (Germany)," *Acier-Stahl-Steel* (English version), No. 3, March 1960.
5. Anon., "Norderelbe Bridge K6: A Welded Steel Motorway Bridge," *Acier-Stahl-Steel* (English version), No. 11, November 1963.
6. Havermann, H. K., "Die Brücke über die Norderelbe im Zuge der Bundesautobahn Südliche Umgehung Hamburg, Tiel I: Ideen-und Bauwettbewerb," *Der Stahlbau*, No. 7, July 1963.
7. Aschenberg, H. and Freudenberg, G., "Die Brücke über die Norderelbe im Zuge der Bundesautobahn Südliche Umgehung Hamburg, Tiel II: Konstruktion des Brückenüberbaus," *Der Stahlbau*, No. 8, August 1963.

8. Aschenberg, H. and Freudenberg, G., "Die Brücke über die Norderelbe im Zube der Bundesautobahn Südliche Umghung Hamburg, Tiel III: Statische Berechnung des Brückenüberbaus," *Der Stahlbau*, No. 9, September 1963.

9. Havemann, H. K. and Freudenberg, G., "Die Brücke über die Norderelbe im Zuge der Bundesautobahn Südliche Umgehung Hamburg, Tiel IV: Bauausführung der stähleren Überbauten," *Der Stahlbau*, No. 10, October 1963.

10. Schöttgen, J. and Wintergerst, L., "Die Strassenbrücke über den Rhein bei Maxau," *Der Stahlbau*, No. 1, January 1968.

11. Thul, H., "Cable-Stayed Bridges in Germany," *Proceedings of the Conference on Structural Steelwork* held at the Institution of Civil Engineers, September 26–28, 1966, The British Constructional Steelwork Association, Ltd., London.

12. Thul, H., "Stählerne Strassenbrücken in der Bundesrepublik," *Der Bauingenieur*, No. 5, May 1966.

13. Freudenberg, G., "Die Stahlhochstrasse über den neuen Hauptbahnhof in Ludwigshafen/Rhein," *Der Stahlbau*, No. 9, September 1970.

14. Naruoka, M. and Sakamoto, T., "Cable-Stayed Bridges in Japan," *Acier-Stahl-Steel* (English version), No. 10, October 1973.

15. Weisskopf, F., "World's Longest-Span Cable-Stayed Girder Bridge over the Rhine near Duisburg (Germany)," *Acier-Stahl-Steel* (English version), July–August 1972.

16. Tamms and Beyer, "Kniebrücke Düsseldorf," Beton-Verlag GmbH, Düsseldorf, 1969.

17. Demers, J. G. and Simonsen, O. F., "Montreal Boasts Cable-Stayed Bridge," *Civil Engineering*, ASCE, August 1971.

18. Taylor, P. R. and Demers, J. G., "Design, Fabrication and Erection of the Papineau-Leblanc Bridge," Canadian Structural Engineering Conference, 1972, Canadian Steel Industries Construction Council, Toronto, Ontario, Canada.

19. Kondo, K., Komatsu, S., Inoue, H. and Matsukawa, A., "Design and Construction of Toyosato-Ohhashi Bridge," *Der Stahlbau*, No. 6, June 1972.

20. Anon., "Erskine Bridge," *Building with Steel*, British Constructional Steelwork Association Limited, Vol. 5, No. 4, June 1969.

21. Kerensky, O. A., Henderson, W. and Brown, W. C., "The Erskine Bridge," *Structural Engineer*, Vol. 50, No. 4, April 1972.

22. Anon., "Opening Batman Bridge 18th May, 1968," Department of Public Works, Tasmania, Australia.

23. Payne, R. J., "The Structural Requirements of the Batman Bridge as They Affect Fabrication of the Steelwork," *Journal of the Institution of Engineers*, Australia, Vol. 39, No. 12, December 1967.

24. Tesár, A., "Das Projekt der neuen Strassen brücke über die Donau in Bratislava/CSSR," *Der Bauingenieur*, No. 6, June 1968.

25. Mohsen, H., "Trends in the Construction of Steel Highway Bridges," *Acier-Stahl-Steel* (English version), June 1974.

26. Gute, W. L., "First Vehicular Cable-Stayed Bridge in the U.S.," *Civil Engineering*, ASCE, November 1973.

27. Borelly, W., "Nordbrücke Mannheim-Ludwigshafen," *Der Bauingenieur*, No. 8, August 1972; No. 9, September 1972.

28. Volke, E., "Die Strombrücke im Zuge der Nordbrücke Mannheim-Ludwigshafen (Kurt-Schumacher-Brücke), Teil I: Konstruktion und Statik," *Der Stahlbau*, No. 4, April 1973; No. 5, May 1973.

29. Rademacher, C. H., "Die Strombrücke im Zuge der Nordbrücke Mannheim-Ludwigshafen (Kurt-Schumacher-Brücke), Teil II: Werkstattfertigung und Montage," *Der Stahlbau*, No. 6, June 1973.

# 6

# Pedestrian Bridges

## 6.1  INTRODUCTION

Previous chapters have emphasized vehicular cable-stayed bridges. In this chapter we intend to present a few examples of cable-stayed pedestrian bridges. Some of these structures, in our opinion, provide architecturally attractive and exciting designs. As the reader might suspect, a number of pedestrian bridges have been built in Germany and in the natural course of evolution have extended to a number of other countries.

There are advantages in reduced superstructure depth, simplicity of erection and aesthetics. The lightness of appearance which js obtainable will become apparent. When conditions are suitable, and they generally are, most pedestrian bridges have been built as asymmetric structures with only one pylon. Superstructures can be quite conventional and, if cable-stayed, can be designed as continuous beams on elastic supports.

In October 1964, the Bureau of Public Roads, U.S. Dept. of Commerce (now the Federal Highway Administration, U.S. Department of Transportation) in the series *Standard Plans For Highway Bridges,* published Volume 5, "Typical Pedestrian Bridges." This document, reproduced in

Appendix A, contains a typical design for a single tower cable-stayed bridge. Considering the date of publication, this represents a very progressive attitude toward this type of construction.

## 6.2   FOOTBRIDGE AT THE GERMAN PAVILION, BRUSSELS EXHIBITION, 1958

This bridge, Fig. 6.1 and 6.2, was the main feature of the German Pavilion at the Brussels Exhibition in 1958. After the exhibition closed it was dismantled and re-erected over a roadway near Duisburg, Fig. 6.3.

**FIGURE 6.1.   German Pavilion, Brussels, 1958. (Courtesy of the British Constructional Steelwork Association, Ltd.)**

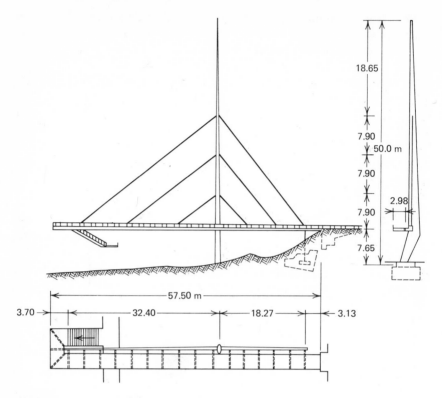

**FIGURE 6.2.    Footbridge, German Pavilion, Brussels, 1958. (Courtesy of the British Constructional Steelwork Association, Ltd.)**

The German Pavilion Bridge is unusual in that it represents a laterally displaced, single vertical plane, transverse stay geometry (see Fig. 2.3). The offset torsionally resistant box girder consists of four $\frac{13}{16}$ in. thick plates, has a constant depth of 4 ft 4 in. and a width that varies from 13 to 28 in. It has a short span of 60 ft, which is anchored to the abutment and cantilevers out from the pylon 120 ft. A 10 ft wide timber deck is supported on 11 ft centers by 10 in. deep transverse wide flange beams which cantilever out 13 ft. from the box girder. A streamlined multicell steel pylon, approximately 165 ft in height (Fig. 6.4), supports the superstructure. The torsion box girder is supported from the pylon by six stays, approximately 2 in. diameter, in a harp configuration, Fig. 6.2.

The girder and pylon were originally painted white. When the structure was re-erected at Duisburg, the girder and pylon were repainted yellow to contrast with the green foliage.[1,2,3]

**FIGURE 6.3.  Duisburg Pedestrian Bridge. (Courtesy of the British Constructional Steelwork Association, Ltd. Ref. 1.)**

## 6.3  VOLTA-STEG BRIDGE AT STUTTGART-MUNSTER

The asymmetric Volta-Steg Bridge, Fig. 6.5, spans the Neckar River at Stuttgart with a major span of 246 ft and a minor span of 69 ft. The pylon height is approximately 40 ft above the pier. Superstructure width is $11\frac{1}{2}$ ft. The two principal girders have a depth of 3 ft 9 in., and the deck is of orthotropic construction with an approximately $\frac{3}{4}$ in. asphalt surface. Total steel weight is 146.6 tons.[1,3]

## 6.4  BRIDGE OVER THE SCHILLERSTRASSE, STUTTGART

A rather spectacular structure, also located in Stuttgart, is the Schillerstrasse pedestrian bridge, Fig. 6.6. This structure was sited in an ancient royal park and aesthetics considerations were very important. A slender superstructure was dictated with gently sloping approaches rather than staircases. As can be seen in Fig. 6.7, the structure forks at the pylon, providing two approaches on that side, the other side has a single straight approach. As seen in elevation, the total length of the structure, from abutment to abutment, is 304 ft. The major span is 225 ft and the minor

FIGURE 6.4.   View of pylon, Duisburg pedestrian bridge. (Courtesy of Beratungsstelle für Stahlverwendung. Ref. 3.)

**FIGURE 6.5.** Volta-Steg Footbridge. (Courtesy of Beratungsstelle für Stahlverwendung. Ref. 3.)

**FIGURE 6.6.** Schillerstrasse Footbridge. (Courtesy of the British Constructional Steelwork Association, Ltd.)

**FIGURE 6.7.    Schillerstrasse Footbridge. (Courtesy of Beratungsstelle für Stahlver-
wendung. Ref. 3.)**

span is 79 ft, measured from the pylon. Cable geometry is a double sloping plane with three fore-stays and two back-stays in each plane. The stays
all converge at the top of the tower.

The superstructure in the long, straight portion, is a very flat box girder,
18 ft in width and 20 in. in depth, Fig. 6.8. Top and bottom flange plates
are longitudinally stiffened by trapezoidal stiffeners. Thickness of the top
and bottom flange plates are $\frac{5}{16}$ in. and $\frac{1}{4}$ in respectively. Web plates
are $\frac{5}{16}$ in. in thickness. Transverse diaphragms are provided at 7 ft $10\frac{1}{2}$
in. intervals, and consist of trusses made of approximately 1 in. diameter
bars forming the diagonals. The transverse diaphragm is very similar in
appearance to the conventional bar joist used in floor construction of
buildings, Fig. 6.8.

The slender pylon has a height of 78 ft 9 in. In cross section it is a
hollow octagonal steel section tapering from 4 ft wide at the base to 1 ft
10 in. wide at the top, Fig. 6.7. Plate thickness varies from 1 in. at the
top to $\frac{13}{16}$ in. at the bottom. The stays and their connection details are discussed in Section 8.14.[1,3,4]

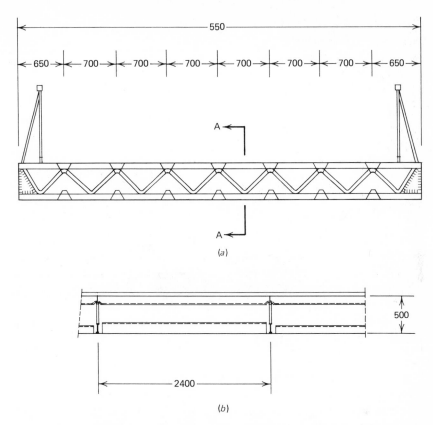

**FIGURE 6.8.** Schillerstrasse Footbridge: (*a*) deck cross section; (*b*) section A-A. (Courtesy of the British Constructional Steelwork Association, Ltd. Ref. 1.)

## 6.5   THE GLACISCHAUSSEE BRIDGE, HAMBURG

The Glacischaussee Bridge, is a single vertical plane structure, Fig. 6.9, with the fore-stays in a fan configuration and the back-stays in a star configuration. The 178 ft long steel superstructure is a trapezoidal box with side cantilevers, Fig. 6.10. The 93 ft 10 in. high triangular cross-section pylon pierces the deck as indicated in Fig. 6.10. The girder was fabricated in five longitudinal sections and field bolted. It took only two hours to erect the girder using mobile cranes. Views of the stay anchorage are shown in Figs. 6.11 and 6.12.[1,3,5,6]

**FIGURE 6.9.** Glacischaussee Footbridge. (Courtesy of the British Constructional Steelwork Association, Ltd. Ref. 1.)

## 6.6   LODEMANN BRIDGE, HANOVER, GERMANY

The Lodemann Bridge has an inverted Y pylon, Fig. 6.13, with a single vertical cable plane that radiates from the peak of the A-frame. It is a single pylon asymmetric structure with a major span of 223 ft and a minor span of 187 ft. The superstructure is a center inverted trapezoidal box girder about 4 ft 8 in. deep, with a top and bottom flange width of 2 ft $7\frac{1}{2}$ in. and 4 ft 4 in., respectively. The box projects 2 ft $7\frac{1}{2}$ in. above the deck, Fig. 6.14, to provide a barrier between pedestrian walk on one side and bicycle track on the other, each of which is 9 ft 4 in. in width. Steel members cantilever out from each side of the center girder to support a $4\frac{3}{4}$ in. thick reinforced concrete deck slab.[3,6]

## 6.7   RAXSTRASSE FOOTBRIDGE, AUSTRIA

This asymmetric structure has a 111 ft 6 in. high inclined A-frame pylon, Fig. 6.15. The superstructure has a clear span of 177 ft between abutments and is free of the pylon. It has a clearance above the roadway of approximately 16 ft 4 in. Walkway width is 13 ft.

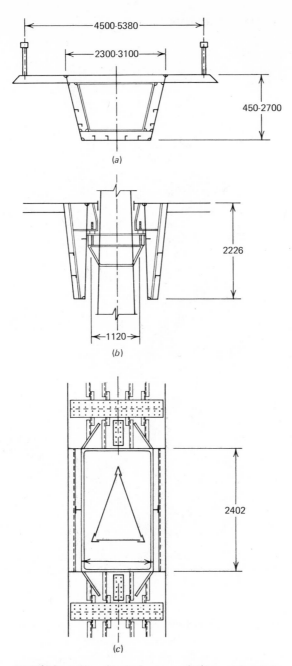

**FIGURE 6.10.** Glacischaussee Bridge: (*a*) girder cross section; (*b*) girder cross section at pylon; (*c*) plan cross section of pylon. (Courtesy of the British Constructional Steelwork Association, Ltd. Ref. 1.)

**FIGURE 6.11.   View of girder anchorage, Glacischaussee Foot-bridge. (Courtesy of the British Constructional Steelwork Association, Ltd. Ref. 6.)**

The superstructure is an orthotropic deck with two longitudinal inverted T plate girders spaced at 9 ft 10 in., Fig. 6.16. The deck plate is about $\frac{5}{16}$ in. thick. Longitudinal stiffeners are $4\frac{3}{4}$ in. by $\frac{5}{16}$ in. flat plates at $19\frac{3}{4}$ in. Transverse stiffeners are $9\frac{27}{32}$ in. by $\frac{5}{16}$ in. plates at 8 ft $10\frac{5}{16}$ in. Longitudinal edge stiffeners are $11\frac{13}{16}$ in. by $\frac{5}{16}$ in. Every fourth transverse stiffener is cantilevered out to pick up a cable-stay anchorage, Fig. 6.17.

**FIGURE 6.12.   View of pylon anchorage, Glacischaussee Foot-bridge. (Courtesy of the British Constructional Steelwork Association, Ltd. Ref. 6.)**

These stiffeners are $17\frac{23}{32}$ in. by $\frac{5}{16}$ in. and are inclined to match the inclination of the stays.

Cable stays support the superstructure at the fifth points and are $\frac{29}{32}$ and $1\frac{1}{16}$ in. in diameter, fixed at the pylon anchorage and adjustable at the deck anchorage. Back-stays are $1\frac{1}{32}$ in. in diameter and anchored to gravity foundations independent of the northern abutment.

**FIGURE 6.13.**   Lodemann Footbridge. (Courtesy of Beratungsstelle für Stahlverwendung. Ref. 3.)

The all-welded, A-frame pylon straddles the deck and is 39 ft $4\frac{7}{16}$ in. wide at its base. Pylon legs are tapered and triangular in cross section with a maximum measurement on a side of 4 ft $7\frac{1}{8}$ in. Plate thickness is about $\frac{3}{8}$ in. The pylon is pin connected at its base. Structural steel weight was approximately 77 tons. Cable weight was 1.65 tons.

## 6.8   PONT DE LA BOURSE, LE HAVRE

This graceful asymmetric structure is located in Le Havre, France. It is a double inclined plane cable arrangement transversely and a radiating configuration in elevation, Figs. 6.18 and 6.19. Total length of superstructure is 344 ft 6 in. with a major span of 240 ft 10 in. and a minor span of 103 ft 8 in.

The superstructure has a depth of 3 ft. 6 in. with a marked camber of 19 ft 8 in. in the 344 ft 6 in. length to provide a clearance of 23 ft at high water and 30 ft 6 in. at mean water level.

In cross section, the superstructure consists of two longitudinal edge girders 19 ft on centers. Cross beams frame-in at the lower flange so that the longitudinal girder forms part of the parapets. The deck is a $3\frac{7}{8}$ in. reinforced concrete slab.

**FIGURE 6.14.   Lodemann Footbridge. (Courtesy of Beratungsstelle für Stahlverwendung. Ref. 3.)**

Pylon height is 114 ft 10 in. with triangular cross-section legs. Outside stays are $3\frac{3}{8}$ in. in diameter, inside stays are $2\frac{1}{4}$ in. in diameter.[7]

## 6.9   CANAL DU CENTRE, OBOURG, BELGIUM

Located in Obourg, Belgium, this concrete pedestrian bridge, Fig. 6.20, consists of eight precast double-T deck sections approximately 55 ft in

**FIGURE 6.15.    Raxstrasse Footbridge. (Courtesy of Waagner-Biro Aktiengesell-schaft, Vienna.)**

length. It is a single pylon symmetric structure with an inclined double-plane stay arrangement transversely and a radiating stay configuration in elevation. The only function of the outer stay is to position the hinged pylon. Erection of the foundation took 24 days; 8 days to erect the deck sections; 2 days for the pylon; 10 days to place and anchor the stays and stress the outer stays; 3 days to stress and adjust the remaining stays to

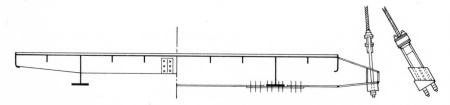

**FIGURE 6.16.    Deck cross section, Raxstrasse Footbridge. (Courtesy of Waagner-Biro Aktiengesellschaft, Vienna.)**

**FIGURE 6.17.** Raxstrasse Footbridge. (Courtesy of Waagner-Biro Aktiengesellschaft, Vienna.)

obtain the proper profile; and 1 day to complete the joints between deck units.[8]

## 6.10  RIVER BARWON FOOTBRIDGE, AUSTRALIA

The solid trapezoidal concrete deck of this structure located in Geelong, Australia, is unusual in that it encases an approximately 3 ft 7 in. diameter PVC (Poly-vinyl chloride) sewer pipe, Fig. 6.21. The end spans and

**FIGURE 6.18.** Pont de la Bourse. (Courtesy of *Acier-Stahl-Steel*. Ref. 7.)

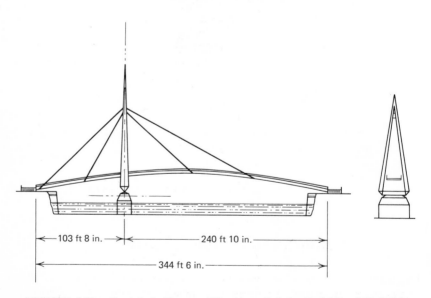

←——103 ft 8 in.——→|←————240 ft 10 in.————→

|←——————————344 ft 6 in.——————————→|

**FIGURE 6.19.** Pont de la Bourse. (Courtesy of *Acier-Stahl-Steel*. Ref. 7.)

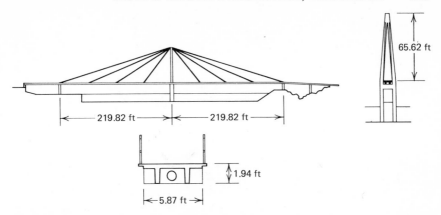

**FIGURE 6.20.    Canal du Centre Footbridge. (Courtesy of Crosby Lockwood & Sons, Ltd. Ref. 8.)**

lower section of the pylon were of cast-in-place construction on falsework, while the center span was cast-in-place cantilever construction in segments of approximately 9 ft 10 in. in length. Legs of the upper portion of the pylon were precast and bolted at the cross beam. The pylon is hinged at the base and has Freyssinet flat jacks in the joint for adjustment that may be required to compensate for distortions arising from creep and shrinkage.[8]

## 6.11    MOUNT STREET FOOTBRIDGE, AUSTRALIA

A single vertical plane pedestrian bridge location in Perth, Western Australia has an unusual cable configuration in elevation which must have

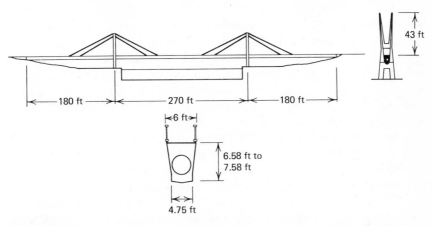

**FIGURE 6.21.    River Barwon Footbridge. (Courtesy of Crosby Lockwood & Sons, Ltd. Ref. 8.)**

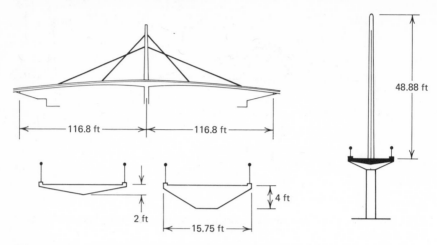

**FIGURE 6.22.    Mount Street Footbridge. (Courtesy of Crosby Lockwood & Sons, Ltd. Ref. 8.)**

been selected for aesthetic reasons, Fig. 6.22. The cast-in-place superstructure is trapezoidal in cross section and has a varying depth, haunching at the pier. The pylon consists of precast units that are prestressed vertically to the deck. Stays are of parallel wire construction with proprietary prestressing anchorages. A dead-end anchorage is used at the deck and a jacking anchorage at the pylon.[8]

## 6.12    MENOMONEE FALLS PEDESTRIAN BRIDGE, U.S.A.

The Menomonee Falls Pedestrian Bridge, Figs. 1.16, 6.23, and 6.24, received an Award of Merit 1971/Special Type in the Prize Bridge Contest sponsored by the American Institute of Steel Construction. It is the first modern cable-stayed bridge, either vehicular or pedestrian, built in the U.S.A. It is a 361 ft, three-span structure with a center span of 217 ft and end spans of 72 ft and was designed by the Wisconsin Division of Highways Bridge Section.[9]

Transverse stay arrangement is two sloping planes. In elevation a single stay emanates from the top of the pylon on each side and in each plane. Each stay consists of a 3 in. diameter structural strand.

The superstructure has two principal longitudinal girders (W 33 × 130) spaced at 8 ft 3 in. and supporting a $5\frac{1}{2}$ in. reinforced concrete slab. The lower flange has a lateral bracing system consisting of 4 × 3 angle diagonals and 12 in. channel diaphragms.

FIGURE 6.23. Menomonee Falls Pedestrian Bridge. (Courtesy of the Wisconsin Division of Highways.)

FIGURE 6.24. Menomonee Falls Pedestrian Bridge. (Courtesy of the Wisconsin Division of Highways.)

FIGURE 6.25.   General view, Prince's Island Pedestrian Bridge. (Courtesy of Joseph A. Chilstrom, FHWA, Washington Division.)

FIGURE 6.26.   Underside view of girder, Prince's Island Pedestrian Bridge. (Courtesy of Joseph A. Chilstrom, FHWA, Washington Division.)

FIGURE 6.27. Longitudinal view of deck, Prince's Island Pedestrian Bridge. (Courtesy of Joseph A. Chilstrom, FHWA, Washington Division.)

Height of the A-frame pylon is approximately 56 ft 6 in. with a 15 ft 2 in. distance between center of legs at the base. The legs are a steel box section 20 by 10 in., with the larger dimension parallel to the longitudinal axis of the bridge. Plates on the 20 in. side are a constant thickness of $\frac{1}{2}$ in., while on the 10 in. side the thickness varies from $\frac{1}{2}$ in. below the superstructure to $\frac{3}{8}$ in. above.

## 6.13   PRINCE'S ISLAND PEDESTRIAN BRIDGE, CANADA

This structure, located in Calgary, Alberta, Canada, is an asymmetrical, cable-stayed pedestrian and cycle bridge with a total length of 600 ft and a 220 ft suspended river span, Fig. 6.25. The box girder has a depth of 3 ft 2 in., a top flange width of 12 ft, and a bottom flange width of 5 ft, Fig. 6.26. An epoxy and silica sand wearing surface is bonded to the orthotropic deck, Fig. 6.27. The A-frame pylon, Fig. 6.25 and 6.27, attains a height of 55 ft above the deck surface and supports four $1\frac{3}{4}$ in. diameter galvanized structural strands. The piers are supported on steel pipe piles driven to bedrock. The foundations also include rock anchors for stability under ice pressures and anchor pier uplift. The structure was designed by Carswell Engineering Ltd. of Calgary.

# REFERENCES

1. Anon., "Steel Footbridges," British Constructional Steelwork Association, Ltd., 16M/623/1266.

2. Fuchs, D., "Der Fussgängersteg auf der Brüsseler Weltausstellung 1958," *Der Stahlbau*, No. 4, April 1958.

3. Feige, A., "Fussgängerbrücken aus Stahl," Merkblatt 251, Beratungsstelle für Sthalverwendung, Düsseldorf.

4. Leonhardt, F. and Andrä, W., "Fussgängersteg über die Schillerstrasse in Stuttgart," *Die Bautechnik*, No. 4, April 1962.

5. Reimers, K., "Fussgängerbrücke über die Glacischaussee in Hamburg für die Internationale Gartenbau-Ausstellung 1963," Schweissen and Schneiden, June 1963.

6. Anon., "Suspended Structures," British Constructional Steelwork Association, Ltd., 16M/842/68.

7. Bachelart, H., "Pont De La Bourse Footbridge over the Bassin du Commerce Le Havre (France)," *Acier-Stahl-Steel* (English version), No. 4, April 1970.

8. Gee, A. F., "Cable-Stayed Concrete Bridges," in *Developments in Bridge Design and Construction*, edited by Rockey, Bannister, and Evans, Crosby Lockwood & Son, Ltd., London, 1971.

9. Woods, S. W., Discussion to "Historical Development of Cable-Stay Bridges," by Podolny and Fleming, *Journal of the Structural Division*, ASCE, Vol. 99, No. ST 4, April 1973, Proc. Paper 9640.

# 7

# Cable Data

## 7.1 INTRODUCTION

Several types of cables are available for use as stays in cable-stayed bridges. The form or configuration of the cable depends on its make-up; it can be composed of parallel wires, parallel strands or ropes, single strands or ropes, locked-coil strands, or solid bars.

This chapter discusses cables made from the basic single wire and describes the manufacturing process and mechanical properties that influence design and construction practices.

A definition of the terms useful in this section is given below. The reader should understand the specific meanings of the terms and their unique application.[1]

Cable—any flexible tension member, consisting of one or more groups of wires, strands, or ropes.

Wire—a single continuous length drawn from a cold rod.

Strand—(with the exception of parallel wire strand) an arrangement of wires helically placed about a center wire to produce a symmetrical section, Fig. 7.1.

**189**

FIGURE 7.1. Structural strand. (Courtesy of the United States Steel Corporation. Ref. 2.)

FIGURE 7.2. Structural rope. (Courtesy of the United States Steel Corporation. Ref. 2.)

Rope—a number of strands helically wound around a core that is composed of a strand or another rope, Fig. 7.2.

Locked-coil strands—resemble strands except the wires in some layers are shaped to lock together when in place around the core, Fig. 7.3.

Parallel wire strand—individual wires arranged in a parallel configuration without the helical twist, Fig. 7.3.

## 7.2  DEVELOPMENT OF CABLE APPLICATIONS[2]

The structural application of a flexible tension member or cable dates back to the period before recorded history. Primitive man constructed cables of tangled grape vines to bridge large ravines and small rivers. The early Chinese built suspension bridges of hemp rope and iron chains. The Incas of Peru constructed their suspension bridges over major rivers with cables of hemp rope as the principal load-supporting member.

Records indicate that copper cables were used in the ancient city of Ninevah near Babylon about 685 B.C. A short piece of such a cable is on display in an English museum.

Another old piece of wire rope made of bronze and used in a treadmill at Pompeii was discovered in the ruins caused by the eruption of Mount Vesuvius in 79 A.D. There is evidence that the Romans manufactured

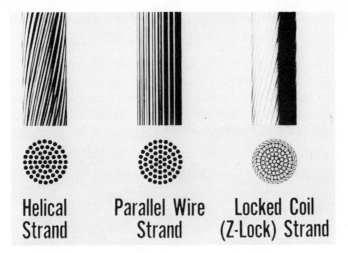

Helical Strand          Parallel Wire Strand          Locked Coil (Z-Lock) Strand

FIGURE 7.3.    Strand types. (Courtesy of the Bethlehem Steel Corporation.)

cables of wire and wire rope; a specimen 1 in. diameter, 15 in. in length, is displayed at the Musio Barbonico at Naples, Italy.

The wires for the early ropes were constructed by hand. In the succeeding centuries the only changes were refinements of the craftsmen's skill and the introduction of new materials and field construction techniques. The art of cable manufacture reached such a height of perfection that only a close examination could reveal that the wires were hand made. Examples of wire made by the Vikings are so uniform that some are of the opinion that they were mechanically drawn.

Machine-drawn wire first appeared in Europe during the fourteenth century, but A. Albert of Germany is credited with producing the first wire rope in 1834 which closely resembles the wire rope of today. Some sources claim that a man named Wilson produced the first rope in England in 1832. These dates may be somewhat in error but indicate the general period of time of the development of the wire rope in Europe.

The first American machine-made rope was used in a service application in 1846. Since then many changes have taken place, such as the introduction of better quality, high-strength steels, efficient manufacturing processes and machines, and new field applications accompanied by new techniques of construction. The technology of structural strand and rope has improved consistently with time.

V. G. Shookhov, a Russian engineer, was one of the first to use cables as a structural load-carrying member in a building. He designed cable-suspended roofs for four pavilions at an exhibition at Nijny-Novgorod in 1896, and used the same design for the Bary Boiler works in Moscow. In 1933, at the Chicago Worlds Fair a pavilion to house a locomotive round-house was constructed with a cable-suspended roof structure.

Since 1933, many buildings have been constructed using cables as sus-pension systems for roof structures, as hangers supporting roofs, and as stays for roof structures. The application of cables to cable-stayed bridge construction has been presented in Chapter 1.

## 7.3  MANUFACTURING PROCESS

One of the most important features of the steel cables is their inherent strength and structural integrity. This strength is a result of the excellent quality control maintained throughout the manufacturing process from the type of iron ore to the finished product.[2]

Wires are produced by cold drawing a rod through a series of successive dies, Fig. 7.4. This process reduces the cross-sectional area of the rod by 65 to 75%. At the same time, cold drawing improves the internal struc-ture of the steel and thus increases the tensile strength.

The rods for the wire drawing operation are reduced from billets, heat-treated and carefully quenched at controlled temperatures, and inspected to ensure that all physical and chemical standards are met. They are then cleaned by dipping in acid to remove mill scale, rinsed in water, neutral-ized, coated with a lubricant which facilities the drawing process, and, finally, coiled for shipment to the wire mill. The heat treatment improves the tensile strength, relieves the stresses caused by the hot rolling, and controls the crystalline structure.

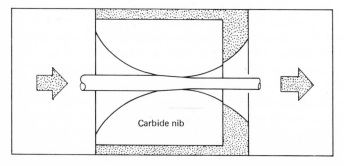

Carbide nib

**FIGURE 7.4.    Cross section of die. (Courtesy of the United States Steel Corporation. Ref. 2.)**

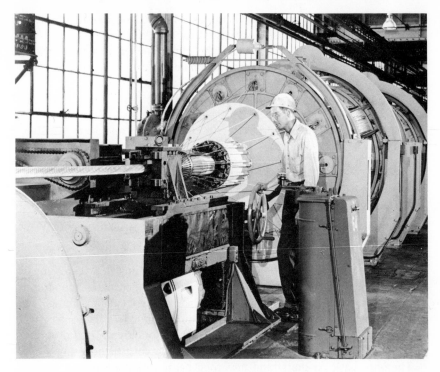

**FIGURE 7.5.    Stranding machine. (Courtesy of the Bethlehem Steel Corporation.)**

After the wire is drawn to a finished size, it undergoes a series of quality control tests. The tests include the usual tensile test for strength, torsion test to determine uniformity and toughness, and careful gauging to verify the diameter. The objective of all the operations of hot rolling, wire drawing, heat treating, and quality inspection is to ensure a wire of proper size, strength, and toughness before it is made into a rope or strand.

Strands and rope are produced from the individual wires that have been wound on steel spools, similar to bobbins of an ordinary sewing machine, before being placed in the cradle of a stranding machine, Fig. 7.5. In the stranding operation the individual wires are led from the spools over sheaves, through bushings along the periphery of the machine, converging through a twister-head into the proper location for the strand, which is guided through a stationary die. As the strand is pulled through the stationary die, the stranding machine and twister-head rotate continuously, forming the strand.

Larger structural strands are made by adding successive layers of wires. As the strand is pulled off the stranding machine it is wound on a reel or spool, depending on its subsequent use, as a finished structural strand or a strand to be used to produce a structural rope.

A structural rope is made in much the same manner as a strand except that strands replace wires in a larger machine.

## 7.4   MECHANICAL PROPERTIES

The static mechanical properties of structural strand and rope are stated in two standard specifications of the American Society for Testing and Materials (ASTM) designated as:

1. Standard Specification for Zinc-Coated Steel Structural Strand, ASTM A586-68.
2. Standard Specification for Zinc-Coated Steel Structural Wire Rope, ASTM A603-70.

These specifications contain information on the physical requirements, tests for zinc coating weight, data on the wires used to make the strands, strength tables, sampling, testing, inspecting, and packaging. Those planning to use the strands in construction should familiarize themselves with the two ASTM Standards in order to know how to design for its properties or handle the material knowledgeably during the construction stages.

The wire basic to the strand and rope has an ultimate tensile strength ranging from a high value of 220,000 psi (155 kilogram-force per square millimeter) for wires with class A zinc coating to a low value of 200,000 psi (141 kgf/mm$^2$) for wires with class C zinc coating.

The breaking strength for strand and rope with class A zinc coated (Section 7.7) wires throughout, range as shown in Table 7.1.

Other combinations of zinc coatings and number of wires will affect the minimum breaking strength to the extent that designers and contractors

TABLE 7.1   Breaking Strength—Strand, Rope

| Item | Strand | | Rope | |
|---|---|---|---|---|
| Minimum size, diameter | $\frac{1}{2}$ inch | 12.7 mm | $\frac{3}{8}$ inch | 9.53 mm |
| Minimum breaking strength | 15.0 tons[a] | 13.6 tons[b] | 6.5 tons[a] | 5.9 tons[b] |
| Maximum size, diameter | 4 inch | 101.6 mm | 4 inch | 101.60 mm |
| Maximum breaking strength | 925 tons[a] | 839.2 tons[b] | 730 tons[a] | 662 tons[b] |

[a] These are in tons of 2000 pounds.
[b] Metric tons

should refer to the ASTM Standards for the specific strength associated with each manufactured rope or strand size.

## 7.5    PRESTRETCHING

Although steel ropes and strands are considered to have safe and satisfactory elastic properties for most conventional service requirements, for certain end uses, such as structural applications, additional stretching of the manufactured product is necessary. True elasticity is required for applications such as main cables for suspension and stayed bridges, hangers or suspenders for arch and suspension bridges, guy ropes for high towers, cable-supported roof structures, and hangers for buildings.

Prestretching may be defined as the application of a predetermined tension force to a finished strand or rope in order to remove the cable looseness (constructional stretch) inherent in the manufacturing process. The prestretched cable becomes an elastic material within the limits of the prestretching operation, which enables a designer or contractor to predict the elongation under load with the high degree of accuracy necessary for structural applications. Another reason for prestretching the cables is that it permits measuring and marking of the proper spacing for the location of suspenders or the center of the towers. Although this marking is not required for cables to be installed in cable-stayed bridges there may be an occasion to locate spacers or vibration tie downs for certain configurations.

The prestretching operation consists of stretching a certain specified length of cable (sometimes as long as 5000 ft), in long successive "bites," on a stretching machine with tension jacks or screws, Fig. 7.6.[3]

Removal of the constructional stretch is effected by repeated applications of a tension load to the cable, which forces the component wires to seat themselves in closer contact. Upon removal from the prestretching bed, the cable is left with well-defined and uniform elastic properties that are similar to the steel itself. The prestretching load applied to the cable does not usually exceed 55% of the rated minimum breaking strength of strand or 50% for rope which essentially eliminates the constructional stretch of the cable or rope.

The prestretching equipment used by the cable manufacturers enables the designer to better predict the elastic behavior of the strand or rope after erection in the structure, because it eliminates the constructional stretch of the cable. Loading curves can be furnished as proof of the results of the operation.

FIGURE 7.6.   Prestretching operation. (Courtesy of the Bethlehem Steel Corporation.)

## 7.6   MODULUS OF ELASTICITY

The magnitude of the elastic elongation of a cable under tension is dependent upon the value of the Young's modulus of elasticity ($E$), which is defined as "the ratio of unit stress in the cable to a corresponding unit strain within a defined stress range."[1] Unlike the usual conventional tension test, the value for the modulus of elasticity for cables is determined from a gauge length of not less than 100 in. (254 cm) and is computed on the basis of the gross metallic area which includes the zinc coating. Experience in prestretching has indicated that stress-strain data taken from 1600 ft (487.7 m) lengths are much more accurate than those taken from a 100 in. (254 cm) gauge length.

The elongation data used to compute the modulus of elasticity are taken when the cable is stressed not less than 10% of the minimum breaking strength and not more than 90% of the prestretching load. The data is presented in the form of a load deflection diagram, Fig. 7.7.

The value for the modulus of elasticity is determined by calculation using the conventional expression for elastic elongation of a specified length of the material such that:

$$E = \frac{Pl}{Ae}$$

where $E$ = Young's Modulus of Elasticity, psi
   $P$ = increment of load, lb
   $l$ = gauge length, in.
   $A$ = gross metallic area, sq in.
   $e$ = elongation caused by load increment, in.

As an example a $2\frac{1}{8}$ inch diameter galvanized structural strand has the the following data:

At 10% breaking strength (58,400 lb) the elongation is 0.102 in. At 90% prestretching load (259,200 lb) the total elongation is 0.395 in. The difference in load and elongation is:

   $P$ = 200,800 lb
   $e$ = 0.293 in.
   $A$ = 2.715 sq in.
   $l$ = 100 in.

Therefore,

$$E = \frac{(200,800)(100)}{(2.715)(0.293)} = 25,242,000 \text{ psi}$$

It is to be noted that the value for $E$ is somewhat less than the usual 29,000,000 psi for structural steel.

In fact, the value of $E$ varies with the type of cable, such as strand, rope, or parallel wires, and is also dependent on the amount of zinc coating applied to the wires. The ASTM Specifications state minimum values to be used for the various sizes and coatings. The minimum modulus of elasticity of prestretched structural strand and rope for class A coating of zinc on the wires is given in Table 7.2.

The ASTM Specifications also state that for heavier zinc coatings, classes B and C, on outside wires, the value of $E$ is to be reduced by 1,000,000 psi; for other combinations of zinc coatings on all wires, the manufacturer should be consulted.

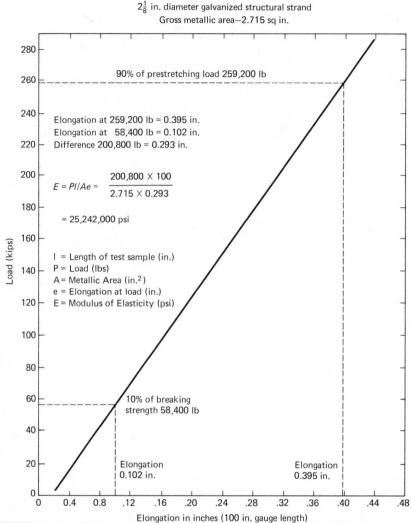

$2\frac{1}{8}$ in. diameter galvanized structural strand
Gross metallic area—2.715 sq in.

90% of prestretching load 259,200 lb

Elongation at 259,200 lb = 0.395 in.
Elongation at   58,400 lb = 0.102 in.
Difference 200,800 lb = 0.293 in.

$$E = Pl/Ae = \frac{200,800 \times 100}{2.715 \times 0.293}$$

= 25,242,000 psi

l = Length of test sample (in.)
P = Load (lbs)
A = Metallic Area (in.²)
e = Elongation at load (in.)
E = Modulus of Elasticity (psi)

10% of breaking
strength 58,400 lb

Elongation
0.102 in.

Elongation
0.395 in.

Load (kips)

Elongation in inches (100 in. gauge length)

FIGURE 7.7. Typical modulus of elasticity chart, based on data recorded during prestretching. (Courtesy of the United States Steel Corporation. Ref. 2.)

TABLE 7.2   Minimum Modulus of Elasticity Class A Coating of Zinc

| Type | Diameter | Modulus of Elasticity |
|---|---|---|
| Strand | $\frac{1}{2}$ to $2\frac{9}{16}$ in. | 24,000,000 psi |
| | $2\frac{5}{8}$ in. and larger | 23,000,000 psi |
| Rope | $\frac{3}{8}$–4 in. | 20,000,000 psi |

## 7.7   PROTECTIVE COATINGS

Protection of structural cables against corrosion is essential because the pitting and/or nicking of the surface of the steel wires creates points of weakness, so they cannot resist stress concentrations at these points. Corrosion affects all steel products in varying degrees, and structural cables are no more or less susceptible to rusting than other steel products. Therefore, protection of the wires is provided by various thicknesses of zinc coatings, depending on the location of the wire in the cable and the degree of atmospheric exposure expected.

The ASTM Specifications (A586 and A603) require that zinc conform to ASTM Specification B6, for zinc metal (slab zinc), high grade or better. The various thicknesses of zinc coatings are classified as class A, B, and C. The class A coating is the basic thickness that varies from 0.40 to 1.00 ounce per square foot of uncoated wire surface, (122.0 to 305.0 $g/m^2$), Class B coatings are twice as heavy, and Class C coatings are three times as heavy, Fig. 7.8.[2] The ASTM Specifications indicate values for breaking strength for three types of cables, depending on the zinc coatings and location of wires in the cable, such as:

1. Class A coating throughout all wires.

2. Class A coating for the inner wires and class B coating for the outer wires.

3. Class A coating for the inner wires and class C coating for the outer wires.

Other coatings and arrangements are possible if the atmospheric conditions warrant more or less protection. The contractor and designer should consult a manufacturer for any special weight of zinc coating desired on the wires.

As a general guideline the *Manual for Structural Applications of Steel Cables for Buildings*,[1] states that a class A coating is adequate for indoor

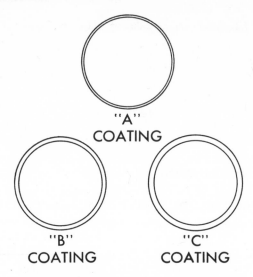

FIGURE 7.8.   Zinc coatings. (Courtesy of the
Bethlehem Steel Corporation.)

use and most outdoor exposures. For more severe exposures, classes B
and C are available as noted in the ASTM Specifications. If the exposure
is extreme, such as conditions of heavy condensation, salt or chemical
atmosphere, added protection should be supplied.

Acceptable methods of added cable protection are painting, plastic
jacketing, and the use of rust-preventive compounds, which are noted in
the recommendations of the Steel Structures Painting Council publication.
The federal specification TT-P-641 suggests that zinc dust–zinc oxide
paints can be used to restore the original zinc coating protection, especially
where there is little bending action in the cables.

It is to be noted, however, that as the zinc coating is increased in thickness it displaces a larger portion of steel area, and as a result, the breaking strength of the same size cable is reduced when using the heavier coatings. Therefore, the contractor and designer must consider the advisability
of a reduced strength when requiring more zinc protection for a given
diameter of cable.

## 7.8   STRAND AND ROPE COMPARED

The difference in the ASTM Specifications for the breaking strengths of
strand and rope for a given diameter is not the only consideration involved

in choosing the cable for a specific application. Because of the difference in the manufacturing process that uses strands instead of wires to make a rope, other characteristics of strand and rope may influence the choice.

The significant differences between strand and rope, which should be considered in making a selection, are summarized below.

1. Strand is stronger than rope of the same diameter.

2. Strand has a higher modulus of elasticity than rope.

3. Strand is less flexible than rope, and is not used with small bend radius curvatures.

4. Strand has larger diameter wires than rope of the same size, consequently, strand of a given class of zinc coating is more resistant to corrosion because of the thicker coating on the larger wires.

5. Strand is specified when flexibility or bending is not a major requirement.

6. Rope is specified when bending of the cable is an important consideration in the application of the cable.

7. The outside surface of strand is smoother than rope; therefore it may be protected more easily with a paint covering.

8. Strand uses smaller accessory fittings, because the strand diameter required for a given load is smaller than rope.

9. Rope is easier to handle in the field because it is more flexible.

10. Saddles for ropes are generally smaller than those for strands because rope can be bent to a shorter radius.

11. Angle changes at bands and clamps may be larger for rope than strand because of the flexibility of the rope.

## 7.9  PARALLEL WIRE STRANDS

Until 1968, structural strands and ropes were manufactured by winding several wires around a core in a helical pattern. Although ropes are still produced in the same manner, strands are now being produced by an alternate process and consist of a group of parallel wires held closely together by end fittings. The number of wires per strand may vary from 19 to 127 for convenience in handling. The number of parallel strand units in a single cable is determined by the magnitude of the design tension force to be supported.

The shop assembly of parallel wire strands for applications in cable-stayed or suspension bridges is a comparatively recent development by the manufacturers of bridge cables.[4,5,6] The in-plant process is an alternative to the old method of "spinning" parallel wires in place for the main suspension cables and also provides an alternate cable to the shop assembled helical strand or rope.

The use of parallel wire strands is not a recent innovation, but an improvement and acceptance of a concept that originated during the late 1920s and early 1930s. At that time the cable consisted of a group of prefabricated, machine-made helical strands. Although this was a beginning, engineers, manufacturers, and contractors were still hoping to develop a completely prefabricated parallel wire strand. In later years several occasions presented an opportunity to continue the development of the parallel wire strand; these were: a small bridge over the Gila River at Globe, Arizona,[7] and the original construction of the Portsmouth, Ohio, bridge over the Ohio River[8] where the parallel wire strands were prefabricated at the site along the river bank. Another variation of the concept was used for the Otto Beit Bridge, Rhodesia, Africa, where British contractors hauled many wires simultaneously in horizontal layers.[9]

The first successfully applied parallel wire cables which were completely shop fabricated were manufactured by Bethlehem Steel Corporation for the Newport Suspension Bridge constructed in May 1968.

The Newport Bridge, built for the Rhode Island Bridge and Turnpike Authority, crosses Narragansett Bay connecting Jamestown Island to Newport Island. The strands for the main cables were prefabricated in the shop and consist of parallel wires which were cut to length, had fittings added, and reeled for transportation. At the site, the strand reel was anchored at one end and the strands were unreeled by pulling the lead fitting over the support points to the other anchorage.

The manufacturing process for the parallel wire strand uses wires of the same diameter in various combinations of 19, 37, 61 etc. to form a perfect hexagonal configuration, Fig. 7.9. Experience has indicated that the perfect hexagon is the most compact grouping of wires and provides a geometry in which equal lengths of individual wires can most easily be maintained and thus achieve uniform stressing in all wires. For a required stress condition, the selection of the proper wire diameter will determine the economical areas for the strand. Wire diameters most often used range from 0.177 to 0.225 in. in diameter. This provides a sufficient range to permit an economical cable selection.

The criteria for an economical cable selection encompasses the following considerations:

1. Minimum number of strands.
2. Largest number of wires per strand.
3. Largest diameter of individual wires.

**FIGURE 7.9.** Parallel wire strand—hexagonal arrangement before compaction into round cable.

Items 2 and 3 are within limits imposed by wire properties, by bending and reeling problems, and by transportation facilities.

Preassembled parallel wire strands, in contrast to structural strands, do not require prestretching because there is no constructional stretch to be eliminated. All the wires are in a straight alignment and the elastic behavior of the strand approaches that of the wire. In final position in the bridge structure the cable is in a circular shape, having been compacted from the as-erected hexagonal configuration, Figs. 7.10 and 7.11.

The quality of steel for the individual wires of a preassembled parallel wire strand is the same as that used for the wires of structural rope and strand, except that most specifications add a provision that for any heat the average tensile strength shall be a minimum of 225,000 psi. The minimum modulus of elasticity of a parallel wire strand is in the range of 27,500,000 to 28,500,000 psi compared to the lower value of 24,000,000 psi, for strand of $2\frac{9}{16}$ inches diameter and smaller. The higher modulus is the result of the wires being placed in a parallel position and, therefore, approaching the elastic characteristics of the individual wire.

Some laboratory tests for the modulus of elasticity for parallel wire strands have indicated a value of 27,500,000 psi, which is slightly lower

FIGURE 7.10. Parallel wire strand after compaction into round cable.

FIGURE 7.11. Parallel wire strand—close-up of compaction.

than the value for the individual wire.[5] As a result, some engineers prefer to use an effective value of 27,500,000 psi for the parallel wire strand to account for the fact that the modulus of elasticity of the individual wires tends to fall over time.

Tension tests to destruction of preassembled parallel wire strands are usually not necessary because of the rigorous tests performed on the individual wire. However, break tests may be made to verify the adequacy of the sockets which are required to develop 100% of the breaking strength of the strand.

The allowable working stress for the parallel wire strands is in the range of 85,000 to 90,000 psi, which is factored down from a yield strength of 160,000 psi. Consequently, working stresses have a factor of safety of 1.88 and 1.78, against yielding of the strands.

The end fittings for the parallel wire strands are the zinc socketed type which are described in Chapter 8.

In designing connections for cable stays, the designer can either terminate the strands at each tower or run them continuously through the towers from the center span to the end span on saddles especially fabricated for the number of strands to be accommodated. The saddles must have a generous radius to minimize the local bending effects and ensure only tension stresses in the cables.

A large angular change may cause differentially stressed lengths among the wires of the strand sufficient to alter the stress condition in the strand. A corollary effect is to be avoided during erection when care must be taken to avoid the longitudinal movement of wires with respect to each other over the saddle.

The cable can be protected after installation by one of two recent developments:

1. A multiple-layer plastic coating impregnated with glass-reinforced acrylic resins.

2. An elastomeric coating made from liquid neoprene, multiple coats of uncured neoprene sheets and top coats of hypolon paint.

The latter system air cures to form a coating that is not only resilient, but also weatherproof. The designers and contractors should familiarize themselves with all the methods available and determine the one technique which is most effective for the particular application and atmospheric conditions.

## 7.10   HANDLING

The handling of structural cables to prevent damage is an important consideration which may save lives and money during the erection and life of the structure.

The strands should not be dragged over obstacles that can cause cuts, nicks, abrasions, or remove the zinc on the wires before they are wrapped. These defects can cause premature failure of the cables because of the local stress concentrations.

Storing unprotected strands in locations subject to corrosive elements can cause pitting and rusting which are detrimental to the wires, thus reducing the tension load capacity and increasing the likelihood of early failure.

The reel should be permitted to rotate as the strand is unwound to avoid kinks or twisting which will damage the wires. The twisting will cause a shortening or lengthening of the strand resulting in stress changes in the wires.

During erection, when lifting a strand into position, the crane or sling attachment should be connected in such a manner as to avoid sharp bends in the cable, and the cable should be kept clear of obstructions in order not to cause abrasion.

Strands are packaged at the manufacturer's plant in accordance with approved practices. The strands are packaged in coils or on reels at the discretion of the manufacturer and in such a manner that no permanent deformation of the wires will occur.

## REFERENCES

1. *Manual For Structural Applications of Steel Cables for Buildings*, American Iron and Steel Institute, Washington, D. C., 1973.
2. Scalzi, J. B., Podolny, W., Jr. and Teng, W. C., "Design Fundamentals of Cable Roof Structures," ADUSS 55-3580-01, United States Steel Corporation, Pittsburgh, Pennsylvania, October 1969.
3. "Bethlehem Wire Rope for Bridges, Towers, Aerial Tramways, and Structures," Catalog 2277-A, Bethlehem Steel Corporation, Bethlehem, Pennsylvania.
4. Scalzi, J. B. and McGrath, W. K., "Mechanical Properties of Structural Cables," *Journal of the Structural Division*, ASCE, Vol. 97, No. ST 12, December 1971, Proc. Paper 8604.
5. Birdsall, B., Discussion to "Mechanical Properties of Structural Cables," by Scalzi and McGrath, *Journal of the Structural Division*, ASCE, August 1972.
6. Birdsall, B., "Main Cables of Newport Suspension Bridge," *Journal of the Structural Division*, ASCE, Vol. 97, No. ST 12, December 1971, Proc. Paper 8606.

7. "Contrasts in Bridge Building," *Suspension Bridges: A Century of Progress*, John A. Roebling's Sons Co. (Engineering Societies' Library, New York No. ESL624.5 R62s).

8. Steinman, D. B., "Ohio River Suspension Bridge at Portsmouth," *Engineering News-Record*, Vol. 99, No. 16, October 20, 1927.

9. Shirley-Smith, H. and Freeman, R., Jr., "The Design and Erection of the Birchenough and Otto Beit Bridges, Rhodesia," *Proceedings of the Institution of Civil Engineers*, London, Vol. 24, 1945.

# 8

# Cable Connections

## 8.1  INTRODUCTION

The structural design and construction details of the individual component members of cable-stayed bridges are similar to those of suspension bridges and/or other conventional type bridges. The one principal difference is the attachment of the special end fittings to the cable itself.

Many types of fittings are available, depending on the size of the cable, the manufacturer producing the fitting, and the unique desires of the designer. Cables may be connected to the towers and the superstructure by pins joining the end fittings of the cables to the attaching fittings on the supporting member or by terminal fittings that transmit their force in bearing. As in suspension bridges, saddles may be used on the towers to permit the use of a continuous cable and allow movement to take place.

**208**

This chapter first discusses the general considerations of these connections, which are applicable to all geometrical types of cable-stayed bridges, and then discusses typical terminal fittings and saddles.

Specific cable anchorage details for selected existing bridge structures are illustrated and discussed as case studies. Other unique construction and erection details that are of interest to the designer and the contractor are also included. We have assumed that all conventional details are familiar to the professionals and contractors and, therefore, have not included them.

American designers and contractors have expertise and experience with cable assemblies in bridge structures and in cable-supported roofs of many varieties. However, because experience with the type of cables and methods of construction of cable-stayed bridges is quite limited, a review of the details of construction by others will be helpful. The experience gained in other countries may not be directly applicable to practice in the United States, but concepts may be adapted and improved by our techniques and ingenuity. As in any new concept and innovation we must evaluate the experience of others carefully. This chapter illustrates the types of connections used in existing bridges in Europe without attempting to evaluate or compare them in any way, either with each other or with an equivalent American practice.

In general cable connections should

1. Provide full transfer of loads.
2. Provide access for inspection.
3. Provide protection against weather.
4. Provide protection against accidental damage to the cable.
5. Provide sufficient space for initial tensioning and later adjustments.
6. Use standard fittings as much as possible.
7. Consider erection procedures when selecting types of connections.

## 8.2 GENERAL CONSIDERATIONS

When choosing the particular geometrical configuration of the cable stays and the number of cables to be used in a specific system, several considerations must be taken into account. Foremost among these considerations is the comparative cost of additional material required in the superstructure to resist the horizontal thrust versus the increased cost of the tower, cables, and their connections to the supporting members of the bridge. In other words, the number of cables used may depend on the economic balance between the distribution of the cables along the span and the number of cables to be connected at the tower and superstructure.

The use of a few cables results in large tensile forces, which require massive and sometimes complicated anchorage systems to transfer the loads to the tower and the bridge deck. The deck structure will have to be heavily reinforced at a few points of attachment as a result of the concentration of loads. When only a few cables are used, the deck structural system must be composed of a relatively deep girder or box in order to span the large distances between the cable connections. In addition to a deeper deck structure, the transverse girders at the cable attachments may also require reinforcement in order to distribute the horizontal thrust of the cable as uniformly as possible throughout the structural system. With only a few cables connected to the tower, large connection details are required, and these become exceptionally bulky, heavy, and cumbersome if all the cables are to converge at one point near the top.

When a large number of cables are used, the connections along the bridge deck are generally uniformly spaced and provide a nearly continuous elastic supporting media for the deck structure. With closely spaced connection points, the principal longitudinal girders or box members can be shallow in depth, thus requiring less material and simplifying the fabrication and erection procedures. The use of a large number of cables automatically implies smaller diameter cables requiring smaller terminal fittings and connection details to the tower and bridge girders. Although the handling of the cables may be increased slightly, the reduced weight and easier fabrication and erection may more than offset this disadvantage by reducing the total time and cost of the project. With a larger number of connection points along the bridge deck, the designer can distribute the cable horizontal thrust more efficiently along the deck, use smaller anchorage details, and, usually, needs no additional reinforcement to the transverse beams. The connections to the towers will be distributed to many locations along the height of the tower, thus simplifying the fitting details and distributing the gravity load almost uniformly along the height instead of concentrating the total load at the top.

In all configurations the designer has the choice of terminating the cables at the tower connections or permitting them to pass through the tower as a continuous member. In the latter case, the cable must be supported on a special fitting, referred to as a saddle, designed to fit the number and size of cables passing through the tower at that location. When only a few cables are to be connected to the tower, it may be more advisable to pass them over a saddle, rather than attempt a costly or impractical terminating connection to the tower. When a large number of smaller cables are distributed throughout the height of the tower, it becomes more practical and sometimes more economical to terminate the cables at the tower.

The cable saddles may be either rigidly connected to the tower or supported on expansion bearings that permit longitudinal movements to take place. When the saddles are fixed to the tower they add stiffness to the structural system, thus increasing the rigidity of the total bridge structure.

On some structures the designer may choose to connect the base of the tower to the supporting structure, thus producing a fixed condition. In this instance the saddle is allowed to move longitudinally, thus reducing the bending moments acting on the tower. Only the gravity load is transferred to the tower. The saddle movement adjusts itself to accommodate the necessary balance of the horizontal forces on each side of the tower. When a large number of cables are used, the saddle for the top cable stay may be fixed, and all or a few of the lower saddles permitted to move. As may be expected, the selection of a geometrical cable arrangement depends on many conditions particular to a specific application.[1]

## 8.3   END FITTINGS

Structural strands and ropes have end fittings by which they can be connected to other parts of a structure. These fittings vary in shape, size, and weight depending on the diameter of the cable to which they are attached. Experience in the shop and field has indicated that fittings should be designed, manufactured, and attached to the cable so that they are capable of transferring the breaking strength of the cable without exceeding the yield strength of the fitting.

Two types of end fittings are generally referred to as sockets: the swaged type and the zinc-poured type. The swaged sockets are used for the small diameters of strand and rope, ranging from $\frac{1}{2}$ to $1\frac{3}{8}$ in. for strand cables and $\frac{3}{8}$ to 2 in. for rope cables, Fig. 8.1.[2] The swaging process consists of carefully pressure squeezing the fitting over the cable in a hydraulic press in order not to damage the wires. The size limitations of the swaged fittings restrict their use and they are not now used for the main stay cables of bridges. The poured-zinc type of fitting is the accepted method for the large cable sizes used in cable-stayed bridge construction in the United States. The standard fittings are illustrated in Fig. 8.2, which denotes the type of fitting and the size of the cable which it can accommodate. The cable diameters range form $\frac{1}{2}$ to 4 in. for strand and $\frac{3}{8}$ to 4 in. for rope.

The poured-zinc method of attaching the end fittings to the cable is a unique technique, which must be performed by experienced technicians. The method involves several operations, and each must be accomplished properly in order to guarantee the full strength of the cable.

| Name | Description | Attach-ment | Sizes Available Diam. in. | |
|---|---|---|---|---|
| | | | Strand | Rope |
| Closed Swaged Socket | 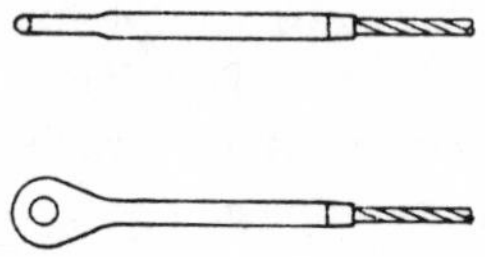 | Swaged | ½–1⅜ | ⅜–2 |
| Open Swaged Socket | 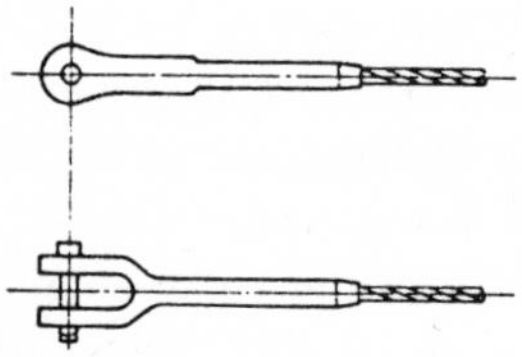 | Swaged | ½–1⅜ | ⅜–2 |
| Threaded Swaged Socket | 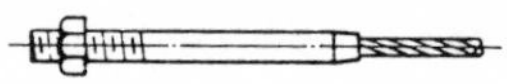 | Swaged | ½–1⅜ | ⅜–2 |

FIGURE 8.1. Swaged end fittings. (Courtesy of the American Iron and Steel Institute. Ref. 2.)

The sequence of operations begins with the "brooming" of the individual wires of the cable for the length sufficient to be inserted in the "basket" of the socket, Fig. 8.3. These broomed ends are carefully cleaned and immersed in a flux solution to prepare them to adhere to the zinc. The wires are then placed in the basket of the socket in a manner that will ensure that every wire will be surrounded by the poured molten zinc, Fig. 8.4. The ASTM Specifications A586 and A603 for strand and rope, respectively, specify that the slab zinc shall conform to ASTM Specification B6 for zinc metal (slab zinc), high grade or better. By careful attention to each phase of the operation, the attached fitting will develop the full breaking strength of the cable.

The Japanese have reported[3] that the pouring temperature of the zinc alloy when filling the socket affects the fatigue strength of the wires at the socket. The results of the study, indicated graphically in Fig. 8.5, show that a casting temperature of 450°C increases the fatigue strength of the wires when compared to a temperature of 480°C. A German report[4] has indicated that at the 450°C casting temperature there is still a decrease in fatigue resistance as compared to the original parent wire material, in or near the cable anchorage. Thul has reported[4] improved results in tensile, fatigue, and creep tests using a cold casting material composed of steel balls, zinc dust, and an epoxy resin mixture. However, no conclusion has been reached with respect to the aging resistance of the resin material in the anchorage.

Another method of attachment for a parallel wire cable is to use a button head on the wires that bear on a steel anchorage plate. This technique is similar to one of the prestressed posttensioned anchorage systems. The anchorage system closes the end of the conical cavity of the steel socket after the voids have been filled with a special mixture of metals and epoxy resin, Fig. 8.6. This method of anchoring the cables reportedly increases the magnitude of strength almost twice that for zinc-filled sockets.[5]

The selection of the type of end fitting to be used in a particular application depends on the details of the connection as determined by the designer. Construction sequence and equipment may influence the decision. In other instances, the type of fitting may be dictated by the structural arrangement of the components—certain fittings may be impractical or impossible because of the inaccessibility of the parts to be connected.

The standard fittings shown in Fig. 8.2 can have many different applications. With the open and closed sockets, a pin can be inserted through the fitting and connect it to other parts of the structure, Fig. 8.2a and b. Several types of bridge sockets are available that meet the needs of the various designs, Fig. 8.2c, d, i. Sockets that transfer their loads by direct bearing on another structural component have various configurations in order to

| Name | Description | Attachment | Sizes Available Diam. in. | |
| --- | --- | --- | --- | --- |
| | | | Strand | Rope |
| Closed Socket (a) | 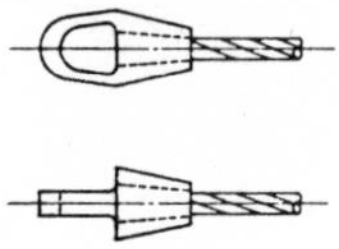 | Poured Zinc | ½–4 | ⅜–4 |
| Open Socket (b) | 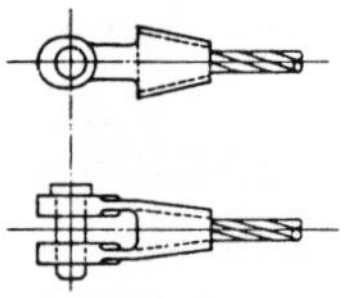 | Poured Zinc | ½–4 | ⅜–4 |
| Open Bridge Socket (c) | 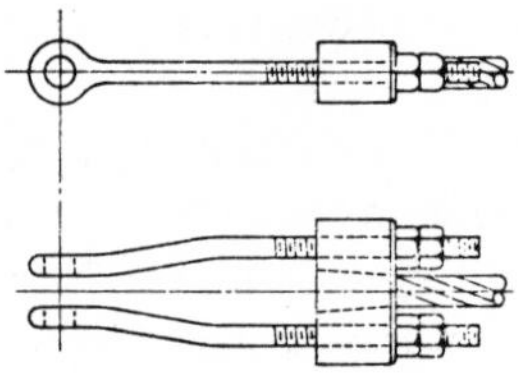 | Poured Zinc | ½–4 | ⅜–4 |
| Closed Bridge Socket (d) | 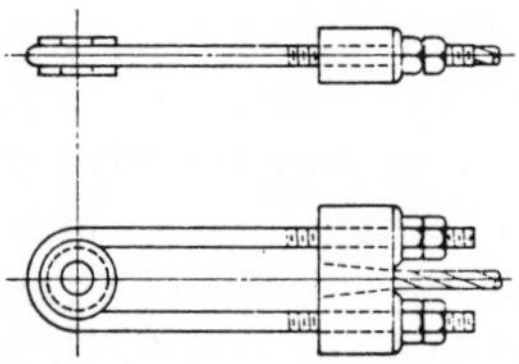 | Poured Zinc | ½–4 | ⅜–4 |

**FIGURE 8.2.** Socketed end fittings. (Courtesy of the American Iron and Steel Institute. Ref. 2.)

| Name | Description | Attach-ment | Sizes Available Diam. in. | |
|---|---|---|---|---|
| | | | Strand | Rope |
| Button Socket (e) | Bearing Surface | Poured Zinc | ½–4 | ⅜–4 |
| Bearing Sockets (f) | (a) Bearing Surface (b) | Poured Zinc | ½–4 | ⅜–4 |
| Threaded Socket (g) | Bearing Surface Internal threads optional | Poured Zinc | ½–4 | ⅜–4 |
| Threaded Stud Socket (h) | Bearing Surface | Poured Zinc | ½–4 | ⅜–4 |
| Bridge Socket Bowl (i) | Anchor rods by others | Poured Zinc | ½–4 | ⅜–4 |

**FIGURE 8.2.** (*continued*)  **215**

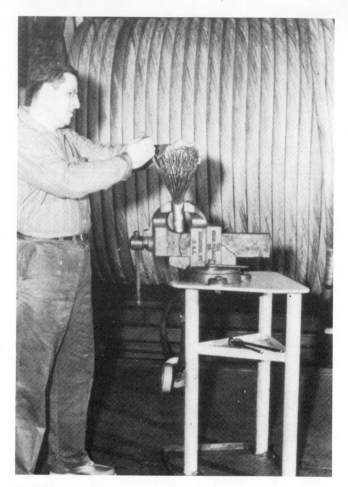

**FIGURE 8.3.  Brooming out. (Courtesy of U.S. Steel Corporation. Ref. 18.)**

suit the specific method of application. Fig. 8.2e, f, g, h. The type with external nuts and internal threads allow for periodic checks and adjustments of the cable tension. Jacks are used to apply tension to the cable through the threaded portions of the fittings. The use of the bearing sockets is analogous to the concept of the dead-end anchorage of a post-tensioned, prestressed concrete member.

In general, the standard fittings have been found satisfactory in all previous applications. Therefore, it is recommended that, where possible, standard, rather than specially designed, fittings be specified. Individually

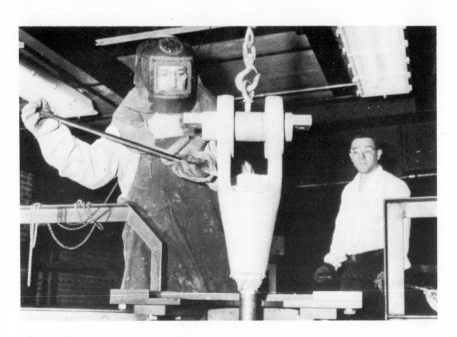

**FIGURE 8.4.** Pouring zinc into socket. (Courtesy of U.S. Steel Corporation. Ref. 18.)

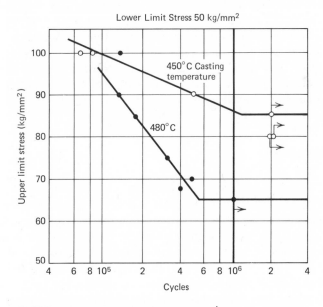

**FIGURE 8.5.** Fatigue text of wire $\phi$ 5 mm with zinc-copper-alloy filled sockets. (Courtesy of *Der Stahlbau*. Ref. 3.)

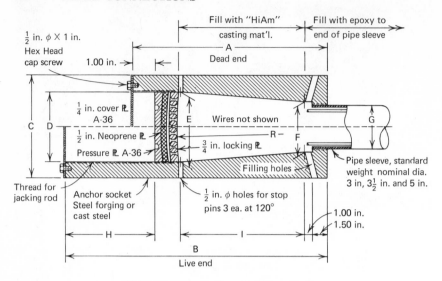

**FIGURE 8.6.    Cable end fitting detail.**

designed fittings must be tested for strength and deformation before acceptance. If other than a standard fitting is to be used, the designer should consult a manufacturer before proceeding with a special fitting design. Normally, a minor change in design of the structure will allow the use of a standard fitting, resulting in a reduction of fabrication and construction time and costs. The BBRV type of terminal hardware as used in posttensioned, prestressed concrete has been used with button-headed parallel wires in some cable-stayed applications, but requires special care to prevent entrance of moisture.

## 8.4    SADDLES

Saddles are grooved cable supports designed with due consideration of the bearing pressures, bend radii, and groove diameters. All surfaces in contact with the cables should be smooth to avoid nicks in the wires. To avoid stress concentrations and minimize excessive bending at the end of the grooves, a generous contour is provided, which eliminates cable chafing.

A bridge engineer must consider both the design features of a saddle and the requirements of the cables before he designs the saddle. Because design conditions vary for each bridge, the saddles must be designed and fabricated expressly for each installation (as is current practice for conventional suspension bridges), Fig. 8.7. The saddles may be produced from

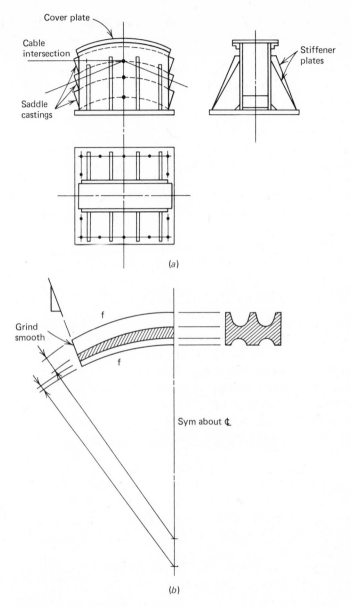

FIGURE 8.7. Typical saddle: (*a*) tower saddle; (*b*) saddle casting.

fabricated plates or steel castings with grooves in the form of an arc for the individual cables to rest on. The profile of the saddle transverse to the direction of the cables is formed to suit the desired cable arrangement.

In order to ensure the proper seating of the individual cables or strands, a zinc or aluminum filler may be used in the groove. These soft materials will flow plastically and provide a smooth surface for the cables to rest upon.

The radius of the saddle grooves must provide a contact area between cable and saddle that results in permissible bearing pressures on the cables and the saddle. The radius must also be selected to maintain the bending stresses in the outer fibers of the cable within allowable limits. When movement of the saddle is not provided in the design and construction, the unbalanced forces at the saddle must be resisted by friction and shear on the plates between layers of strands or cables. Additional clamping force may be provided by a clamp over the top of the cables that holds them in a fixed position.

In view of the many special features involved in the design and fabrication of saddles, engineers and contractors should consult a manufacturer or cable specialist. There is needed expertise from experience in determining:

1. Adequate tolerances for the saddle grooves.

2. Values for the coefficient of friction to prevent sliding of the cables on the grooves.

3. Maximum allowable bearing pressures.

4. The percentage reduction in breaking strength for various ratios of saddle radius to cable diameter.

5. The suitable deflection angles for live loads.

6. The best method of supporting a bundle of cables.

Saddles are an important construction detail, and engineers should consult with a contractor or cable specialist on methods of erection and a manufacturer or cable specialist on the best composition and geometry to suit the load conditions of each application. The best bridge design on paper is of no value to anyone unless it can be built easily and economically to perform its function.

In the absence of experience and for preliminary design by nonspecialists the AISI suggests certain limits that should be considered in the design of saddles, Fig. 8.8. Before specifying them, it is recommended that the saddles be discussed in greater detail with a specialist.

## 8.5  CLAMPS

A clamp is defined as "a fitting which develops its load-carrying capacity by friction." They are used to attach a small cable to a larger cable such

## SADDLES

1. Saddle groove diameters for individual wire ropes and strands shall be the nominal diameter of the cable plus the following tolerances:

| Wire Rope | | | Strand | | |
|---|---|---|---|---|---|
| | Tolerance, in. | | | Tolerance, in. | |
| Wire Rope Diam. in. | Min. | Max. | Strand Diam. in. | Min. | Max. |
| $\frac{3}{8}$–$1\frac{1}{8}$ | $+\frac{1}{32}$ | $+\frac{3}{32}$ | $\frac{1}{2}$–$1\frac{1}{2}$ | $+\frac{1}{32}$ | $+\frac{3}{32}$ |
| $1\frac{1}{16}$–$1\frac{1}{2}$ | $+\frac{1}{16}$ | $+\frac{1}{8}$ | $1\frac{9}{16}$–$2\frac{1}{2}$ | $+\frac{1}{16}$ | $+\frac{1}{8}$ |
| $1\frac{9}{16}$–3 | $+\frac{1}{8}$ | $+\frac{5}{16}$ | Over $2\frac{1}{2}$ | $+\frac{1}{6}$ | $+\frac{3}{16}$ |
| Over 3 | $+\frac{3}{16}$ | $+\frac{5}{16}$ | | | |

2. The maximum design coefficient of friction to prevent sliding in a saddle shall be 7%. This provides a factor of safety with respect to the actual coefficient. It must be recognized that the actual coefficient is a variable and can be much higher than 7%, perhaps as high as 12 to 15%.

3. The maximum allowable projected bearing pressure exerted on the strand or rope by the saddle shall be 4000 psi for 3 in. diameters and larger, increasing linearly to 6000 psi for 1 in. diameters.

4. Cast saddles shall have a smooth surface in groove. All rough and high spots shall be removed.

5. For individual strands or wire ropes, the effective design breaking strength as a function of the ratio of saddle radius (base of groove) to cable diameter is as follows:

| Strand Ratio | Wire Rope Ratio | Effective Design Breaking Strength as % of Specified Minimum Breaking Strength |
|---|---|---|
| 20 & over | 15 & over | 100 |
| 19 | 14 | 95 |
| 18 | 13 | 90 |
| 17 | 12 | 85 |
| 16 | 11 | 80 |
| 15-min. | 10-min. | 75 |

6. The above criteria apply only to saddles where the live load changes the deflection angle of the cable less than 2° for strand and 4° for wire rope per saddle end, or where a larger change happens so infrequently that there is no damage to the cable or enlargement of the saddle groove. If the criteria do not apply, it is suggested that a cable manufacturer be consulted.

7. If it is desired to support in a saddle a cable consisting of a bundle of strands or ropes, it is suggested that a specialist be consulted.

**FIGURE 8.8. Saddles. (Courtesy of the American Iron and Steel Institute. Ref. 2.)**

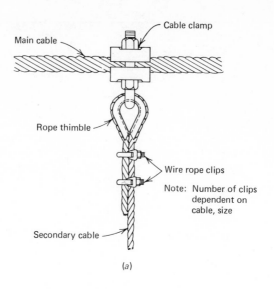

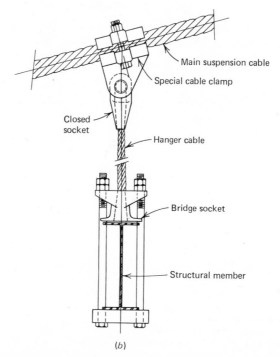

FIGURE 8.9. Cable clamp connection: (*a*) cable-to-cable connection; (*b*) cable-to-cable and cable-hanger-to-beam connection. (Courtesy of U.S. Steel Corporation. Ref. 18.)

as a hanger. As in all fittings, clamps, Fig. 8.9, must be designed to suit specific conditions and requirements. The engineer who designs the clamp must take into account the permissible bearing pressures, the limiting deflection angles for bends or change in direction of the cable, the holding power to be developed, and the groove diameters. When all these factors have been considered and smooth contact surfaces have been provided, the clamp should prove to be adequately designed. As in saddle design, edges of the fitting that contact the cable should be sufficiently rounded to avoid cutting or nicking the cable during erection and under the action of service load conditions.

Similar to saddle design, the AISI has suggested factors and data that should be considered in the design of a clamp, Fig. 8.10.[2] However, before a final design is attempted, a specialist should be consulted for the latest information and data.

## 8.6  MAIN RIVER BRIDGE, GERMANY

A new type anchorage system developed by Dyckerhoff and Widman was installed for the first time on the Main River Bridge constructed in 1971, near Hoechst, a suburb of Frankfurt.[6] The anchorage and assembly system (referred to as Dywidag) is a unique method of using 25 parallel threaded bars, 16 mm ($\frac{5}{8}$ in.) in diameter, arranged and anchored to resist dynamic and static loads, Fig. 8.11. The bars are maintained in a parallel condition by a polyethelene spacer encased in a steel conduit. For the dynamic load anchorage zone, an end piece connected to the conduit has rivets on its surface to improve the bond when it is embedded in the concrete. This end piece is closed with a guide cap from which the individual bars with their own conduits are led to the end anchorages.

The advantages claimed for the Dywidag parallel bar system are:[7,8]

1. Full use of the stress range of the bar by eliminating weakness in the anchorage.

2. Complete corrosion protection by concrete encasement.

3. Accurate and simple erection by tensioning individual bars.

4. Elimination of cable elongation and slip in the anchorage so that later adjustments of the bars are unnecessary.

5. Simple assembly of the system on the construction site.

6. Favorable behavior with respect to wind excited vibrations.

## CLAMPS

1. The groove diameter of clamps shall be the nominal diameter of the cable plus the following tolerances:

| | Wire Rope | | | Strand | | |
|---|---|---|---|---|---|---|
| | | Tolerance, in. | | | | Tolerance, in. |
| Wire Rope Diam. in. | | Min. | Max. | Strand Diam. in. | Min. | Max. |
| $\frac{3}{8}-1\frac{1}{8}$ | | $+\frac{1}{32}$ | $+\frac{3}{32}$ | $\frac{1}{2}-1\frac{1}{2}$ | $+\frac{1}{32}$ | $+\frac{3}{32}$ |
| $1\frac{3}{16}-1\frac{1}{2}$ | | $+\frac{1}{16}$ | $+\frac{1}{8}$ | $1\frac{9}{16}-2\frac{1}{2}$ | $+\frac{1}{16}$ | $+\frac{1}{8}$ |
| $1\frac{9}{16}-2\frac{1}{4}$ | | $+\frac{3}{32}$ | $+\frac{5}{32}$ | Over $2\frac{1}{2}$ | $+\frac{1}{6}$ | $+\frac{3}{16}$ |
| $2\frac{5}{16}-3$ | | $+\frac{1}{8}$ | $+\frac{3}{16}$ | | | |
| Over 3 | | $+\frac{5}{32}$ | $+\frac{1}{4}$ | | | |

2. The depth of the groove in clamps shall be such that when the clamps are fully tightened their faces will not be in contact with each other.

3. The maximum design coefficient of friction to prevent slipping of the clamps on the strand or rope shall be 7%.

4. The maximum allowable projected bearing pressure exerted on the strand or rope by the clamp shall be 4000 psi for 3 in. diameters and larger, increasing linearly to 6000 psi for 1 in. diameters.

5. Cast clamps shall have a smooth surface in the grooves. All rough and high spots shall be removed.

6. All edges that contact the strand or wire rope shall be sufficiently rounded to suit all conditions of design and erection.

7. The angular change in direction of the cable at a clamp of the type shown in the sketch shall not exceed $1\frac{1}{2}°$ for strand or 3° for wire rope, as shown in the sketch below.

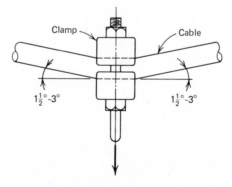

8. The nuts on the bolts of the clamps shall be alternately and gradually tightened to avoid any excessive unequal stressing of the clamp components or the cable to which the clamp is being attached.

**FIGURE 8.10.   Clamps. (Courtesy of the American Iron and Steel Institute. Ref. 2.)**

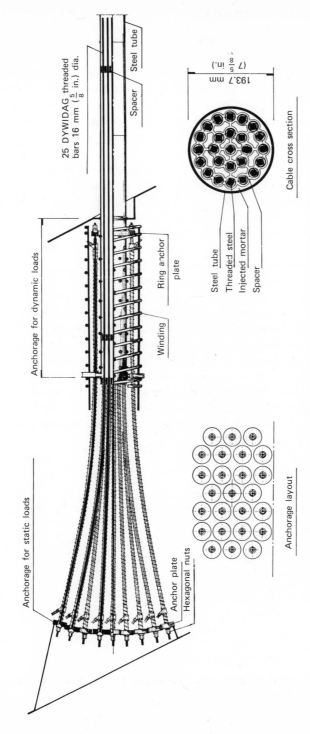

25 DYWIDAG threaded bars 16 mm ($\frac{5}{8}$ in.) dia.

Steel tube

Spacer

Anchorage for dynamic loads

Ring anchor plate

Winding

Anchorage for static loads

Anchor plate

Hexagonal nuts

193.7 mm ($7\frac{5}{8}$ in.)

Cable cross section

Steel tube
Threaded steel
Injected mortar
Spacer

Anchorage layout

FIGURE 8.11. Anchorage for Main River Bridge. (Courtesy of Dickerhoff & Widman.)

## 8.7    SEVERIN BRIDGE, GERMANY

The Severin Bridge at Cologne, Germany, has one tower and six cables that radiate in two directions from the top of the tower to the deck structure. Fig. 8.12. The size of the cables vary with the location of the intersection point along the deck span. The cables closest to the tower are the smallest in size and the ones extending further away are the largest as indicated by an increased cross-section, Fig. 8.12. Each of the cables includes 4 to 16 individual strands. Most of the cables are continuous through the tower, resting on saddles and clamped rigidly to them, Fig. 8.13. All cables intersect the center line of the tower in elevation; viewed transversely, they are slightly offset from the center line of each leg toward the longitudinal center line of the bridge, Fig. 8.13. However, the planes of the cables intersect at the center line of the bridge at a theoretical connection point above the tower.

The cable sag caused by cable dead weight is vertical, which offsets the inclined plane in space. Because the cable planes are inclined, the saddles are also inclined to receive the cables more efficiently.[9,10] When the number of strands in a cable is different on each side of the tower at a common intersection point (see cables 1–6 and 2–5 in Fig. 8.13), the additional unbalanced strands are anchored directly to the tower, Figs. 8.13 and 8.14. Cables 3 and 4 are anchored and terminated at the tower because the angle of intersection is too acute for a saddle. A radius of curvature appropriate for the tower structure would have been too small for an effective saddle, and a saddle with proper radius would have been large, unsightly, and uneconomical.

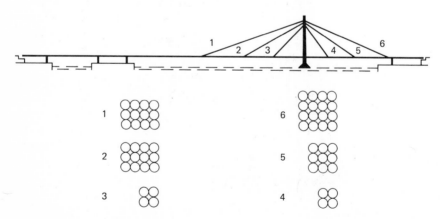

FIGURE 8.12.    Cable arrangement, Severin Bridge. (Courtesy of *Acier-Stahl-Steel.* Ref. 9.)

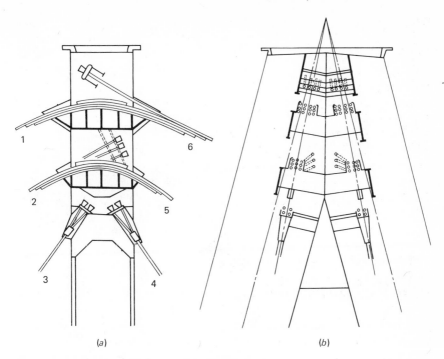

(a)    (b)

FIGURE 8.13.   Anchorage of strands to tower, Severin Bridge: (*a*) elevation; (*b*) transverse section. (Courtesy of *Acier-Stahl-Steel*. Ref. 9.)

The unbalanced forces from the two cables meeting at a common intersection point are resisted by frictional forces. Normally sufficient frictional forces are developed from the gravity pressure contact between the cables and the saddle and the additional friction as a result of the clamped bearing lid holding the cables fixed in position, Fig. 8.15. However, in this application the total frictional force developed was insufficient to provide the factor of safety of 2.5 which was required to resist the unbalanced forces. Therefore, the necessary additional frictional capacity was furnished by horizontal plates inserted between the individual layers of strands. These plates are tapered to provide a gradual transition from the pressure region to the nonpressure region, Fig. 8.15.[9,10] The plates are fixed to the seat and cover saddle. The plates increase the number of friction planes from two to six, and, therefore, the frictional stress is increased three times for the same cable pressure.

The connection of a cable to a diaphragm of a deck girder that transfers the loads to the superstructure is usually accomplished by separating the cable into its component strands and anchoring them individually, Fig. 8.16.

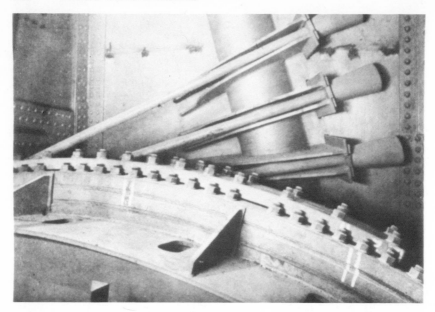

**FIGURE 8.14.**   Unbalanced strands anchored to tower, Severin Bridge. (Courtesy of *Acier-Stahl-Steel.* **Ref. 9.**)

The rectangular configuration for the stays is formed by placing the strands in tiered layers. The sides of the rectangular cable assembly are tilted with respect to the vertical and strand socket planes, Fig. 8.17, as they are anchored at an abutment. Up to the point of strand spreading, the strand center lines are parallel to the tangent plane of each cable, which is slightly distorted because of the cable gravity load sag.

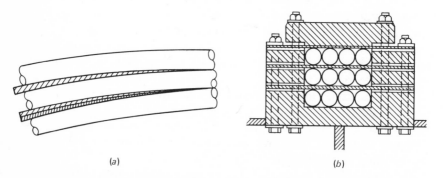

(a)                                        (b)

**FIGURE 8.15.**   Friction plates, Severin Bridge: (*a*) elevation; (*b*) cross section. (Courtesy of *Der Stahlbau.* **Ref. 10.**)

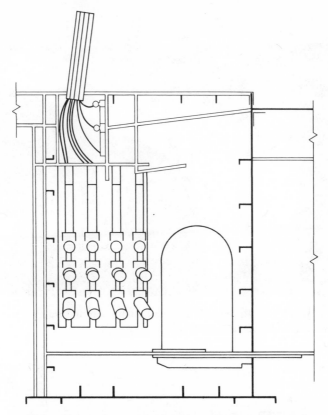

**FIGURE 8.16.** Connection of cable to girder, Severin Bridge. (Courtesy of *Acier-Stahl-Steel*. Ref. 9.)

Where the cable enters the girder diaphragm, the individual strands are flared in three directions, upward, downward, and inward, before they reach their connection bearing points. As a result of this spatial geometry, a special guide saddle is required to lead the strands in the proper direction, Figs. 8.18 and 8.19. The connection of the cable strand sockets is made along a circular arc inside the main girder box, Fig. 8.20. The strand anchorages are closely spaced and in the same layered arrangement as the strands in the cable.[10]

## 8.8 STRÖMSUND BRIDGE, SWEDEN

The cable anchorages for the Strömsund Bridge in Sweden are more conventional because of the typical standard fittings for the individual strands.

**FIGURE 8.17.** Rectangular cable assembly, Severin Bridge. (Courtesy of *Der Stahlbau*. Ref. 10.)

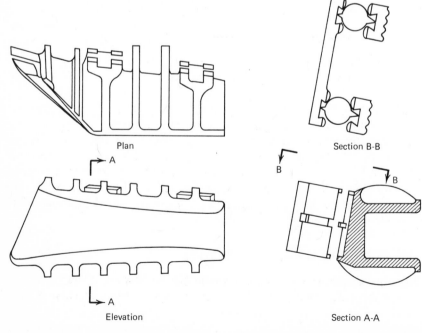

Plan

Section B-B

Elevation

Section A-A

**FIGURE 8.18.** Special guide saddle, Severin Bridge. (Courtesy of *Der Stahlbau*. Ref. 10.)

**FIGURE 8.19.  Top view of guide saddle, Severin Bridge. (Courtesy of** *Acier-Stahl-Steel.* **Ref. 9.)**

As a result, the installation is simpler than that of the Severin Bridge. At the top of the tower the cables are connected with open strand sockets, thus terminating them there, Fig. 8.21. At the girder connection the cables are terminated with standard bearing sockets, Fig. 8.22. Shims are inserted under the sockets to provide for adjustment against a bearing block that rests against an inclined transverse box beam.[11]

## 8.9   GALECOP BRIDGE, HOLLAND

The Galecop Bridge, Holland,[12] is a twin, single-plane cable-stayed structure which is skewed 39 degrees, thus producing an offset of the towers and cables as seen in an elevation view, Fig. 8.23a. Each orthotropic deck structure has a width of 113 ft 6 in., with six main girders, Fig. 8.23b.

The two center girders of each bridge are stiffened by the centrally located cable system and tower. The towers pass through openings in the roadway deck and are fixed at the base to the piers. At the top of the tower the twelve 75 mm ($2\frac{61}{64}$ in.) diameter strands are divided into two groupings of six strands each and supported on two saddles, Fig. 8.23 c and d. The girder connection is made to a transverse diaphragm by flaring the strands into two layers of six each, terminating them with standard

**FIGURE 8.20.** Anchorage at main girder, Severin Bridge. (Courtesy of *Der Stahlbau*. Ref. 10.)

bearing sockets and shims for adjustment. The inclined transverse diaphragm lies between the two main center line longitudinal girders and transfers the cable loads to them. The cables are flared horizontally and vertically in order to space them sufficiently to fit the sockets on the diaphragm, Fig. 8.23 *c* and *d*.

## 8.10 GEORGE STREET BRIDGE, ENGLAND

The George Street Bridge crossing the Usk River at Newport, Monmouthshire, England, has main towers of rectangular hollow concrete, Fig. 8.24a. The towers are 170 ft above the caission and have a base dimension of

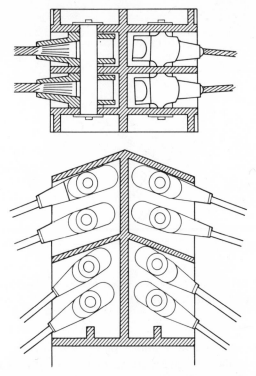

**FIGURE 8.21.   Tower connection, Strömsund Bridge. (Courtesy of *Der Stahlbau*. Ref. 11.)**

13 ft 6 in. by 9 ft 10 in., tapering to 9 ft $11\frac{1}{2}$ in. by 7 ft at the top. The wall thickness of the towers varies from 18 in. at the base to 12 in. below the uppermost saddle.

The cables are supported in pairs by cast steel saddles located within the hollow sections of the tower. To achieve fixity of the cables at the towers, they are securely held to saddles by a clamp over the top, Fig. 2.24b. In order to compensate for the longitudinal movement of the cables from traffic loadings and differential temperature effects, the saddles are placed on high-load steel roller bearings, which permit the necessary movements to take place. The steel rollers are $6\frac{1}{2}$ in. in diameter and are subjected to a maximum load of approximately 33 tons per lineal inch.[13]

## 8.11   PAPINEAU-LEBLANC BRIDGE, CANADA

The Papineau-Leblanc Bridge, located in Montreal, Canada, is a two-tower arrangement with four stays attached to each tower, Fig. 8.25a.

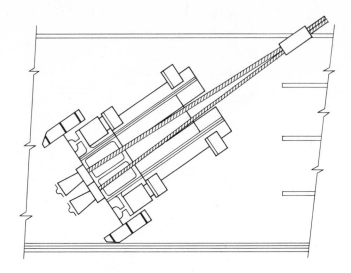

**FIGURE 8 22.    Girder connection, Strömsund Bridge. (Courtesy of *Der Stahlbau*. Ref. 11.)**

Each stay is divided into two bundles of 12 strands each, which facilitates the anchorage to the box girder. Each bundle of strands is connected to one side of the center web of the two cell box girder, Fig. 8.25$b$. The longer stays have $2\frac{5}{16}$ in. diameter strands, and the shorter stays have strands of $1\frac{5}{8}$ in. diameter.

An analysis of the displacements and internal forces in the cables at the towers indicates that slippage will occur if the saddles for the end and outer stays are fixed in position. Therefore, it was determined that the longer stays of the end and center spans be continuous and rest on a saddle that is allowed to move on a plastic base.

The connection of the cables to the girder is accomplished by terminating the individual strands in a socket with internal threads. The threads accommodate ASTM A354 threaded rods, which pass through and are anchored to a curved bearing plate attached to the web of the girder. The bearing plate is finished on the back face to provide proper seating of the washers. As the strand is pulled to the correct tension, adjustments are made by the thickness of washers and positioning of the nuts on the back face of the bearing plate. The bearing plate must be designed to accept the tension loads and transfer them to the adjacent webs by stiff diaphragms.[14,15]

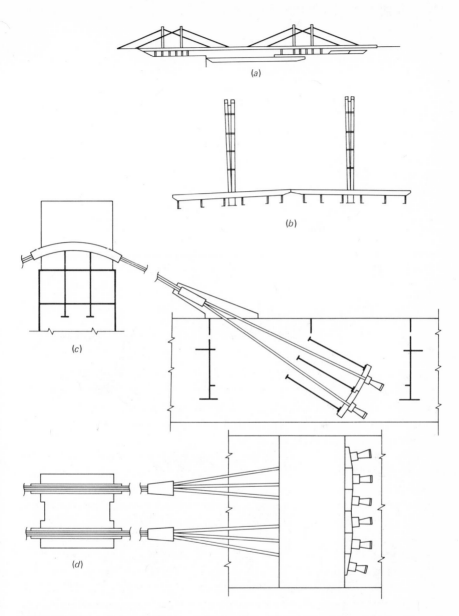

FIGURE 8.23. Galecop Bridge: (*a*) elevation; (*b*) cross section; (*c*) elevation of connections; (*d*) plan of connections. (Courtesy of the British Constructional Steelwork Association, Ltd.)

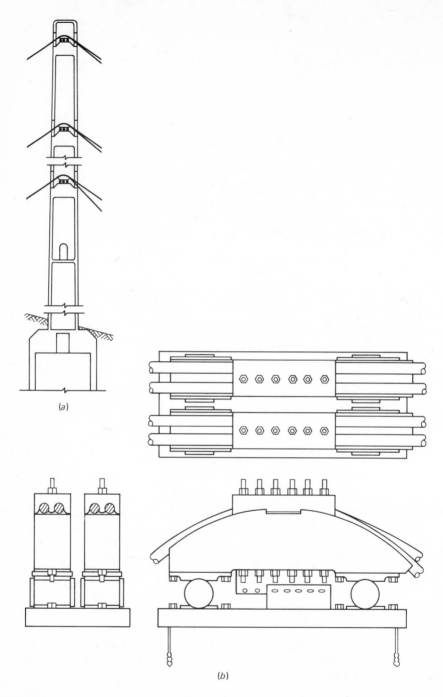

**FIGURE 8.24.** George Street Bridge: (*a*) cross section of pylon; (*b*) detail of saddle. (Courtesy of the Institution of Civil Engineers. Ref. 13.)

**236**

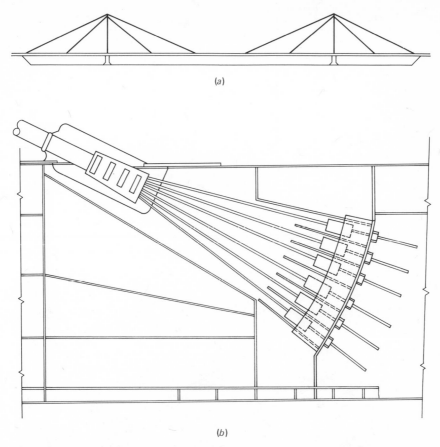

(a)

(b)

**FIGURE 8.25.   Papineau-Leblanc Bridge: (*a*) elevation; (*b*) anchorage to web of box girder. (Courtesy of the Canadian Steel Industries Construction Council. Ref. 15.)**

## 8.12   SITKA HARBOR BRIDGE, U.S.A.

The Sitka Harbor Bridge, which has a 450 ft center span, was designed by the Alaska Department of Highways to permit vehicular traffic to cross the narrow portion of the harbor. The bridge is noteworthy because of its unusual cable connections to the deck structure, Fig. 8.26. The uniqueness of the design is the fact that the cables are anchored to a transverse tube, 5 ft in diameter and 47 ft long. The tube passes through the longitudinal box girders as a transverse diaphragm and cantilevers outward on each side of the deck, Fig. 8.27.

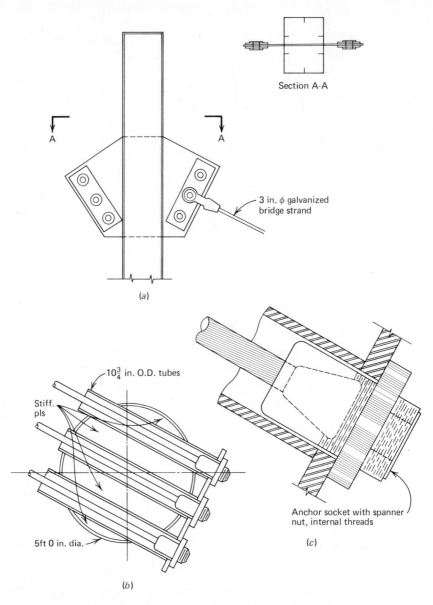

**FIGURE 8.26.** Sitka Harbor Bridge: (*a*) elevation of pylon; (*b*) cable anchor tube; (*c*) cable connection at anchor.

FIGURE 8.27.  Cable anchorage, Sitka Harbor Bridge.

The stay cables are 3 in. diameter galvanized structural strands which terminate at the towers and the transverse tubes. The tower connection is a standard open socket for each strand and is pinned to a gusset plate which is continuous through the center of the tower, Fig. 8.26. The connection of the cables to the transverse tube makes use of a $10\frac{3}{4}$ in. pipe sleeve which permits the cables to pass through and become anchored at a bearing plate attached to the pipes. The bearing plate is 20 in. by 5 ft.

The cable fittings are standard bearing sockets with internal threads for jacking and external threads to hold the spanner nut. The $10\frac{3}{4}$ in. pipes transfer the cable tension load to the transverse tube by welds along the pipes on the stiffener plates attached to the tube, Fig. 8.26. The space between the cable and the pipe is fitted with a polymer sealer for protection against the severe climatic conditions.

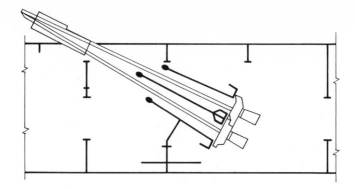

FIGURE 8.28.   Cable anchorage at girder, Rhine River Bridge at Maxau. (Courtesy of *Der Stahlbau*. Ref. 16.)

## 8.13   RHINE RIVER BRIDGE AT MAXAU, GERMANY

The Rhine River Bridge at Maxau, Germany, has a single-plane, fan type cable arrangement located within the median strip.[10] The superstructure is a single rectangular box girder with side cantilevers. The cable forces are transmitted to the box girder by inclined transverse diaphragms, Fig. 8.28.

FIGURE 8.29.   Opening in top flange for cables, Rhine River Bridge at Maxau. (Courtesy of *Der Stahlbau*. Ref. 16.)

As a result of the opening in the top flange of the box girder, the longitudinal component of the cable is not immediately distributed uniformly to the total cross section of the box girder, Fig. 8.29. Reinforcement is necessary for strength and stiffness around the opening to compensate for the removed flange material.

The cables lie in two horizontal planes and pass continuously through the tower over a saddle near the top of the tower, Fig. 8.30. The saddle has two friction plates held in place by a cover that is bolted securely to the flanges of the saddle, Fig. 8.31. The additional plates are installed to develop the total frictional capacity required to accommodate the differential force between the cables on each side of the tower.

## 8.14  SCHILLERSTRASSE FOOTBRIDGE, GERMANY

The Schillerstrasse Footbridge is a single-tower, five-cable bridge; its slim octagonal tower has a slight taper that narrows to limiting dimensions for the cable anchorages, Figs. 8.32 and 8.33. The cables are terminated at the tower in a space of only 1000 mm (3 ft. $3\frac{7}{8}$ in.) high by 490 mm (1 ft $7\frac{5}{16}$ in.) wide. Since this bridge is the first cable-stayed bridge to use

**FIGURE 8.30.  View of cables at tower saddle, Rhine River Bridge at Maxau. (Courtesy of *Der Stahlbau*. Ref. 16.)**

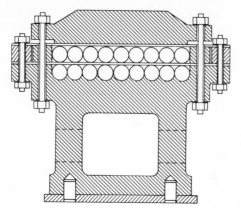

FIGURE 8.31.   Tower saddle, Rhine River Bridge at Maxau. (Courtesy of *Der Stahlbau*. Ref. 16.)

parallel wire strands, it is quite unique. The cables consist of a varying number of 6 mm ($\frac{1}{4}$ in.) diameter wires, Table 8.1, of the type common to posttensioned, prestressed tendons using the BBRV anchorages.

TABLE 8.1   Number of Wires Per Stay

|  | Cable Number | | | | |
|---|---|---|---|---|---|
| Stay number | 1 | 2 | 3 | 4 | 5 |
| No. of wires per cable | 44 | 28 | 20 | 22 | 90 |

The use of the smaller diameter BBRV anchorages resolves the difficult problem of fitting the 10 cable anchorages in the restricted top portion of the tower. The anchorage is accomplished by installing a pipe or trumpet into the tower and welding it so that it is aligned with the direction of the cable. After the cable and anchorages are installed, the cavity in the tower is filled with high-strength concrete. The concrete provides a bearing medium for the cable anchorages and at the same time increases the mass of the octagonal cross-sectional shape, Fig. 8.33.

As the cable approaches its anchor in the superstructure, it follows the line of inclination of the cable axis as seen in elevation, but is curved in plan view by a built-up saddle, Fig. 8.34, which anchors it to the transverse girder.

Weather protection for the cable is provided by a polyethylene tube with sufficient inside diameter to allow pressure grouting. The cement mortar grout is introduced under pressure at the lower anchorage and

FIGURE 8.32.   View of anchorage at tower, Schiller-strasse Footbridge. (Courtesy of *Die Bautechnik*. Ref. 17.)

pumped the entire length of the cable. An air vent at the top of the cable permits an observer to see when the grout has filled the cable.[17]

## 8.15   RAXSTRASSE FOOTBRIDGE, AUSTRIA

The Raxstrasse Footbridge in Vienna has an inclined A-frame tower with anchored back-stays and eight cable stays supporting the deck on either side at the fifth points, Fig. 6.15. Connection of the cables to the deck is accomplished by means of cantilevering the transverse floor beams out-wardly in a bracket. The anchorage is achieved by a standard bearing socket, and adjustment is by threaded rods, Figs. 6.16 and 6.17.

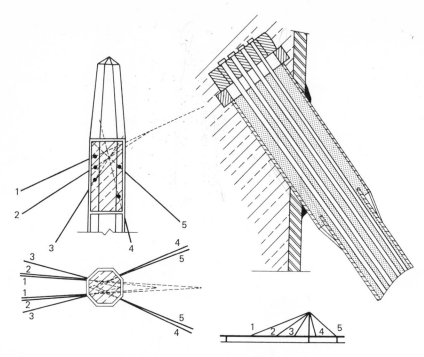

**FIGURE 8.33.** Tower anchorage details, Schillerstrasse Footbridge. (Courtesy of *Die Bautechnik*. Ref. 17.)

## 8.16   LUDWIGSHAFEN BRIDGE, GERMANY

In 1968, at Ludwigshafen, Germany, an old dead-end railway station was converted into a modern through station which now routes both rail and road traffic at five different levels. The highest level is a vehicular roadway supported by a cable-stayed bridge with a converging cable arrangement, Fig. 5.20. The four-legged tower, A-shaped on all sides, is approximately 250 ft in height. The superstructure is of the orthotropic type and is approximately 80 ft in width.

The cables are anchored to the deck structure by cantilever brackets along the span, Fig. 8.35a. The terminal fittings on the cable are standard bearing sockets, Fig. 8.35b, connected to stiffened built-up attachments. The A-frame converges at the top to a rather small structure to support the large number of cables; an architecturally distinctive solution to the termination and type of connection of all the cables, Fig. 5.22.

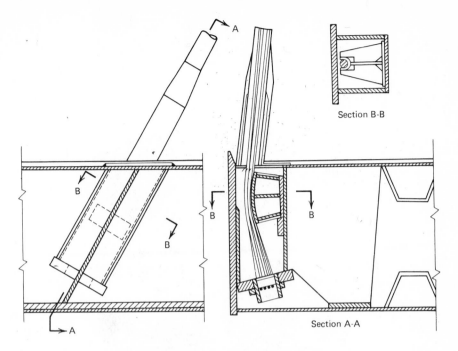

Section B-B

Section A-A

**FIGURE 8.34.** Cable anchorage detail at girder, Schillerstrasse Footbridge. (Courtesy of *Die Bautechnik*. Ref. 17.)

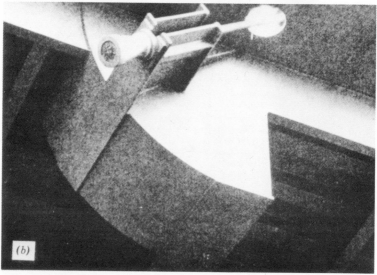

FIGURE 8.35. Ludwigshafen Bridge: (*a*) Bracketed cable attachment;
(*b*) bearing socket. (Courtesy of *Der Stahlbau*. Ref. 19.)

# REFERENCES

1. Simpson, C. V. J., "Modern Long Span Steel Bridge Construction in Western Europe," *Proceedings of the Institution of Civil Engineers*, 1970 Supplement (ii).

2. *Manual for Structural Applications of Steel Cables for Buildings*, American Iron and Steel Institute, Washington, D. C., 1973.

3. Kondo, K., Komatsu, S., Inoue, H. and Matsukawa, A., "Design and Construction of Toyosata-Ohhashi Bridge," *Der Stahlbau*, No. 6, June 1972.

4. Thul, H., "Schragseilbrücken," Preliminary Report, Ninth IABSE Congress, Amsterdam, May 1972.

5. Andrä, W. and Zellner, W., "Zugglieder aus Paralleldrahtbündeln und ihre Verankerung bei hoher Dauerschweilbelastung," *Die Bautechnik*, No. 8, August 1969 and No. 9, September 1969.

6. Anon., "Der Bau der 2. Mainbrückeder Farbverke Hoechst AG," Dickerhoff & Widmann, Inc., New York, N. Y.

7. Anon., "Dywidag—Spannverfahren, Paralleldrahtseil," Bericht Nr. 14, Herausgegeben von der Aubteilung für Entwicklung, June 1972, Dickerhoff & Widmann, Inc., New York, N. Y.

8. Finsterwalder, U. and Finsterwalder, K., "Neue Entwicklung von Paralleldrahtseilen für Schrägseil und Spannbandbrücken," Preliminary Report, Ninth IABSE Congress, Amsterdam, May 1972.

9. Fischer, G., "The Severin Bridge at Cologne (Germany)," *Acier-Stahl-Steel* (English version), No. 3, March 1960.

10. Hess, H., "Die Severinsbrücke Köln," *Der Stahlbau*, No. 8, August 1960.

11. Wenk, H., "Die Strömsundbrücke," *Der Stahlbau*, No. 4, April 1954.

12. Schor, R. J., "Steel Bridges in Holland," *Proceedings of the Conference on Steel Bridges*, British Constructural Steelwork Association, Ltd., London, 1968.

13. Brown, C. D., "Design and Construction of the George Street Bridge over the River Usk, at Newport, Monmouthshire," *Proceedings of the Institution of Civil Engineers*, Vol. 32, September 1965.

14. Demers, J. G. and Marquis, P., "Le Pont a Haubans de la Riviere-des-Prairies," *L'Ingenieur*, June 1968.

15. Taylor, P. R. and Demers, J. G., "Design, Fabrication and Erection of the Papineau-Leblanc Bridge," Canadian Structural Engineers Conference, 1972, Canadian Steel Industries Construction Council, Toronto, Ontario, Canada.

16. Schöttgen, J. and Wintergerst, L., "Die Strassenbrücke über den Rhein bei Maxau," *Der Stahlbau*, No. 2, February 1968.

17. Leonhardt, F. and Andrä, W., "Fussgängersteg über die Schillerstrasse in Stuttgart," *Die Bautechink*, No. 4, April 1962.

18. Scalzi, J. B., Podolny, W., Jr., and Teng, W. C., "Design Fundamentals of Cable Roof Structures," ADUSS 55-3580-01, United States Steel Corporation, Pittsburgh, Pennsylvania, October 1969.

19. Freudenberg, G., "Die Stahlhochstrasse über den neuen Hauptbahnhof in Ludwigshafen/Rhein," *Der Stahlbau*, No. 9, September 1970.

# 9

# Erection and Fabrication

## 9.1   INTRODUCTION

Fabrication and erection costs are very significant to project cost esti-mates, and, as a result, present trends are to fabricate components as large as possible for simplified erection. In this manner larger components of the project are assembled in the shop in contrast to assembling many smaller units in dangerously elevated, exposed positions on the project site. The vagaries of inclement weather conditions are avoided to a certain extent because fewer components must be erected.

The techniques and methods of erecting cable-stayed bridges are as varied and numerous as the ingenuity and number of erector contractors. It is common practice for the design engineer to suggest a method of erection in the bid document, because of his knowledge of the specific design details. The contractor has the option of accepting the suggested construction method or submitting an alternate method of his choice. Alternate methods are normally subject to the approval of the design engineer. The required approval is considered necessary because the erection method not only affects the stresses in the structure during erection but may also have an effect on the final stresses of the completed structure. The design engineer must satisfy himself that the final stress distribution and geometry of the completed structure is in accord with his concept and calculations.

The methods of erection for cable-stayed bridges are broadly described by three general methods; namely, the staging method, the push-out method, and the cantilever method. These methods are generally described and then discussed on a case-study basis of completed bridges.

## 9.2   METHODS OF ERECTION

A general description of the three erection methods is provided in this section; more specific details are provided in the following sections.

The staging method of erection is most often used where there is a low clearance requirement to the underside of the structure and temporary bents will not interfere with any traffic below the bridge. Its advantage is its accuracy in maintaining required geometry and grade and its relatively low cost for low clearance.

The push-out technique has been used successfully on a number of occasions in Europe but is relatively new to American construction. This method is commonly used in Europe where care must be taken not to interfere with traffic below the bridge and where cantilever construction is impractical. In this method, large sections of bridge deck are pushed out over the piers on rollers or sliding teflon bearings. The deck is pushed out from both abutments toward the center, or, in some instances, from one abutment all the way to the other abutment. Assembling the components in an erection bay at one or both ends of the structure and progressively pushing the components out into the span as they are completed can simplify construction and reduce costs. With this method as much as 1500 tons of steel, spanning a number of supports, have been pushed out and, in some instances, it has been used where a horizontal curvature is required.

The cantilever erection method is very often employed in cable-stayed bridge construction where temporary supports are necessary. It may increase the steel requirements over that required for final positioning to accommodate the increased moments and shears during the erection process. The principal advantage is that it does not interfere with traffic below the bridge.

## 9.3    STAGING METHOD

### 9.3.1    Rhine River Bridge at Maxau, Germany

The superstructure erection for the Rhine Bridge began at an abutment on two temporary land piers and then proceeded by short cantilevers to rest on a temporary river pier and the permanent tower pier, Fig. 9.1. Gen-

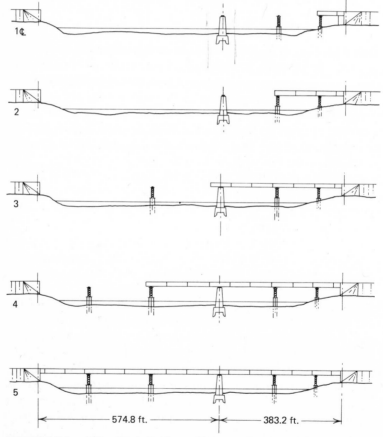

**FIGURE 9.1.    Rhine River Bridge at Maxau. (Courtesy of *Der Stahlbau*. Ref. 1.)**

erally, the units were approximately 65 ft in length and each weighed up to 27.5 tons. They were placed by a derrick mounted on rails on the bridge deck. In the navigation channel, two temporary piers located at the third points of the navigation channel were utilized, Fig. 9.1. Upon completion of the suspended structure the tower erection was begun. The tower was erected in nine units, each weighing up to 4.4 tons with cross-sectional dimensions of 6.5 by 9.8 ft.

The erection method for this bridge consisted simply of erecting the entire suspended structure on temporary piers, followed by the tower erection, and cable connections. Finally, the tower saddles were jacked to stress the cables to the desired tensile load to obtain profile and then the temporary piers were removed.[1]

### 9.3.2   Toyosato-Ohhashi Bridge, Japan

The Toyosato Bridge structure is a three-span, continuous orthotropic box girder with a single plane fan configuration of stays and A-frame towers. The box girder was erected by the staging method, Fig. 9.2. Field welding was used for the longitudinal joints of the deck plates, and high-strength bolts were used for the transverse joints.[2] The legs of the A-frame towers were erected independently and then joined by the lower saddle support portal member, Fig. 9.3. The main girder was jacked into position to en-

**FIGURE 9.2.   Girder erection by staging, Toyosato-Ohhashi Bridge. (Courtesy of A. Matsukawa.)**

**FIGURE 9.3.    Tower erection, Toyosato-Ohhashi Bridge. (Courtesy of A. Matsukawa.)**

able all the cables to be installed and then the jacks were released and the temporary bents were removed, Fig. 9.4.

The stays on this project were prefabricated parallel wire strands (PPWS). To facilitate their erection, temporary bents were erected on the deck at approximately 65 ft (20 m) intervals, and a catwalk was installed from the bridge deck to the towers, Fig. 9.5. Rollers were installed on the catwalk to temporarily support the strands of the cable, while a carrier that pulls the strands was also installed. The carrier grips the socket of the strand, draws it out of its reel, and pulls the strand over the rollers on the catwalk.

Each shop fabricated hexagonal strand is composed of bundled parallel wires. The top stay consists of 16 strands of 154 wires each, and the bot-

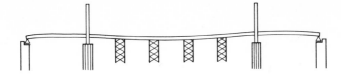

1. Installation of main girder and tower

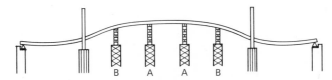

2. Jack up (Point A = 140 cm, point B = 85 cm)

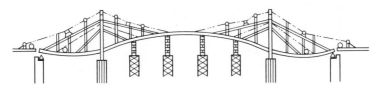

3. Installation of cables

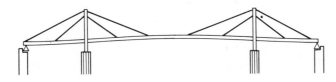

4. Jacks were released

**FIGURE 9.4.    Erection procedure. (Courtesy of *Der Stahlbau*. Ref. 2.)**

tom stay is composed of 12 strands of 127 wires each. To avoid bending stresses in the strands as they pass over the saddle, the strands are prefabricated with a curvature to fit that of the saddle (referred to as "curved strands"). To preclude stretching or deforming during installation that part of the strand called "curved strand" required an additional strong-back support rigging as it was pulled up the catwalk, Fig. 9.6. Figure 9.7 shows the strands in position in the saddles. After adjustment for sag, the cable made up of strands is squeezed (Fig. 9.8) into a circular shape using a technique similar to that employed with conventional suspension bridges.

FIGURE 9.5. Cable erection, Toyosato-Ohhashi Bridge. (Courtesy of
*A. Matsukawa*.)

FIGURE 9.6. Erection of curved strand, Toyosato-Ohhashi Bridge. (Courtesy of
*Der Stahlbau*. Ref. 2.)

254

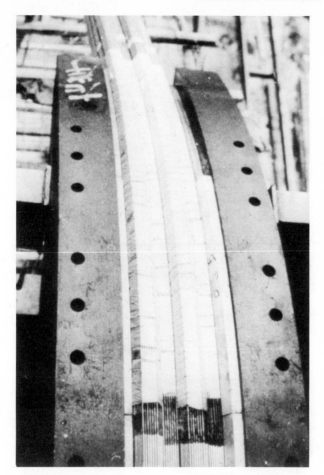

**FIGURE 9.7.  Strand in position on saddle, Toyosato-Ohhashi Bridge. (Courtesy of *A. Matsukawa*.)**

In the autumn, when a favorable temperature of 10°C was obtained, the cables were wrapped in an acrylic resin, Fig. 9.9.

## 9.4  PUSH-OUT METHOD

### 9.4.1  Julicher Strasse Bridge, Germany

The Jülicher Strasse Bridge is a highway overpass crossing a railroad installation in an urban area of Düsseldorf, Fig. 9.10. It is a three-span

FIGURE 9.8. Cable compaction, Toyosato-Ohhashi Bridge. (Courtesy of *Der Stahlbau*. Ref. 2.)

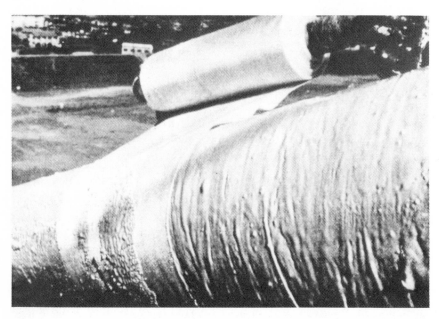

FIGURE 9.9. Plastic cable wrapping, Toyosato-Ohhashi Bridge. (Courtesy of *Der Stahlbau*. Ref. 2.)

**FIGURE 9.10.    Jülicher Strasse Bridge.**

structure with a center span of 324 ft and equal side spans of 104 ft, Fig. 9.11*a*. In cross section it consists of a 6 ft 6 in. median, two roadways of 24 ft 7 in., two walkways of 10 ft 2 in., and two safety strips of 5 ft 3 in.; the overall width is 86 ft 6 in., Fig. 9.11*b*. These features are combined in an orthotropic deck, three cell center box girder with overhangs. The torsionally stiff center box girder is divided into three cells of 13 ft $1\frac{1}{2}$ in., 6 ft 1 in., and 13 ft $1\frac{1}{2}$ in. width and a constant depth of 4 ft 7 in., 1/70 of the span. The transverse girders cantilever out from the box girder and are spaced at intervals varying from 6 ft $10\frac{1}{2}$ in. to 7 ft $6\frac{1}{2}$ in. The tower is 54 ft high and is an externally smooth box. Structural steel weight was $62\frac{1}{2}$ lb per sq ft.[3]

The erection problem was that the federal railway operation, which consisted of six electrified tracks under the eastern side span and the marshalling yard under the center span, could not be interrupted. The push-out concept was selected as the most feasible for the site conditions. The concept was not entirely new, but had never before been used on a cable-stayed bridge. In addition, flat teflon bearings were used for the first time.[4]

An area behind the west abutment of approximately 200 by 130 ft was utilized as the assembly shop. It was possible to assemble the entire bridge

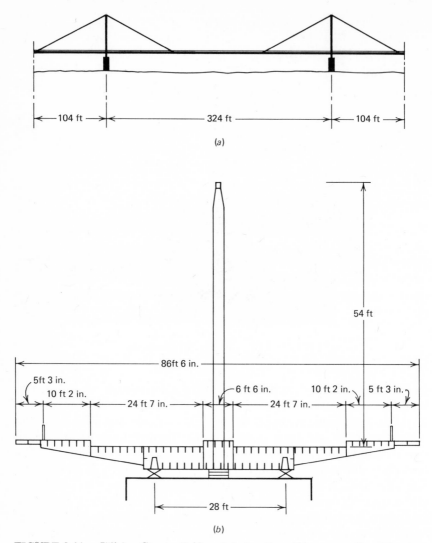

**FIGURE 9.11.** Jülicher Strasse Bridge: (*a*) elevation; (*b*) cross section. (Courtesy of *Acier-Steel-Steel*. Ref. 3.)

cross section to an approximate length of 165 ft in this area. A portal crane with a 100 ton capacity was able to handle the largest components of the deck. Erection units were approximately 53 ft in length and were assembled from six subunits and, as much as possible, were automatically welded at the assembly site. As a result of the length of the assembly work area and the performance range of the crane, it was possible to have three units in various stages of assembly at one time. In this manner, relatively large assembly units were fabricated and welded in a concentrated area under "workshop" conditions, thus providing the greatest possible accuracy in transverse and longitudinal alignments. The cable stays could be laid out on the deck next to the pylon and were erected along with the saddles to the top of the tower by the portal crane.

The erection procedure is shown in Fig. 9.12. It should be noted that in the final position the reaction load of the towers is borne by the permanent piers VIII and XI. However, during the push-out operation the tower reaction must be resisted by a lateral-beam diaphragm which in turn transmits the load to the longitudinal box girders. For this reason the cable stays are only partially tensioned. The jacking mechanism at the saddle is used to compensate for the cantilever deflection of the leading edge of the pushed out section.

When the leading edge of the bridge reaches pier VIII, Fig. 9.12d, the bearing is elevated approximately 4 in. by jacks. As a result of this action the bearing pressure at pier VII is relieved. As the structure is pushed out farther, the bearing pressure at pier VIII will increase. It was determined that the allowable bearing pressure was reached when the leading edge extended approximately $24\frac{1}{2}$ ft past pier IX. At this point the bearing at pier VIII is lowered to its original position. This procedure is then repeated until the structure is in its final position. The erection condition at pier IX is illustrated in Fig. 9.13. Pier IX is at a skew to the longitudinal axis of the bridge as a result of the railway trackage. The final bridge profile is obtained by simultaneously jacking the bearings at piers IX and X and the saddle bearings until the proper elevation is reached and the required tensile force is developed in the cable stays.

No special reinforcement of the box girder was required other than the accommodation of the bearing pressure from the sliding bearing at the outside cells of the box girders.[4] The teflon sliding bearings are illustrated in Fig. 9.14.

### 9.4.2   Paris-Massena Bridge, France

The Masséna Bridge spans 73 tracks from the Austerlitz station. Included in the crossing are the four main lines from Paris to Orleans, shunting

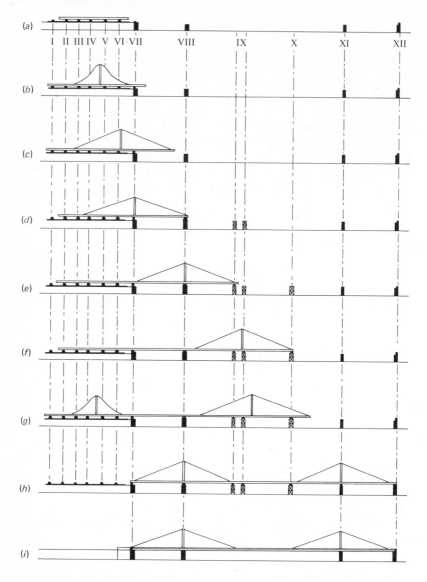

FIGURE 9.12. Jülicher Strasse Bridge, erection procedure. (Courtesy of *Der Bauingenieur*. Ref. 4.)

FIGURE 9.13.   Erection at pier IX, Jülicher Strasse Bridge. (Courtesy of *Der Bauingenieur*. Ref. 4.)

yards, sidings for marshalling main and suburban passenger train lines, and the Paris-Masséna Shop. A site plan is shown in Fig. 9.15.[5,6] The bridge has an overall length of 1615 ft 9 in., divided into six spans of 137 ft 6 in., 182 ft 3 in., 265 ft 7 in., 529 ft 7 in., 264 ft 7 in., and 236 ft 3 in., Fig. 9.16. The deck has an overall width of 118 ft, consisting of a 13 ft median, two roadways of 46 ft, and two sidewalks of 6 ft 6 in. The cable stays are in a single vertical plane and transmit the cable thrust to two main longitudinal box girders, Fig. 9.17.

The box girders are 13 ft $1\frac{1}{2}$ in. deep and 16 ft 5 in. wide, spaced 59 ft center to center. Clearance from top of rail is 28 ft. In the suspended spans, the deck is composed of an orthotropic plate composite with a 4 in. reinforced concrete deck slab. In the approach spans, which do not take a cable longitudinal force, the girders are connected by lateral beams with an $8\frac{5}{8}$ in. thick concrete slab, which takes the place of the orthotropic deck.

The towers are tapered steel boxes 108 ft 3 in. above the deck with cable saddles at 59 and 82 ft, Fig. 9.18. The top saddles are articulated and the lower ones fixed to obtain a balance between shear forces and buckling strength of the towers. The towers are rigidly connected to transverse girders between the main box girders and are supported on the pier

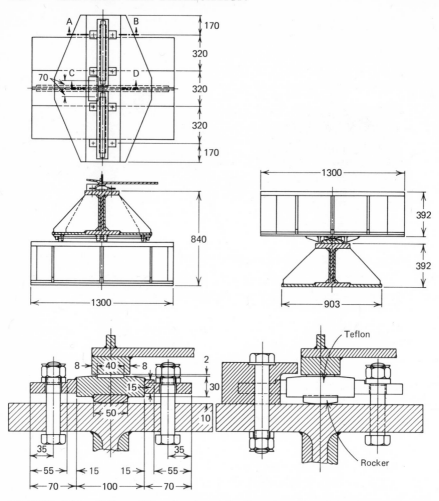

**FIGURE 9.14. Teflon sliding bearing (all dimensions in millimeters), Jülicher Strasse Bridge. (Courtesy of *Der Bauingenieur*. Ref. 4.)**

by spherical bearings. The cables are anchored to the deck 105 and 210 ft from the tower.

This structure was essentially erected by the push-out technique. The sections were pushed out from both ends of the bridge until they cantilevered into the center span approximately 148 ft past the towers. The center portion was then erected by the cantilever method, Fig. 9.19.[5] The box girders were assembled on working platforms at either end of the structure by tower cranes. Each segment of a box girder consisted of five

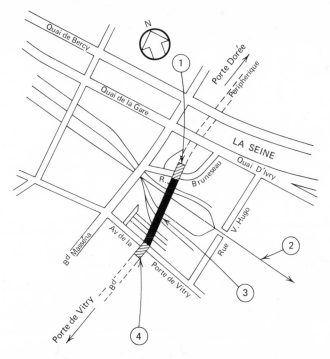

**FIGURE 9.15.** Paris-Masséna Bridge. Key: (*1*) access structure over Rue Bruneseau; (*2*) main railway line to Bordeaux; (*3*) cable-stayed bridge; (*4*) access structure over Avenue de la Porte de Vitry. (Courtesy of *Acier-Stahl-Steel*. Ref. 5 and 6.)

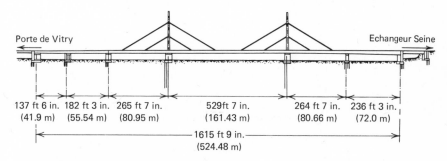

**FIGURE 9.16.** Elevation, Paris-Masséna Bridge. (Courtesy of *Acier-Stahl-Steel*. Ref. 5 and 6.)

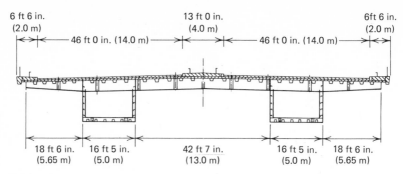

**FIGURE 9.17.  Deck cross section, Paris-Masséna Bridge. (Courtesy of** *Acier-Stahl-Steel.* **Ref. 5 and 6.)**

shop fabricated components, Fig. 9.20. Most of the girder segments were either 52 ft 6 in. or 65 ft $7\frac{1}{2}$ in. (16 or 20 m) in length and had an average weight of 15 tons. Twenty segments exceeded 29.5 tons, and the lower section of the tower weighed 61 tons.

After three girder segments were assembled on the working platform, they were launched by means of cradles fitted with rollers hung from steel wire strands, Fig. 9.21. The two box girders were launched separately. When the launching operation was completed, the girders were transferred to permanent bearings. Two cranes on the Vitry end traveling on girders lifted and erected the deck plates, box section stay anchorage units, box

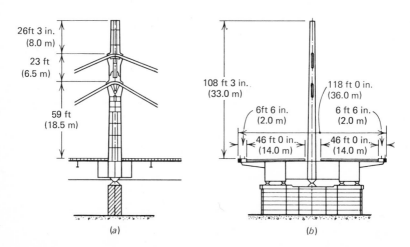

**FIGURE 9.18.   Tower, Paris-Masséna Bridge:** (*a*) **elevation;** (*b*) **cross section. (Courtesy of** *Acier-Stahl-Steel.* **Ref. 5 and 6.)**

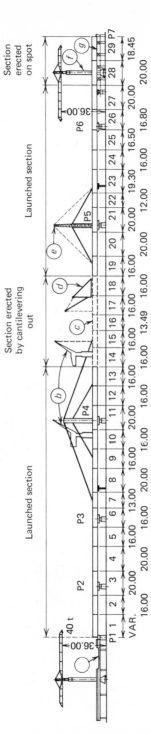

FIGURE 9.19. Paris-Masséna Bridge: Erection procedure (dimensions in meters): (*a*) working platform at Porte de Vitry approach; (*b*) 15 ton tower crane traveling on box-girders; (*c*) center of main span; (*d*) double-jib crane, each jib of 1065 ft ton capacity; (*e*) gantry for erection of towers and cable stays. Maximum capacity, 69 tons; (*f*) tower crane of 30 ton capacity, for handling units into position; (*g*) section numbers 28 and 29 assembled in position. (Courtesy of *Acier-Stahl-Steel*. Ref. 5.)

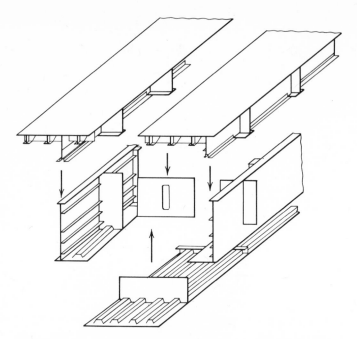

FIGURE 9.20. Isometric diagram of the five components of the box girder section, Paris-Messéna Bridge. (Courtesy of *Acier-Stahl-Steel*. Refs. 5 and 6.)

FIGURE 9.21. Launching a box girder, Paris-Messéna Bridge. (Courtesy of *Acier-Stahl-Steel*. Refs. 5 and 6.)

section units under the tower, the lateral cantilevers, and steel tower. From the Seine end, a double-jib derrick crane was mounted on the two box girders and traveled from the end to the center of the bridge. These cranes were used to erect the center portion of the structure, by the cantilever method, after the lower stays were in place.[5,6]

The stays were erected strand by strand. When the upper stay was anchored to the deck, the tension was adjusted to allow the closure segment to be erected. The stays that were continuous over the saddles were anchored to the deck with an initial tension of 10 tons. The saddles, which

**FIGURE 9.22.    Interior of tower showing telescoping saddle support, Paris-Masséna Bridge. (Courtesy of *Acier-Stahl-Steel*. Ref. 5.)**

FIGURE 9.23.  Box girders after erection by launching, view from Vitry end, Paris-Messéna Bridge. (Courtesy of *Acier-Stahl-Steel*. Ref. 5.)

FIGURE 9.24.  Longitudinal box girders before erection of transverse girders and decking, Paris-Messéna Bridge. (Courtesy of *Acier-Stahl-Steel*. Ref. 5.)

268

were on telescoping supports sliding in the tower, were jacked until the stays were under full load, Fig. 9.22. The lifting force in the lower stays was obtained by six 295 ton jacks with a travel of 1 ft $7\frac{3}{4}$ in., while the upper stays were lifted by four 492 ton jacks with a possible 3 ft $3\frac{1}{2}$ in. travel. Various stages of erection are indicated in Figs. 9.23 through 9.25.

## 9.5    CANTILEVER METHOD

### 9.5.1    Strömsund Bridge, Sweden

The Strömsund Bridge described in Chapter 5 is notable because it was the first modern cable-stayed bridge. It is a three-span structure with end spans of 245 ft each and a center span of 590 ft, Fig. 9.26. The cable geometry

FIGURE 9.25.    General view before closure of center span and after erection of upper stays, Paris-Messéna Bridge. (Courtesy of *Acier-Stahl-Steel*. Ref. 5.)

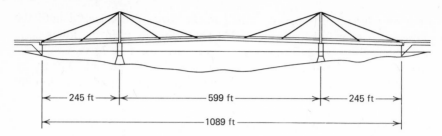

**FIGURE 9.26.** Elevation, Strömsund Bridge. (Courtesy of *Der Stahlbau*. Ref. 7.)

is of the converging type with two vertical planes, one at each side of the roadway deck structure. The portal frame towers are supported at the pier independently of the deck structure, Fig. 9.27. The erection procedure is illustrated in Fig. 9.28.[7] Erection proceeded from both abutments independently of each other with final closure at midspan producing a completed structure.

The end spans were erected to the tower pier with the help of two falsework bents. After erection of the tower, which was braced against the

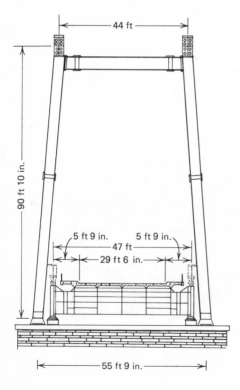

**FIGURE 9.27.** Portal frame tower, Strömsund Bridge. (Courtesy of *Der Stahlbau*. Ref. 8.)

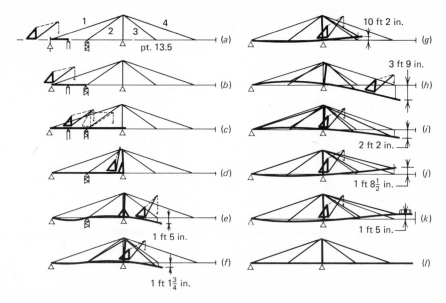

**FIGURE 9.28.** Erection sequence, Strömsund Bridge: (*a*) erection up to the first temporary support; (*b*) erection up to the second temporary support; (*c*) erection up to the tower pier; (*d*) erection of the tower and brace against the stiffening girder; (*e*) erection past point 13.5, the connection of the inside stay 3; (*f*) derrick moved back to the tower pier. Girder jacked at tower. Erection of the inside stays 2 and 3; (*g*) inside stays 2 and 3 tensioned; (*h*) erection past point 18.5, the connection of the outside stay 4; (*i*) derrick moved back to the tower pier. Stay 4 installation: (*j*) stay 4 initially tensioned; (*k*) closure and final tensioning of stay 4; (*l*) completion of superstructure, concrete slab, and so forth, and final position of the bridge geometry. (Courtesy of *Der Stahlbau* Ref. 7.)

girder, cantilever erection proceeded to a point just beyond position 13.5, the anchorage point of stay 3. Because of the negative moments developed in the girder at the main pier, the girder is braced against an erection bracket and bearing mounted to the tower leg, Fig. 9.27. Lateral forces are transmitted from the floor beam through the girder bearings which are in lateral contact with the tower bearings.

To decrease the bending in the girder at the tower and also the deflections of the cantilevered girder, the crane is moved back to the tower pier. While at this location, the crane is used to erect the cables of stays 2 and 3 to the top of the tower. The cables are about $4\frac{3}{4}$ in. shorter than the required lengths under total load to compensate for the tensile elongation under total load. In order to facilitate the installation of the stays, the girder at the tower pier is jacked about 21 in. After installation of the second and third stays, the girder is lowered to its previous position. During this phase, with stays 2 and 3 in position, the top of the tower moves approximately 7 in. toward shore. This change in geometry is used to advantage to install stay 1, which had been preshortened by $8\frac{1}{2}$ in. At this stage the erected portion of the suspended structure is relatively stiff because stay 1 is rigidly anchored and supports the force transmitted by stay 3. The derrick is then moved to its previous position at the end of the cantilever and the girder is erected to a point just beyond station 18.5, which is the connection of stay 4 to the girder. The derrick crane is returned to the tower until stay 4 is installed and initially tensioned. Cantilever erection of the girder is then continued until closure is accomplished at midspan. After closure, final tensioning of stay 4 is accomplished. At this time the steel erection is completed. After the concrete deck is poured, final adjustments are made to obtain the desired profile, Fig. 9.29. The deflected positions of the girder and tower at various stages of erection[8] are illustrated in Fig. 9.29.

### 9.5.2   Papineau-Leblanc Bridge, Canada

The Papineau-Leblanc Bridge has an orthotropic deck structure with a single plane converging cable geometry and consists of a 790 ft center span and 295 ft end spans. Because a greater portion of the bridge is approximately 35 ft above water, an erection method was developed that utilized a 110 ton stiff-leg derrick mounted on two 60 by 120 ft barges. Limited temporary supports were required. The supports were in the form of a single pile supported bent, driven through fill, and located 87 ft from each abutment. After erection of the temporary bents on the pile caps, the first two sections of the girder, from abutment to pile cap, were erected in one piece by two crawler cranes on the river banks. The barge-mounted derrick then erected two more sections that cantilevered out approximately 90 ft past the pile bent. In this position, the stresses in the deck would not allow any further cantilevering; therefore, the capacity of the barge-mounted derrick was utilized to erect the next three units as a single closure unit to the tower pier, Fig. 9.30. The three units comprising the closure lift were spliced together on the barge before being erected. A "pin" connection was made by inserting a few bolts at the bottom of the web splice at the previously erected deck sections. After alignment of all three webs of the two cell box girder, the side span was jacked down at the temporary pile bent and flange splice bolts were inserted as the holes became aligned. All field splices were bolted using ASTM A325 hex head

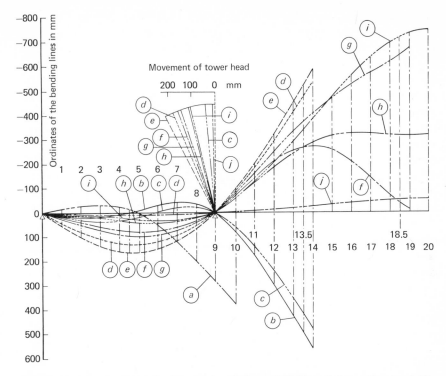

**FIGURE 9.29.** Erection displacements, Strömsund Bridge: (*a*) erection to station point 9; (*b*) erection to station point 13.5; (*c*) installation of stays 2 and 3; (*d*) tensioning of stays 2 and 3; (*e*) tensioning of stay 1; (*f*) erection to station point 18.5; (*g*) preliminary tensioning of stay 4; (*h*) erection to station point 20; (*i*) final tensioning of stay 4; (*j*) final erection condition. (Courtesy of *Der Stahlbau*. Ref. 8.)

bolts. The remaining bolted splice connections in the bottom flange were accomplished by the use of a traveling platform that was suspended from the deck and traveled beneath the deck,[9,10] Fig. 9.31.

When the side spans were completed, the temporary bents were removed and cantilever erection proceeded to the first main span cable-stay attachment. The sections were placed at the front end of the progressive erection by truck and lifted into position by the barge-mounted derrick. The barge alternated between the north and south side erection every two days. In the intervening time the sections already erected were bolted.

The towers are 126 ft in height and taper from 6 sq ft to 5 sq ft and were erected in two major lifts by a pair of 110 ton mobile cranes resting on the deck, Fig. 9.31. The individual cables of the stays were erected one

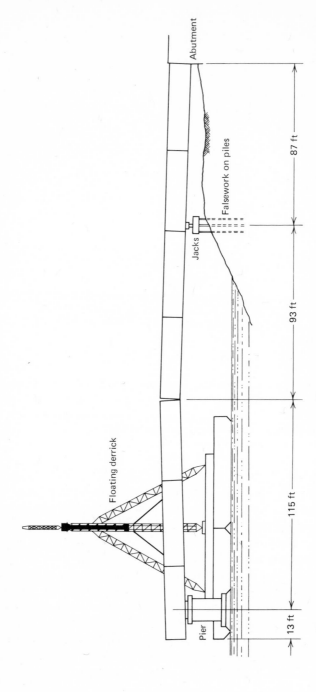

FIGURE 9.30. Erection of sidespan closure segment, Papineau-Leblanc Bridge. (Courtesy of Canadian Steel Industries Council. Ref. 9.)

274

FIGURE 9.31.  Tower and inside stay erection, Papineau-Leblanc Bridge. (Courtesy of Paul Marquis, Gendron Lefebvre and Associates.)

at a time using the mobile cranes. Tensioning of the stays was performed simultaneously from both ends at the deck anchorage.[9]

Erection of the deck continued by the cantilever method, to the outside stay connection, Fig. 9.32. After erection of the outside stay, cantilever erection continued to closure at midspan, Fig. 9.33.

The sequence of erection of the deck structure was performed in several steps. The center web section weighing 30 tons (except at the cable stay anchorage location where they weighed 45 tons) was erected first. The two outside webs, weighing 20 to 25 tons, followed. The bottom panels were placed linking the three webs, and the top orthotropic deck panels were

FIGURE 9.32.    Deck erection, Papineau-Leblanc Bridge. (Courtesy of Paul Marquis, Gendron Lefebvre and Associates.)

FIGURE 9.33.    Closure and midspan, Papineau-Leblanc Bridge. (Courtesy of Paul Marquis, Gendron Lefebvre and Associates.)

276

installed completing the box. Finally, the 11 to 14 ton cantilever overhang of the deck was installed which required careful slope adjustment before final bolting was completed. A computer analysis was performed at various stages of erection to verify tower and girder deflections as well as cable tension values. In this manner the proper distribution of dead load was proportioned to all elements: tower, girder, and stays.[9]

### 9.5.3    Severin Bridge, Germany

Erection of the Severin Bridge was divided into three principal suberection sequences:[12]

1. Erection of the right-hand Rhine side from abutment to point 18, Fig. 9.34*a–n*.
2. Erection of the left-hand Rhine side from abutment to point 17, Fig. 9.35*a–e*.
3. Closure, Fig. 9.36.

Erection was essentially by the cantilever construction method with the principal box girders erected in successive lengths of 52 ft 6 in. and connected by field riveting. The joints between girder segments were prepared during fabrication so that when segments of the girders were joined in the field the proper camber would result. Erection of the center roadway deck between girders and the sidewalk sections, which cantilevered out from the main girders, followed closely behind the erection of the principal girder. The deck sections were shop fabricated in units 52 ft 6 in. in length and approximately 21 ft in width. Three of these units were then field assembled at a preassembly area on a portion of the bridge that had been completed to form a work space of 52 ft 6 in. by 62 ft 6 in. The two longitudinal joints were field welded and the transverse joints at the cross girders were field riveted. The entire deck unit was then erected between the previously erected principal longitudinal girders. The transverse joint between successively erected deck units were field welded and the longitudinal connection to the girders was field riveted. The longitudinal orthotropic ribs of the deck plate had been fabricated 10 in. short of both ends of the deck plate, leaving a 20 in. gap in the ribs after the transverse joints in the successive deck plates had been welded. These gaps were filled by field welding into position short portions of ribs that were accurately cut to the required length. Ten working days were required to install each 52 ft 6 in. length of deck.[13]

Erection of the bridge structure began in April 1958, when pier 5 and three falsework bents were constructed on the right bank of the Rhine

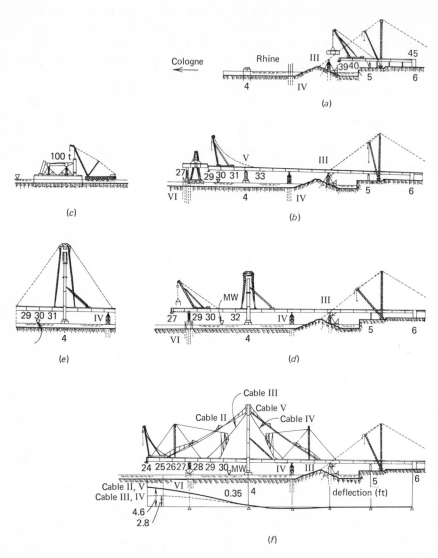

**FIGURE 9.34.** Erection sequence, Severin Bridge: (*a*) erection on temporary piers, sections 41–42, 42–43, 43–44, and 44–45. Cantilever erection toward river from panel 40–41; (*b*) cantilever erection over temporary piers III, IV, and V. Floating crane erection of preassembled sections 27–28–29; (*c*) erection of the first pylon lift with a floating crane (cross section); (*d*) cantilever erection over temporary pier VI and pylon erection to 130 ft height; (*e*) pylon erection to 230 ft height; (*f*) installation of cables III, IV, and II, V; (*g*) cantilever erection to temporary pier VII; (*h*) installation of cables I, VI. (Courtesy of *Der Stahlbau.* Ref. 12.)

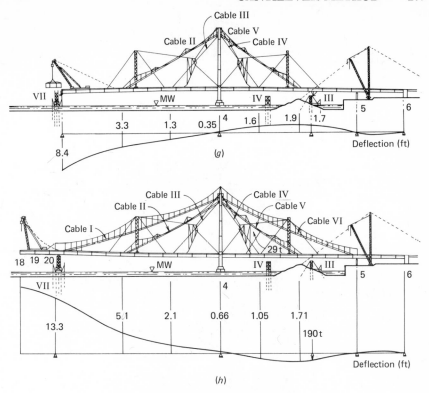

**FIGURE 9.34** (*continued*)

River. With the aid of a cable-supported derrick, the suspended structure from point 41 to abutment 6 was then erected, Fig. 9.34a. At this point, transverse bracing was installed between the two shafts of pier 5 to resist lateral wind forces. The K bracing between the pier 5 shafts had its node point connected to a concrete beam that joined the foundations of the two shafts. A diagonal bracing was also installed from the footing of the shafts to the lower flange of the principal girders to stiffen the structure in the longitudinal direction, Fig. 9.37. Cantilever construction, with the assistance of temporary piers, was employed from this position to erect the deck structure to point 18. Temporary pier III consists of four vertical piles cross braced for wind forces and eight battered piles to resist other horizontal forces, Fig. 9.38.[12]

Cantilever erection of the suspended structure proceeded from the right bank of the Rhine. Upon reaching permanent pier 4, the pylon pier, temporary support piers held the two principal box girders while cantilever erection of the roadway structure continued and erection of the pylon com-

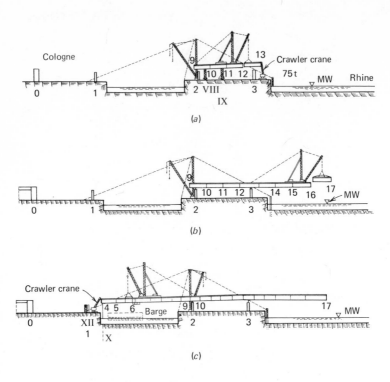

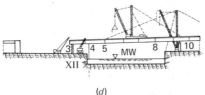

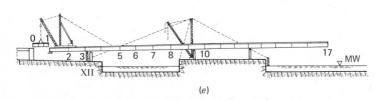

FIGURE 9.35. Erection sequence, Severin Bridge: (*a*) erection of section 9–10–11 with derrick, cantilever erection toward river, erection of section 12–13 with aid of crawler crane; (*b*) cantilever erection to point 17; (*c*) cantilever erection from point 9 to point 6, section 4–5–6 erected with aid of crawler crane; (*d*) erection section 3–4 with crawler crane and truck crane; (*e*) cantilever erection to abutment 0. (Courtesy of *Der Stahlbau*. Ref. 12.)

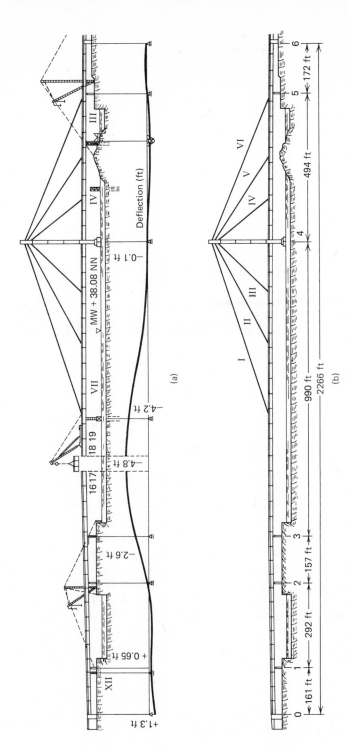

**FIGURE 9.36.** Severin Bridge: (*a*) closure; (*b*) completed structure. (Courtesy of *Der Stahlbau*. Ref. 12.)

**FIGURE 9.37.    Cantilever erection from right bank, Severin Bridge. (Courtesy of *Der Stahlbau*. Ref. 12.)**

menced, Fig. 9.34*b*. A floating crane was used to erect the preassembled girder sections 27–28, 28–29 in order not to overstress the previously erected long cantilever superstructure, Fig. 9.34*b*. The crane was also used to erect the first two sections of the pylon. Each section weighed 90 tons, Fig. 9.34*c*. The balance of the pylon lifts, which weighed somewhat less, were erected by means of a deck-mounted derrick, Fig. 9.34*d* and *e*. The struts between the pylon legs and the deck that stiffen and support the pylon during erection are illustrated in Figs. 9.39, 9.40, and 9.41. The struts were adjustable such that at the time of erection of the pylon head the legs could be adjusted to compensate for any inaccuracies during fabrication and erection. Provision was also made for a torsional moment to be introduced into the pylon legs to correct for fabrication and erection errors that might occur.[12,13] The cable-stay erection sequence was III–IV, II–V, and I–VI. The cable stays are considerably shorter in an unstressed condition than they are in the final stressed condition. To minimize the reduction in cable chord length caused by the cable sag under its own weight, it became necessary to install them as straight as possible into their final alignment. The individual strands of each cable were pulled up along a suspended walkway, Figs. 9.42, 9.43, and 9.44, and hung in pulley blocks such that they were as straight as possible and approximately in their final geometrical position. A force of approximately 10 tons was required to draw the

**FIGURE 9.38.    Bracing at temporary pier III, Severin Bridge.
(Courtesy of *Der Stahlbau*. Ref. 12.)**

strands over the deflection bearing (see Chapter 8) to the cross members
to which they were anchored. To compensate for the shortened cable
length, the structure had to be elevated at the points of cable attachment.
For example, the installation of cables I and VI required that the suspended structure be jacked approximately 16 in. (40 cm) and, concur-
rently, the top of the pylon be displaced toward the abutment by tensioning cable V which had been installed previously.

When erection of the cables was completed and prior to closure, the
right-hand portion of the structure had all temporary supports freed. At

**FIGURE 9.39.** Pylon erection, Severin Bridge. (Courtesy of *Acier-Stahl-Steel*. Ref. 13.)

this time, a check was made to verify that the position of the superstructure was as required by design. This freeing of temporary supports required some minor corrections only at the cable anchorages.[12,13]

Erection on the left-hand side commenced with the erection of that segment of the superstructure between piers 2 and 3, Fig. 9.35. The shafts of pier 2 were cross braced similar to that of pier 5. The superstructure was then cantilevered out over the Rhine approximately 200 ft. Erection then proceeded in the opposite direction until abutment 0 was reached. Erection of the structure was completed in September 1959 with the installation of the closure girders 17–18, Fig. 9.36*a*.

### 9.5.4   Batman Bridge, Australia

The intended erection sequence for the Batman Bridge was as follows,[14] Fig. 9.45:

1. Erect the truss superstructure from the west abutment to the intersection of the superstructure with the pylon.

**FIGURE 9.40. A-frame pylon erection, Severin Bridge. (Courtesy of *Der Stahlbau*. Ref. 12.)**

2. From the east abutment, erect the truss superstructure westward to an expansion joint 56 ft 3 in. west of the first pier on the east shore.

3. While the trusses are being erected on the east shore, erect the pylon to its full height and install the permanent back-stays.

4. Upon completion of the erection of the east shore and the pylon, extend the trusses eastward from the pylon to the expansion joint.

5. Erect and weld the deck and handrails following the truss erection from each side of the river.

Completed sections of the stiffening truss, which are 45 ft in length including all lateral bracing and secondary members, were assembled and bolted with friction grip bolts prior to erection. Each section weighed approximately 55 tons. The first section was erected in place spanning from the west abutment to a temporary steel pier. A special erection frame was mounted on the top of the truss section to erect subsequent 45 ft preas-

**FIGURE 9.41.** A-frame pylon erection, Severin Bridge. (Courtesy of *Der Stahlbau*. Ref. 12.)

sembled sections, Fig. 9.46. The next truss section was cantilevered from the first temporary pier. The subsequent section was erected to extend from the cantilever end to a second temporary pier. The first temporary pier was then removed, producing a 135 ft span from the abutment to the second temporary pier. The second temporary pier was such that its legs straddled the width of the pier. In the same manner subsequent truss sec-

**FIGURE 9.42.** Cantilever jacked up for installation of stay I, Severin Bridge. (Courtesy of *Acier-Stahl-Steel*. Ref. 13.)

tions were transported down the bank on a rail-mounted carriage, passed through the second temporary pier, and were lifted by the erection frame and cantilevered outwardly. An additional 45 ft section was cantilevered from temporary pier 2 to produce a section 180 ft in length from the abutment to the pylon intersection.[15]

The pylon consists of 40 segments, each approximately 15 ft in length. For the first nine sections of each pylon leg the cross beam and cross bracing were erected from a crane mounted on the trusses, Fig. 9.47. When the pylon reached truss elevation, the trusses were raised from the temporary piers by jacks placed on the pylon cross beam. When the temporary pier was removed, the truss was anchored to the abutment and the other end was lowered onto bearings on the pylon cross beam. As the pylon construction rose above the main trusses, the crane used for pylon erection was mounted on a platform on rails attached to the west face of the pylon. The crane in this position proceeded to climb up the pylon erecting two sections on each leg ahead as it climbed, Fig. 9.48. Each pylon unit was lowered down the west bank on temporary rail tracks, taken through the pylon legs, and raised into position. Temporary back-stays, connected to anchorages that were held down by prestressed rock-anchored cables grouted

FIGURE 9.43.    Strand being pulled over the saddle, Severin Bridge. (Courtesy of *Der Stahlbau*. Ref. 12.)

into the bed rock, were installed at various positions to stabilize the leaning pylon until the permanent back-stays were installed at the completion of the pylon erection.

While the pylon was under construction, the bridge truss erection proceeded westward from the east abutment in a similar manner to that of the west side. After assembling an erection frame on the first truss section spanning from the east abutment to a temporary pier, complete bridge sections were alternately cantilevered forward. Additional sections were added to span onto temporary piers, Fig. 9.49. At this time erection of deck units proceeded from the east abutment, and each unit was fully site welded to the preceding unit.[14,15]

**FIGURE 9.44.** Strand being pulled over the saddle and down the catwalk, Severin Bridge. (Courtesy of *Der Stahlbau*. Ref. 12.)

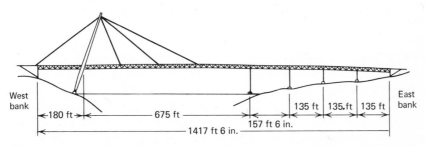

**FIGURE 9.45.** Batman Bridge. (Courtesy of Department of Public Works, Tasmania. Ref. 14.)

**FIGURE 9.46.** Erection frame on bridge section 1, Batman Bridge. (Courtesy of Department of Public Works, Tasmania. Ref. 14.)

Because some pylon units arrived late at the job site, the truss on the west side was erected 90 ft past the pylon. After the back-stays were installed, the erection of the trusses proceeded across the river from the west side. Assembled bridge sections were lowered down the west bank, passed through the pylon legs, lifted to the underside of the truss and suspended from a monorail system attached to the underside of the trusses. The assembly was moved out to the erection face and positioned by the erection frame, Fig. 9.50.[14,15] The bridge sections were cantilevered 90 ft. A temporary fore-stay from the pylon top was installed and the elevation of the leading end of the truss raised to a predetermined level. Two additional truss sections were cantilevered from the temporary fore-stay, and a permanent fore-stay was attached. This procedure was repeated with temporary fore-stays attached at three points between the permanent fore-stays.

Throughout erection a careful check of the stresses in the fore-stays and principal truss members was maintained. Deck erection followed the truss erection in a prescribed pattern to avoid overstressing the structure.

**FIGURE 9.47.   Erection of pylon, Batman Bridge. (Courtesy of Department of Public Works, Tasmania. Ref. 14.)**

Subsequent to the installation of handrails and road surfacing the tension in all fore-stays and reactions of the truss at the expansion joint and pylon were adjusted to obtain the required load distribution.

### 9.5.5   Kniebrücke Bridge, Germany

The Kniebrücke Bridge was erected by the cantilever method starting at the left bank of the Rhine River, Fig. 9.51. Unlike the Severin Bridge, the 1050 ft river span was erected without resorting to the use of temporary supports, which would have impeded navigation in the channel. The erection procedure is summarized in Fig. 9.51.[16]

The roadway superstructure is divided into 34 longitudinal sections varying in length from 46 ft 6 in. to 63 ft 6 in. The pylon was divided into 15 units for the inside cell adjacent to the deck and 10 units for the outside cell. The inside cells varied in length from 20 ft 4 in. to 30 ft 10 in.

FIGURE 9.48. Climbing crane on pylon, Batman Bridge. (Courtesy of Department of Public Works, Tasmania. Ref. 14.)

The joints of the inside and outside cells are staggered in relation to each other as a result of the overlap of 3 ft $3\frac{1}{2}$ in., Fig. 9.52.[17]

The superstructure was delivered to the site in seven subassemblies: two main girder elements consisting of the bottom flange, web, roadway strip adjacent to the girder, and walkway strip; the three center roadway sections; and the two walkway assemblies. Additional secondary units consisted of the diaphragms and cantilever diaphragm sections, Fig. 9.53.

A preassembly and storage area was located on the left bank of the river, Fig. 9.54. A transfer derrick was used to move the materials from

**FIGURE 9.49.  Truss erection, east side, Batman Bridge. (Courtesy of Department of Public Works, Tasmania. Ref. 14.)**

the preassembly area to the bridge deck. The same derrick was used to erect the first two deck units, which were supported on falsework, and install on this portion of the superstructure an erecting crane for the cantilever erection procedure. This portion of the deck also served, throughout the entire erection procedure, as a work area to assemble the three center roadway units.

In the preassembly area, the main girder units were assembled to the walkway units and necessary field adjustments were made at the main splices for the cantilever erection. The main splices, referred to as universal splices, extended across the full cross-sectional width of the decks. During this operation, rivet holes were carefully reamed at the splices to compensate for errors in the bottom flange length. Necessary camber adjustments were made relative to the adjacent unit. Web and flange cover plates were cut to the proper length at the front cantilever end, and units were prepared for splicing. The service walkway, supports for electrical lines, and stormwater and gas main pipes were installed in the units at the assembly area.

**FIGURE 9.50.    Cantilever erection of truss from pylon, Batman Bridge. (Courtesy of Department of Public Works, Tasmania. Ref. 14.)**

On the left bank the superstructure was supported on tension-pendulum piers, which stiffened the river span by supporting the component from the cable stays, Fig. 9.55. Cantilever erection of the principal girder units in this area of the structure was assisted by a pair of movable auxiliary supports, Fig. 9.56. These units, successively placed approximately 40 ft in front of the piers and the pylon, shortened the cantilever and provided the clearance whereby the sections could be adjusted for level, Fig. 9.51 and 9.57. A similar procedure was used in erecting the river span with a movable auxiliary cable-stay system supported by a pair of masts, Fig. 9.51 and 9.58. The masts were successively located at the cable attachment points to the deck.[16,17]

In the cantilever erection procedure, the two main girder units were hoisted by the transfer derrick at the preassembly area to a rail-mounted transporter carriage on the deck. The transporter carriage then traveled down the completed portion of the superstructure to the cantilever end. The girder units were then installed by the erecting crane, Fig. 9.59. In a

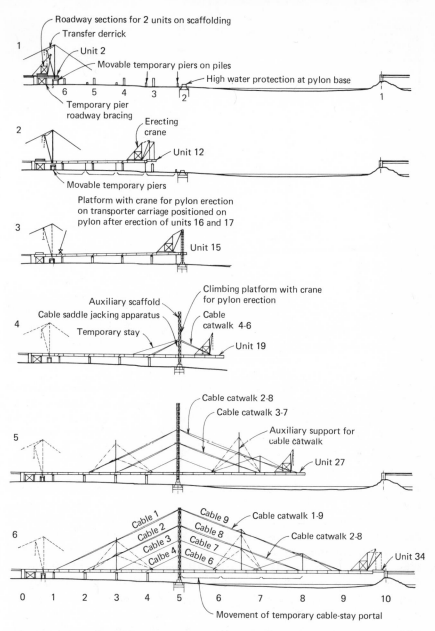

**FIGURE 9.51.** Erection sequence, Kniebrücke, Düsseldorf. (Courtesy of Beton-Verlag GmbH, Düsseldorf. Ref. 16.)

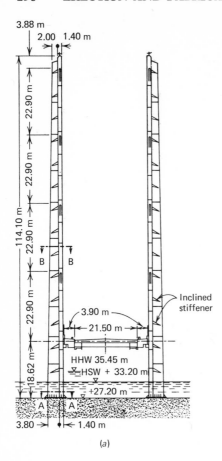

FIGURE 9.52. Elevation and sections of pylon, Kniebrücke, Düsseldorf: (*a*) elevation; (*b*) Section B-B; (*c*) Section A-A. (Courtesy of *Acier-Stahl-Steel.* Ref. 17.)

similar manner the three assembled center deck units were installed, Fig. 9.60.

After the superstructure had been erected to point 14 (just behind the pylon), the pylon cell units were hoisted to the deck by the transfer derrick and transported to the cantilever end by the transporter carriage. When the pylon base units were installed, they were posttensioned to the foundation, Fig. 9.61. The bases were adjusted by jacks, previously installed beneath them; when the bases were at the proper elevation, they were filled with concrete and all the tendons were tensioned. The derrick erected the pylon to a height of 150 ft. After the deck structure unit was cantilevered beyond the pylon, the erecting crane traveled out upon it and proceeded with the remaining superstructure erection. At this point erection of the pylon proceeded by a crane mounted on a climbing platform

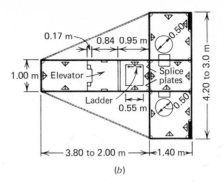

(b)

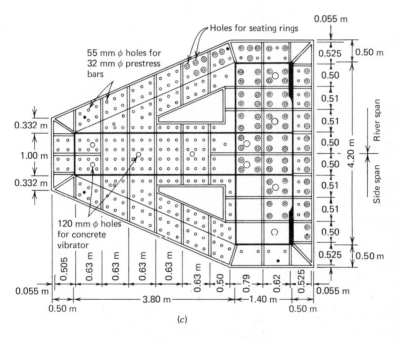

(c)

**FIGURE 9.52.** (*continued*)

that traveled up the pylon as it increased in height, Fig. 9.58 and 9.62. The pylon erection to full height was completed prior to the installation of the cable stay from the bottom.[16,17]

To facilitate the installation of the cable stays, the saddles were initially positioned lower than their final position and were also displaced in a longitudinal direction. Portable jacks located under the end beam of the bearing girder, which cantilevered on each side beyond the pylon, allowed

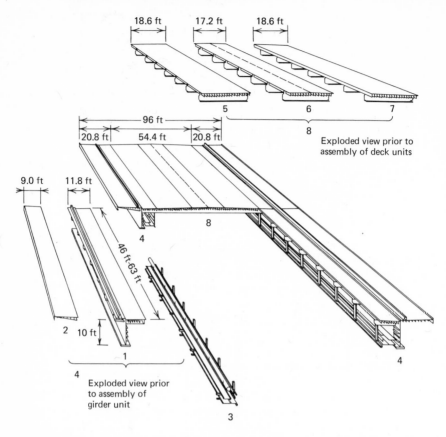

**FIGURE 9.53.** Isometric view of erection components of the superstructure, Knie-brücke, Düsseldorf. (Courtesy of Beton-Verlag GmbH, Düsseldorf. Ref. 16.)

the saddles to be jacked to their specified elevation after the stays were installed, Fig. 9.63. Jacks placed between the saddles and bearing girder controlled the sliding or longitudinal movement of the saddle. Cables were erected using suspended catwalks, Fig. 9.64. With the exception of the bottom pair of catwalks and the second one in the side-span, all catwalks received additional support from a stayed portal frame resting on rocker supports. The top cable in the river span received additional support from auxiliary cable-stays. The first strand of the 13 strands comprising each stay was carefully positioned to serve as a guide for the following 12 strands.

**FIGURE 9.54.** Erection site on left bank, Kniebrücke, Düsseldorf. (Courtesy of *Acier-Stahl-Steel*. Ref. 17.)

### 9.5.6    Lake Maracaibo Bridge, Venezuela

The general structural configuration of the Lake Maracaibo Bridge (also called the General Rafael Urdaneta Bridge) was presented in Chapter 4. This section discusses the construction features of the 771 ft (235 m) main span, Fig. 9.65. The material and illustrations presented have been extracted from ref. 18 through the courtesy of Julius Berger-Bauboag Aktiengesellschaft, Wiesbaden. During construction, structural analyses were performed for the various stages of the erection process. It is interesting to note that the erection analysis involved five times more effort than was required for the design of the structure.

Approximately 13,080 cu yd (10,000 m³) of concrete and 882 tons (800 mt) of reinforcing steel were required to construct a pier, tower, and a continuous cantilever girder. The materials had to be transported 303.5 ft (92.5 m) vertically for the tower and 308 ft (94 m) horizontally at the end of the 131 ft (40 m) jib with a 226 ft (69 m) hook height. For construction of the upper half of the pylon, the main tower crane was extended to a hook height of 321 ft (98 m). The second tower crane was located on the opposite side of the pile cap and was used for placing and

FIGURE 9.55.   Tension-pendulum pier, Kniebrücke, Düsseldorf. (Courtesy of Beton-Verlag, GmbH, Düsseldorf. Ref. 16.)

removing formwork, placing reinforcement and concrete for the X-frames and the lower half of the A-frame pylon. Additional tower cranes were positioned on the service trusses of the cantilever portion of the deck structure. In Fig. 9.66, the main tower crane is extended to its full height. The crane and the next pylon has the service girder in position with a tower crane mounted on it. Also illustrated is the erection of the rear service girder, a pylon with the pier cap completed, a small tower crane ready for dismantling, and two tower cranes pouring concrete for the X-frames and the lower half of the next A-frame pylon. All tower cranes were assembled on shore and positioned as a unit by floating cranes.

Temporary bracing was required during the various erection stages to maintain deformations and stresses within allowable limits. Concrete for the X-frame and the lower half of the A-frame pylon (see Chapter 4) was poured simultaneously. The concrete pouring sequence for the X-frame is

**FIGURE 9.56.** Cantilever erection of units, Kniebrücke, Düsseldorf. (Courtesy of Beton-Verlag, GmbH, Düsseldorf. Ref. 16.)

indicated in Fig. 9.67. Before concreting section 4, inclined braces were installed for each leg until section 8 was completed. After completion of section 13 and prior to pouring section 15, the X-frame outer legs were tied together by six prestressed 1 in. diameter (26 mm) high-tensile steel bars, which were tensioned to the required load. Stresses in the rods were continuously monitored and adjusted by jacks as required. For each X-

**FIGURE 9.57.   Erection to pylon, Kniebrücke, Düsseldorf. (Courtesy of Beton-Verlag, GmbH, Düsseldorf. Ref. 16.)**

frame interior leg, two braces were required; each brace had 18 tons of jack pressure before placement of the transverse cross beams and 35 tons after placement.

The legs of the A-frame tower are of varying dimensions in cross section, decreasing from bottom to top. Because of this shape the formwork and reinforcement required a close check of alignment. Therefore, the formwork for the A-frame pylon also required bracing similar to the legs of the X-frames. The legs of the X-frames were erected before the legs of the A-frame and could be used to brace the A-frame. Erection sequence of the A-frame is illustrated in Fig. 9.68. After erection of section 5, a longitudinal brace and transverse braces were installed between the A- and X-frames. Upon completion of section 9, transverse braces were installed at the section. After section 15 was completed and permanent longitudinal and transverse beams were in place, transverse bracing was installed at section 15. After completion of section 18, additional transverse bracing was installed at this elevation.

The pier cap is a three cell box section 16.4 ft (5 m) in depth, 46.7 ft (14.22 m) in width and 159.3 ft (48.55 m) in length, Figs. 4.4 and 4.5. The X-frame legs were continued into the pier cap to act as a transverse

FIGURE 9.58.    Erection of pylon and lower stays, Kniebrücke,
Düsseldorf. (Courtesy of Beton-Verlag, GmbH, Düsseldorf.
Ref. 16.)

diaphragm. Upon completion of the pier cap, the service girders for the
cantilever portion of the deck structure were hoisted into position. As a
result of the additional moment produced during this stage of erection
additional concentric prestressing was required in the pier cap, Fig. 4.5.
Additionally, because the wet weight of the cantilever and dead load could
overstress the X-frames during construction before cable-stay was in-

**FIGURE 9.59.    Cantilever erection of girder units, Kniebrücke, Düsseldorf. (Courtesy of Beton-Verlag, GmbH, Düsseldorf. Ref. 16.)**

stalled, temporary horizontal ties, tensioned by hydraulic jacks, were required, Figs. 4.5 and 4.6.

To form the 236 ft (72 m) long cantilever girders, special steel trusses were used to support the formwork and wet weight of concrete. These service girders were supported on one end by the completed pier cap and on the other end by temporary rocking piers supported on auxiliary foundations, Fig. 4.7. The cantilever girder is a four cell box, 16.4 ft

FIGURE 9.60.  Isometric model of erection, Kniebrücke,
Düsseldorf. (Courtesy of Beton-Verlag, GmbH, Düsseldorf.
Ref. 16.)

(5 m) deep with 9.8 in. (25 cm) thick webs. It was poured in six sections
in the numerical sequence indicated in Fig. 9.69 to equalize deformations
during the concreting operations. The sequence of pouring was established
to avoid overstressing the pier and cap. Joints of 2 ft 6 in. (75 cm) were
left open between the sections. Upon completion of the last section, the
service girder was raised or lowered to its proper elevation by hydraulic
jacks located under the rocking piers, and the joints between the cantilever
girder sections were poured. A floating concrete mixing plant was anchored alongside the pier where concrete was raised by a hoist and con-
veyed by power buggies to the tower crane and poured into place, Fig. 9.69.

The transverse cable-stay anchorage girder is 73.8 ft (22.5 m) in length
and has its cross section oriented to the inclination of the stays. The 60

FIGURE 9.61. Installation of pylon base units, Kniebrücke, Düsseldorf. (Courtesy of Beton-Verlag, GmbH, Düsseldorf. Ref. 16.)

ton reinforcing cage was fabricated on shore in its proper orientation, Fig. 4.8, and contained the 70 prestress tendons, mild reinforcement, and thick-walled steel pipes for housing the strands of the cable stay, Fig. 4.9. A steel spreader bar was used to erect the prefabricated cage into its position in the structure to be ready for concrete placement, Fig. 4.10.

Catwalks, which were used to facilitate the installation of the strands of the cable-stay, were prefabricated on shore, barged to piers, and placed on the pier caps by a floating crane. One end of the catwalk was raised by

**FIGURE 9.62. Free cantilever erection, August 1968, Kniebrücke, Düsseldorf. (Courtesy of Beton-Verlag, GmbH, Düsseldorf. Ref. 16.)**

winches to the top of the pylon and anchored; the other end was pulled to the anchorage girder and fixed at that location, Fig. 9.70. The strands were delivered to the erection site in 13 ft (4 m) to 16 ft (5 m) coils, unreeled on the surface of the roadway girder, checked for length, cleaned, and given a prime coat, Fig. 9.71. The strands were threaded into pipes in an anchorage girder, pulled up the catwalk to a roller saddle support provided for mounting purposes on the top of the pylon, pulled down the catwalk on the other side, and threaded into the pipes of the opposite anchor girder, Fig. 9.72 and 9.73a.

The strands at the anchorage girder are arranged in a grid of 4 × 4 while at the pylon saddle they are arranged in two layers of eight each, Fig. 9.74. Two to three strands were erected during the day and stretched at night to minimize the effect of thermal expansion of the strands. Each strand was initially tensioned by a hand winch and jacked to an amount which caused the strand socket to project out of the anchorage girder approximately 6 in. (150 mm), Fig. 9.73b. The sockets have internal

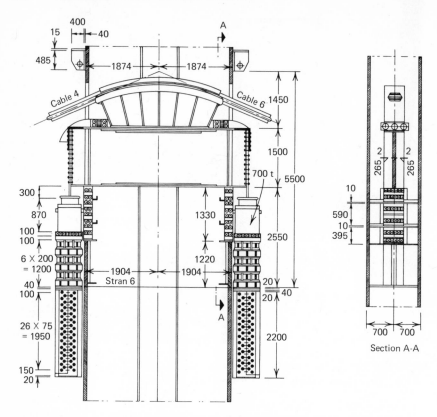

**FIGURE 9.63. Jacking mechanism at the cable saddle bearing, Kniebrücke, Düsseldorf. (Courtesy of Beton-Verlag, GmbH, Düsseldorf. Ref. 16.)**

threads to accept the jacking screws. Two 250-ton jacks, Fig. 9.75, pull the strand out enough that a $3\frac{1}{8}$ in. (80 mm) thick washer can be inserted between the anchorage girder and the strand socket, Fig. 9.73c, and is welded to the anchor plate in the anchorage girder. The strand was tensioned to 106 tons. A second jacking operation tensioned the strands to the theoretically required tension of 161 to 172 tons, when additional washers were inserted and the jacking screws removed, Fig. 9.73d. Finally, the cables were fixed in the saddles by high-strength bolts and clamping plates, Fig. 9.76. The roller saddle used for placing the cables can be seen at the left of the Fig. 9.76.

While the strands were being tensioned, the anchorage girder was post-tensioned in stages. By following a strict sequence of tensioning the

**FIGURE 9.64.    Cable erection, Kniebrücke, Düsseldorf. (Courtesy of Beton-Verlag, GmbH, Düsseldorf. Ref. 16.)**

**FIGURE 9.65.    Main span pylon and cantilever girders, Lake Maracaibo Bridge. (Courtesy of Julius Berger-Bauboag Aktiengesellschaft. Ref. 10.)**

FIGURE 9.66. Various stages of erection, Lake Maracaibo Bridge. (Courtesy of Julius Berger-Bauboag Aktiengesellschaft. Ref. 18.)

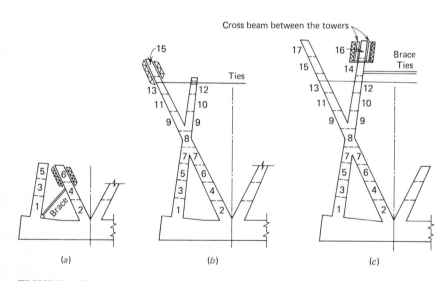

FIGURE 9.67. (a)–(c) Erection sequence at X-frames, Lake Maracaibo Bridge. (Courtesy of Julius Berger-Bauboag Aktiengesellschaft. Ref. 18.)

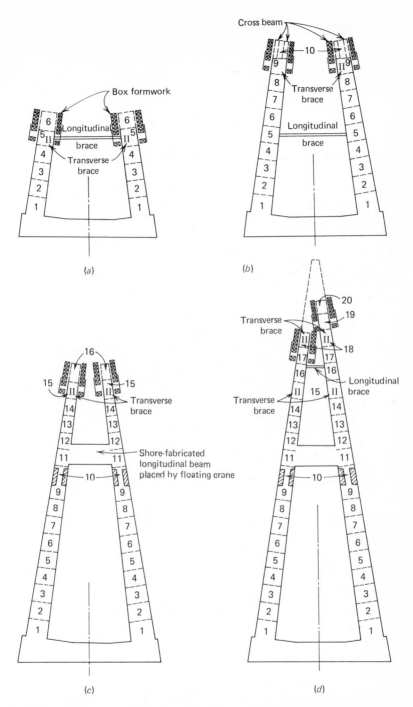

**FIGURE 9.68.** (*a*)–(*d*) Erection sequence of A-frames, Lake Maracaibo Bridge. (Courtesy of Julius Berger-Bauboag Aktiengesellschaft. Ref. 18.)

FIGURE 9.70. Catwalk for cable-stayed installation, Lake Maracaibo Bridge. (Courtesy of Julius Berger-Bauboag Aktiengesellschaft. Ref. 18.)

strands and prestressing the anchorage girder it was possible to avoid tension stresses detrimental to the concrete.

Before tensioning the strands in the stay, the force in the jacks under the rocking piers was measured, and the required tension in the strands was thus determined. Any unintentional difference in dead weight could therefore be accommodated. As the strands in the stay were tensioned, load in the rocking pier jacks was gradually relieved. In this manner, vertical displacement of the transverse anchorage girder was eliminated. This was accomplished by lowering the jacks under the rocking piers in stages such that elastic rebound of the rocking piers and their foundation were simultaneously equalized.

When the stays were finally tensioned to their full value, the rocking piers were relieved of load and the jacks under the rocking piers removed. This prevented the rocking piers from accepting any load resulting from

FIGURE 9.71.  Strand laid out on deck, Lake Maracaibo Bridge. (Courtesy of Julius Berger-Bauboag Aktiengessellschaft. Ref. 18.)

FIGURE 9.72.  Strand erected in position, Lake Maracaibo Bridge. (Courtesy of Julius Berger-Bauboag Aktiengessellschaft. Ref. 18.)

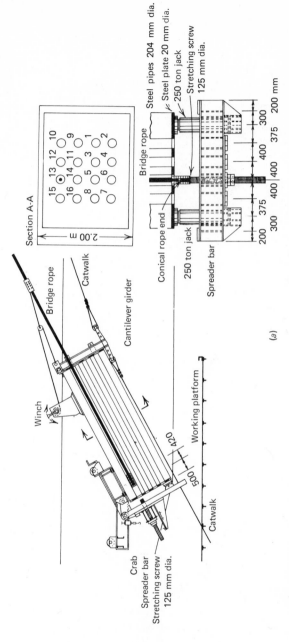

FIGURE 9.73. (a)–(d) Tensioning of bridge strands, Lake Maracaibo Bridge. (Courtesy of Julius Berger-Bauboag Aktiengessellschaft. Ref. 18.)

314

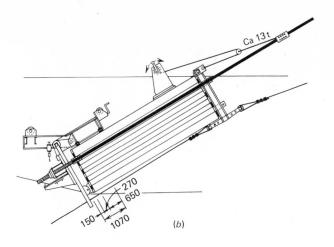

(b)

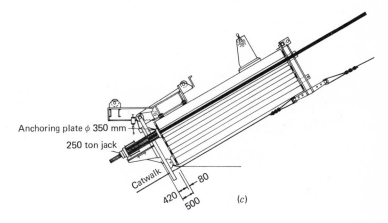

Anchoring plate $\phi$ 350 mm

250 ton jack

Catwalk

(c)

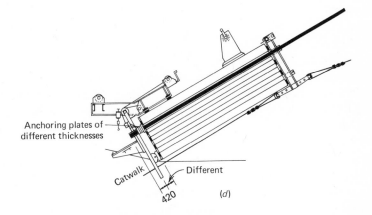

Anchoring plates of different thicknesses

Catwalk

Different

(d)

**FIGURE 9.73.** *(continued)*

**FIGURE 9.74.    Strand position at anchor girder, Lake Maracaibo Bridge. (Courtesy of Julius Berger-Bauboag Aktiengessellschaft. Ref. 18.)**

deflection of the cantilever girder as a result of thermal elongation of the stays. After a few hours, the service girders were removed with the rocking pier and its foundation.

After the service girder was removed, the tension in each strand was reduced approximately 30 tons. To avoid redistribution of the forces, concrete weights totaling 550 tons were stacked on the end of the cantilever girders. The weights were distributed symmetrically to avoid any warping of the cantilever arm resulting from creep. Before the drop-in suspended girders between the ends of the cantilever girders were erected, all but 150 tons of the weight was removed from each cantilever end. The remaining weight represented the roadway surface and sidewalk, and was subsequently removed at a rate corresponding to the rate of progress of the installation of the finishing work. The suspended drop-in spans were made up of four precast T sections, Fig. 4.2.

FIGURE 9.75. Two 250 ton jacks used for tensioning, Lake Maracaibo Bridge. (Courtesy of Julius Berger-Bauboag Aktiengessellschaft. Ref. 18.)

FIGURE 9.76. Installation of clamping plates at pylon saddle, Lake Maracaibo Bridge. (Courtesy of Julius Berger-Bauboag Aktiengessellschaft. Ref. 18.)

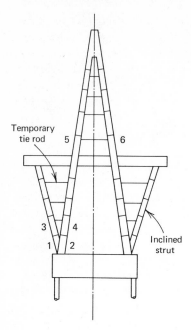

**FIGURE 9.77.   Erection sequences of pylon, Chaco/Corrientes Bridge. (Courtesy of *Civil Engineering*, ASCE. Ref. 20.)**

### 9.5.7   Chaco/Corrientes Bridge, Argentina

The superstructure of the Chaco/Corrientes Bridge (also referred to as the General Manuel Belgrano Bridge) consists of two concrete A-frame pylons connected by transverse beams at the apex of the A-frame and at deck level, Figs. 4.23 and 4.24. The deck system is comprised of two lines of post tensioned concrete box girders supported by two stays radiating from each side of the pylon in two vertical planes, Fig. 4.25. The deck is also supported by inclined struts flanking the pylon legs.

To eliminate the need for falsework, the inclined struts and pylon legs were supported by horizontal ties at successive levels as construction proceeded, Fig. 9.77. The legs were poured in segments by cantilevering the formwork from previously constructed segments. Upon reaching deck level, the girder section between the extremities of the inclined ties was cast on formwork. To further stiffen the pylon structure, a slab was cast between box girders at the level of the girder bottom flanges. This slab is within the limits of the cast-in-place box girders and inclined struts and serves as an additional element to accept the horizontal thrust from the cable-stays. The upper portion of the pylon was then completed using horizontal struts to brace the legs until they were connected at the apex. Fig. 9.77.[19,20]

The precast box girder units, with the exception of those at the cable-stay anchorage were cast 13 ft $1\frac{1}{2}$ in. (4 m) in length by the long-line, match-cast procedure. The soffit bed of the casting form had the required camber built in. Alignment keys were cast into both webs and the top flange. Match casting and alignment keys were required to ensure a precise fit during erection. Each 44 ton unit was transported by barge to the construction site and erected by a traveling crane operating on the erected portion of the deck. Since each box was lifted by a balance beam, four heavy vertical bolts had to be cast into the top flange of each box, Fig. 9.78. The lifting crane at deck level allowed longitudinal movement of the suspended box. Upon erection to the proper elevation, the unit was held to within 6 in. of the mating unit while epoxy joint material was applied. Bearing surfaces of the unit were sand blasted and water soaked prior to erection. The water film was removed before erection and application of the epoxy joint material. The traveling deck crane held the unit in position against its mating unit until it could be posttensioned into position. The crane was slacked off without waiting for the joint material to cure.[19,20]

To minimize overturning forces and stresses in the pylon, it was necessary to erect the precast box units in a balanced cantilever method on both sides of the center line of the pylon. The erection schedule demanded simultaneous erection at each pylon, although the pylons are independent of each other. When four precast box units were erected in the cantilever on each side of the pylon, temporary stays were installed from the top of the pylon to their respective connections at deck level. After installation of the temporary stays, cantilever erection proceeded to the positions of the permanent stays and the procedure was repeated to completion of the installation of the precast box units.[19]

The cable stays are composed of locked-coil strands (Chapter 7) $3\frac{5}{8}$ in. (92 mm) in diameter. Because of the cable length and weight, and the acute angle of inclination of the pylon, saddles were impractical. Each strand was anchored at the top of the pylon and the deck using individual pipe sleeves, Figs. 9.78 and 9.79. To prolong strand life, zinc liners were used at the end of the pipe sleeve opposite the strand bearing socket. The zinc liner holds the strand rigid and at deck level separates the point of maximum vibration from the point of potentially maximum corrosion. The strand sockets were first attached at the pylon, then at the deck, and subsequently tensioning of the cable was performed from the deck level. To minimize the effect of aeolian vibration of the strands, separators were installed to ensure nodes at selected locations.[20]

A summary of the erection sequence used is outlined as follows:

1. Erect precast boxes and posttension successively.

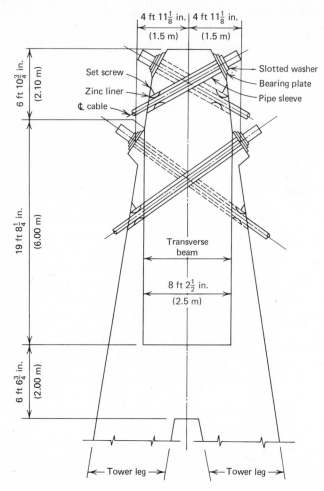

**FIGURE 9.78.    Cable anchorage at top of pylon, Chaco/ Corrientes Bridge. (Courtesy of *Civil Engineering*, ASCE. Ref. 20.)**

2. Erect diaphragms between lines of boxes and posttension.
3. Place temporary and permanent stays as erection proceeds.
4. Remove temporary stays.
5. Remove temporary posttensioning in the cantilevered sections.
6. Place precast deck slabs between box girders.
7. Concrete the three 65 ft 8 in. (20 m) drop-in spans.
8. Place asphalt pavement, curbs, and railings.

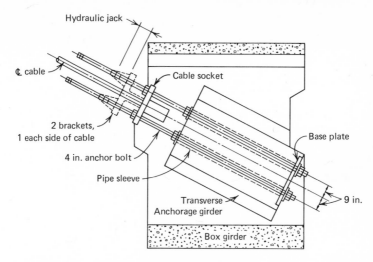

**FIGURE 9.79.   Cable anchorage at girder, Chaco/Corrientes Bridge.
(Courtesy of *Civil Engineering*, ASCE. Ref. 20.)**

### 9.5.8   Pasco-Kennewick Intercity Bridge, U.S.A.

A suggested methodology for the erection of the Pasco-Kennewick Intercity Bridge, by the consultants Arvid Grant and Associates, Inc. and Leonhardt and Andrä, is illustrated in Fig. 9.80, as extracted from the design
drawings.

Phase 1.   Abutment 1 and pier 2 are completed and the cofferdam is
erected for pylon 3.

Phase 2.   Foundation of pylon 3 is completed, and the cofferdam is
erected for pier 5.

Phase 3.   Piers 5 through 8 and abutment 9 are completed. Formwork
for span I and the cantilever section is erected. The caisson is erected for
pylon 4. The casting of pylon 3 begins and contemplates 15 lifts of
approximately 15 ft each.

Phase 4.   Span I and cantilever are cast and ready for prestressing.
Formwork for spans V through VIII are erected (this operation can be
accomplished after the prestressing of span I). Foundation for pylon 4
is completed.

Phase 5.   Span I is prestressed. The auxiliary pier at the end of the
cantilever is left in place to be utilized for adjustment of structure elevation and forces. Spans V through VIII and the cantilever section are cast
and ready for prestressing. Pylon 3, including the portal strut, is completed. Derricks are erected on the strut. Casting of pylon 4 is begun using

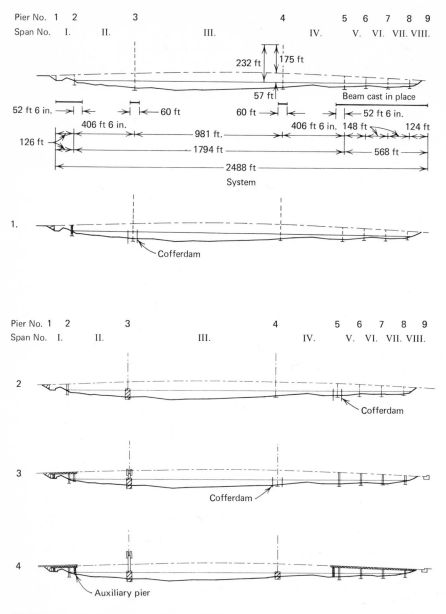

**FIGURE 9.80. Erection sequence, Pasco-Kennewick Intercity Bridge. (Courtesy of Arvid Grant, Arvid Grant and Associates, Inc.)**

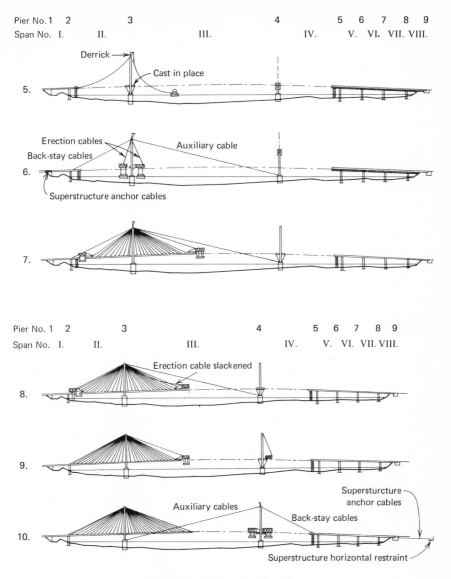

**FIGURE 9.80.** (*continued*)

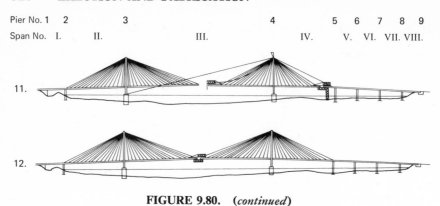

**FIGURE 9.80.** (*continued*)

the same formwork as for pylon 3, which consists of 16 lifts, 14 ft each. The cast-in-place portion of the deck structure at pylon 3 is completed. Auxiliary cables on reels arrive at the construction site. The top socket is hoisted to the top of the pylon and anchored at the position for permanent stay 1. The reels are unrolled and the bottom socket is pulled to the anchorage point by grip hoists.

Phase 6. The back-stay and auxiliary cables are connected at the anchorage for the permanent stay 1. The fore-stay auxiliary cables are anchored at the bottom of pylon 4. Superstructure anchor cables between the end girder of span I and abutment 1 are installed. Back-stay, auxiliary, and superstructure anchor cables are stressed simultaneously by means of pull-rods and center hole jacks. The vertical position of the pylon is checked during stressing and adjusted if required. Floating cranes lift the erection trusses in sections onto the cast-in-place girder and the trusses are then assembled. The erection truss is supported on one end by the completed deck and at the other end by erection cables from the pylon portal strut. Two corresponding precast elements are shipped to the site by barges, connected with lift-slab pull-rods, and lifted simultaneously. They are then connected to the erected portion of the deck and the permanent stay is installed (see precast element erection sequence). The trusses are then moved forward and the sequence repeated.

Phase 7. Spans V through VIII and cantilever are prestressed. The auxiliary pier is left in place for future adjustments. Pylon 4 erection is completed along with the cast-in-place portion of the deck superstructure. The last precast element in the side span, along with its corresponding element in the center span, is erected.

Phase 8. Pylon 3 moments are adjusted with the back-stay and auxiliary cables. The cast-in-place joint in the outer span between approaches and main span is formed. Fore- and back-stay erection cables are slackened. Erection trusses are resting on the girder. The back-stay erection cable is removed and the back-stay erection truss is positioned over the joint and rigidly attached to preclude any movement in the joint resulting from thermal elongation of the cables and overstressing the green concrete. After the closure joint has attained sufficient age the erection truss is dismantled and relocated to the cast-in-place girder at pylon 4. During this stage, the auxiliary pier is used to adjust the elevation of the cantilever, if required. The longitudinal restraint of the deck at the pylon as well as the transverse erection wind bracing are removed.

Phase 9.    To erect the last five precast elements in the Pasco half of the center span, the fore-stay erection cable and the corresponding permanent back-stay cable are stressed simultaneously. For the last element the fore-stay auxiliary cable is removed to allow for the anchorage of the permanent fore-stay cable 1. Superstructure anchor cables at abutment 1 are released.

Phase 10.    Auxiliary cables are relocated to pylon 4 and anchored along with the back-stay cables at the anchorage for permanent cables no. 1. Superstructure anchor cable between the end of girder VIII and the superstructure horizontal restraint is installed. Back-stay, auxiliary, and superstructure anchor cables are stressed simultaneously. Erection trusses are anchored to the cast-in-place portion of the girder at pylon 4, and the first two elements are erected.

Phase 11.    Pylon 4 moments are adjusted with the auxiliary and back-stay cables. The cast-in-place joint in the side span is formed and cast. The back-stay erection truss is rigidly positioned over the joint. An auxiliary pier is used for any necessary adjustment. The back-stay erection cable is removed. Auxiliary cable is removed prior to erection of last element. Longitudinal girder restraint at pylon 4 and transverse erection and wind bracing is removed.

Phase 12.    The last precast element is erected. The fore-stay erection cable has been removed. The superstructure anchor cables at abutment 9 are released. An erection truss is rigidly located over the gap in the center of span III. The cast-in-place joint is formed and cast.

Phase 13.    The membrane and asphalt wearing surface are placed. Railings and lighting fixtures are installed. Final adjustments in all stays are made. Cables are grouted and remaining work completed.

The precast erection sequence is as follows:

1. Connect lift-slab pull-rods and lift precast element to deck elevation.
2. Couple the longitudinal stress bars and shift the element against the erected superstructure to check the matching of the joints.

3. Pull back the element, trowel on the epoxy, and shift the element back to its final position.

4. Stress bars for initial joint pressure. Jack load increment on last installed cables for control of erection bending moments in the structure.

5. Pull in permanent cables and stress simultaneously with the releasing of the erection cables.

6. Complete the stressing of the longitudinal bars. Grout all stressed bars.

7. Weld the top layer of reinforcement and grout the remaining joint.

8. Shift the erection truss to a new position and prepare to lift the next element. This is accomplished by adding additional length to the erection cables, anchoring the erection truss, and stressing the erection cables until the erection truss is resting on its back support.

Installation of the permanent cables is illustrated in Fig. 9.81. The sequence of operations for the cable erection is briefly outlined below. After fabrication, each cable is rolled on a reel and transported to the site. When the cable is due to be erected, it is shipped to the bridge site on a barge. The cable is lifted onto an auxiliary platform at deck level outside the plane of the permanent cables. It is taken up by a trolley, shifted onto the beam, and moved forward to the erection truss. In the meantime, the sky line (guide rope) has been installed at its new position. Its upper end is hinge connected to a trestle on top of the tower head, which permits the rope to follow the different positions of the erection truss. On the trestle the rope can slide in a transverse direction so that it is always on top of the saddle to be installed. The top anchorage remains unchanged during the entire erection process. The lower end of the sky line is anchored at the erection truss by a clamp whose position is adjusted for each new cable.

From the working platform the permanent cable is inspected and hung into sliding hangers that run on the guide rope and are interconnected by the pull rope. The first two hangers are adjustable in length to facilitate threading the cable at the tower head after it has been pulled up by the pull cable. It is finally drawn into the tower head from the inside by an auxiliary tackle. There the head is supported by cross bars 'and horseshoe washers. While the permanent cable is unwound, brakes (installed at the trolley to control the unwinding operation) are applied to the reel.

After the upper end is installed, the bottom anchor head is pulled with grip hoists to the end of the steel pipe at the permanent anchorage. There the pull rod of a center hole jack is screwed into the inside thread of the anchor head and the cable is pulled into its final position where it is anchored with horseshoe washers.

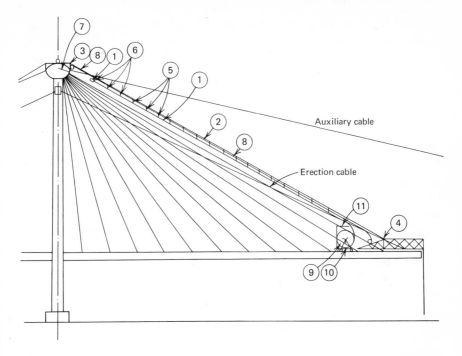

Auxiliary cable

Erection cable

**FIGURE 9.81.    Installation of permanent cables, Pasco-Kennewick Intercity Bridge. List of components: (*1*) permanent cable to be erected; (*2*) guide rope; (*3*) top anchorage of the guide cable at towerhead; (*4*) bottom anchorage of the guide cable at erection truss; (*5*) sliding hangers with fixed length; (*6*) sliding hangers with adjustable length; (*7*) auxiliary tackle; (*8*) pull rope, (*9*) reel for transport of permanent cable; (*10*) trolley for supporting the reel; (*11*) working platform. (Courtesy of Arvid Grant, Arvid Grant and Associates, Inc.)**

The clamps at the lower end of the hangers by which the permanent cable is supported are bigger than the outside diameter of the polyethylene pipe (the pull force is working at the anchor head only), so that after the permanent cable has been installed and partly stressed, they come free of the polyethylene pipe and can be pulled back with the pull cable.

## 9.6   FABRICATION

Aside from technical considerations the factor that most affects bridge design is economics, especially the ratio of material cost to fabrication cost. Therefore, a discussion of fabrication is closely related to economics and is included here for a better understanding of the influence of fabrication on total costs. In the United States the cost of rolled steel members has been relatively low compared to fabrication costs, especially labor costs. This condition has not generally been favorable to least-weight design. Generally, bridge designs in the United States stress simplicity and minimize fabrication, thus producing a heavier structure than is required from a purely analysis point of view. Conversely, until the mid 1950s the European philosophy had been diametrically opposite; relatively elaborate fabrication procedures had been used to produce a least-weight design. Understandably, the material shortages of the post-World War II era created an environment conducive to the development of such innovative concepts as orthotropic plate girders and cable-stay designs in bridges.

However, economic conditions in postwar Europe have changed gradually in some areas. The costs of rolled steel products have remained relatively stable since 1952 but the costs of labor and/or fabrication have been steadily rising. Thus the ratio of material costs to fabrication costs in Europe has been approaching that which exists in the United States. As a result, there has been a great interest in Europe in reducing fabrication costs. Simplified design, mass production, and automation have become increasingly more important than material savings alone. With respect to automation, there has been an increasing trend toward the use of machine programmed numerically controlled fabrication plants. This has led to the development of drilling and milling machines controlled by means of punched cards, punched or magnetic tape, and automatic welding machines.

This convergence of European and American economic philosophy, at least in construction activity, will undoubtedly lead to an increased interchange of ideas and concepts of building and bridge designs. Americans will no longer be able to shrug off European concepts with a simplistic statement such as: "We cannot do that here, the economic situation is different."

It is difficult to predict future trends, but it appears that there may be a merging of the least-weight concept with the automation and simplified design concept. This will undoubtedly result from increased pressures to conserve natural resources and energy and from the necessity to reduce the labor costs.

Cable-stay bridges are, in general, a statically indeterminate system, but may become highly hyperstatic because of the geometrical configuration, number of stays, and built-in redundancies. The completed structural geometry must agree with the specified design geometry for dead loads. One consideration of this design condition is that the elevation of the deck structure must be consistent with the required roadway grade elevations, the cable alignments, and the girder and pylon position dimensions under dead load, such that the pylons are vertical and the bearings are in a no-load position. To satisfy these requirements, structural components must be fabricated to a length that not only meets the required grade, but is properly cambered for the live load deflection. The length must be sufficient to compensate for the elastic shortening caused by axial load and long-term creep effects. The same considerations are true for the pylons and the cable-stays. If these requirements are carefully considered, the desired results will be obtained, irrespective of the method or sequence of erection.

The successful completion of a cable-stayed bridge requires, if not demands, the transfer of design concepts into acceptable fabrication techniques. Thus, a high degree of cooperation is required of the designer, fabricator, and erector. The design engineer must be willing to modify the concept, if necessary, for ease of fabrication and erection. The fabricator and erector must be able to adapt their procedures to the design concept wherever possible.

It is not the intent of this section to describe fabrication procedures in detail. These procedures are the responsibility of the fabricator and may vary from one fabricator to another, depending on equipment and experience. This section will present only those reasons for fabrication decisions as presented by several projects which were reviewed. The reader should bear in mind that what may be a valid decision for one fabricator may not be for another. Furthermore, the reader is reminded that, as a result of the rapidly changing fabrication technology, what may have been economically and technically valid in 1955 is not necessarily applicable in 1975. However, on the premise that one can always learn from previous experiences, a few specific comments appear to be justified.

### 9.6.1   General Fabrication and Welding Comments

Bridge structures are subject to fatigue stresses, and this is particularly true of cable-stayed bridges because of their inherent flexibility. Therefore, the possibility of brittle fracture and fatigue requires special attention to ensure that stress concentrations will not develop from notches, poor edge preparation, welding defects, poor weld shapes, poor welding methods, and, in general, low-quality fabrication techniques.

To avoid fatigue and fracture failures, the fabrication requirements of the Batman Bridge specified that shearing or flame cutting of plates have an allowance that would permit the removal of at least $\frac{1}{8}$ in. of metal by machining for plates up to $\frac{1}{2}$ in. in thickness and at least $\frac{1}{4}$ in. of metal for plates in excess of $\frac{1}{2}$ in. in thickness. Machine flame cut edges were accepted, at the discretion of the engineer, if the edge thus produced was essentially as straight, smooth, and regular as that produced by the finishing cut of a planing machine. Where some doubt existed, a light grinding of the edge was required. Shearing in a direction perpendicular to the direction of main stresses was permitted for minor gusset plates and splice plates. However, shearing was not permitted on plates in main structural members.[15]

With respect to possible stress raisers in welds, the fabricator of the Batman Bridge was not permitted to finish welds with a bad profile. Welds had to be produced with a minimum amount of smooth over fill (i.e., no over-roll), and, if required, fillet welds were to be ground to a smooth finish. Wherever possible, intersection of weld lines were to be avoided. Where this was impossible, the first weld laid down was ground smooth and flush in the region of the intersection, in a direction parallel to the direction of stress. Weld spatter and arc strikes outside of the welds were avoided. Tack welds were required to be kept to a minimum and where possible to be kept in the weld area so that the subsequent automatic welding procedure would tend to "float" them out. Automatic or semiautomatic welding procedures were required for all main longitudinal welds and transverse butt welds in the truss chords. In general, for good fabrication practice, automatic or semiautomatic procedures are encouraged. A general requirement for the Batman Bridge specifications was the use of properly prepared run-on and run-off plates so that butt welds could be carried beyond the edges of the plate. In addition, electrodes were required to be stored in heated boxes or ovens.[15]

### 9.6.2  Superstructure

From the standpoint of material handling during fabrication and erection, it is generally preferred to reduce the bridge cross section into several manageable components. This technique is illustrated for the Kniebrücke Bridge, Fig. 9.53; the Paris-Masséna Bridge, Fig. 9.20; and the Papineau-Leblanc Bridge, Fig. 9.82.

In the fabrication process of the Papineau-Leblanc Bridge, considerable effort was exercised to maximize repetition of the structural components. The cross section was divided into 11 components along longitudinal splice lines indicated in Fig. 9.82. Girder webs were assembled from three panels. Thus a total of 17 stiffened panels compose the cross section. A typical cross section is indicated in Fig. 5.33. Longitudinally the superstructure was divided into 31 sections, requiring a total of 527 stiffened panels (excluding those required for the pylons). As a result of this large amount of repetitive fabrication, cost savings were realized in material orders, jigs, welding, and drilling operations. Standardization of the center to center dimension of the orthotropic stiffeners in the top and bottom

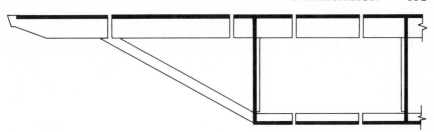

**FIGURE 9.82.   Breakdown of cross section into stiffened panels, Papineau-Leblanc Bridge. (Courtesy of the Canadian Steel Industries Construction Council. Ref. 9.)**

flanges further simplified fabrication. Where the bottom flange increased in thickness and fewer stiffeners were required, alternate stiffeners were omitted and a multiple of the basic spacing was maintained. Thus, the jigs remained constant and no additional fixture layout was required.[9]

The top flange of the box girder also acts as the orthotropic roadway deck. The thickness of the deck plate is 0.437 in. The longitudinal through stiffeners, Fig. 5.33, are $\frac{1}{4}$ in. thick, 13 in. deep, and are spaced on 2 ft centers. Transverse floor beams are spaced 15 ft on centers. The longitudinal through stiffeners were produced in straight 45 ft lengths, kink free, on a patented hydraulic machine designed and built by Dominion Bridge. Since the through stiffener to deck weld was critical, specifications demanded a minimum of 90% penetration weld at the connection. The straight edge produced by the stiffener bending machine made it possible to produce high-quality welds at these critical locations. A rolling pressure head on top of the through stiffener forced the plates into close contact at the welding head and prevented weld blow-through. All panel welding was done by a fully automatic, twin tandem, submerged arc rig mounted on a gantry spanning the welding bed, Fig. 9.83. The subsequent operation position and the transverse floor beams, which were profile burned to fit over the through stiffeners, and were jacked down flat and then tack welded. At another location the floor beams were welded by semiautomatic $CO_2$ machines.

In the fabrication of the West Gate Bridge,[21] the deck plates for the orthotropic deck panels were initially passed through a set of heavy flattening rolls to ensure their flatness. They were then machine flame cut to size with a tolerance for shrinkage in both directions, and bevels for the field welds were also cut at this time. A simultaneous operation was the cold rolling of the longitudinal trapezoidal stiffeners from $\frac{1}{4}$ in. and $\frac{5}{16}$ in. coiled strip and accurately cut to length with an allowance for longitudinal shrinkage. Edges of the stiffeners were machined to proper dimensions and

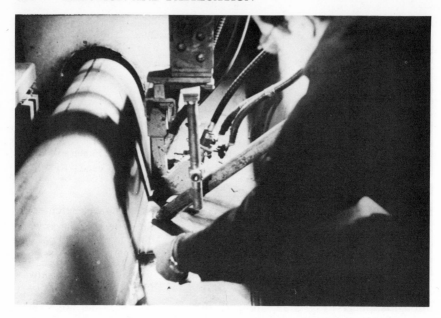

**FIGURE 9.83.** Welding longitudinal stiffener, Papineau-Leblanc Bridge. (Courtesy of Paul Marquis, Gendron Lefebvre and Associates.)

prepared for the partial penetration butt weld that joined them to the deck plate. Transverse stiffeners were cut and profiled to the required shape and spacing of the longitudinal stiffeners.

The trapezoidal longitudinal stiffeners were clamped and tacked to the deck plate at 1000 mm (39.37 in.). The maximum allowable gap between the contact edges of the stiffeners and the plate was 0.5 mm (0.02 in.), Fig. 9.84. A problem was encountered in the partial penetration butt weld joining the longitudinal stiffener to the deck plate. The minimum manual tack weld compatible with the required hardness level had a tendency to over-roll in the continuous automatic submerged arc weld and was rejected. Acceptable solutions were to preheat and use a smaller tack weld or to arc air gouge and grind approximately 50% of the metal from the larger tack weld. On the basis of cost economies, the fabricator chose the latter solution. The longitudinal butt welds were accomplished by a unit that completely welds two stiffeners at each pass.

To correct for the transverse camber in the panels produced by the longitudinal welding, the panels were turned right side up and the correction was made by the controlled application of heat. During this operation, the temperature of the metal was not permitted to exceed 600°C. The transverse stiffeners were then fitted and welded, Fig. 9.85. Upon

FIGURE 9.84.   Clamping and tack welding, West Gate Bridge. (Courtesy of *Acier-Stahl-Steel*. Ref. 21.)

FIGURE 9.85.   Transverse stiffener attachment, West Gate Bridge. (Courtesy of *Acier-Stahl-Steel*. Ref. 21.)

completion of all welding, the panels were inspected and corrected for flatness. Panels were required to be flat within 3 mm (0.118 in.) in 3000 mm (approx. 10 ft) measured in any direction.

### 9.6.3   Pylons

Typically, pylons are designed as box sections with longitudinal and transverse stiffeners. The transverse stiffeners are normally notched to allow the longitudinal stiffeners to pass through. At each end of a box section of pylon, heavy internal flanges are butt welded to the four sides of the box. In this manner the box sections are bolted together without the need for external splices (see Fig. 9.22).

In the Batman Bridge, a submerged arc welding process was used and a fully automatic machine was used for the longitudinal butt welds in the sides of the boxes and for the partial penetration welds for the box corners. For the flange to side wall welds a semiautomatic machine was used. Square butt welds were used for the longitudinal welds with either a $\frac{7}{32}$ in. or $\frac{5}{32}$ in. gap, depending on the plate thickness. Corner welds were prepared for an inclusive angle of 70 degrees, with either two or four runs depending on thickness. Preparation for the flange to side plate weld consisted of a double J for thick plates and a partial penetration double V for thinner plates.[15]

The Papineau-Leblanc Bridge pylon was assembled from four unstiffened plates, with a partial penetration U weld at each corner that was ground flush for appearance. A welding head on a long-reach boom was used to place a fillet weld on the inside of each corner.[9]

The principal problem in the fabrication of the larger sections is the difficulty in achieving a good fit. If in a nominally vertical cable plane geometry the cable plane is not truly vertical, the pylon may be required to be cambered in the transverse direction of the bridge to accommodate the bending forces that may be induced. Therefore, the ends of the section will require milling at a very small angle.

Requirements for the Batman Bridge were that the sections butt evenly together at least 80% on any side and 90% over the entire perimeter. This meant that a three-thousandths feeler gauge could not be entered into the joint from the outside.[15]

Strict mating restrictions were placed on the pylon sections in order to transfer the large compressive loads carried by the pylon from one section to another by direct bearing. To minimize the potential for eccentricity of the load caused by misalignment of the pylon wall plates, specifications required the sections to be fabricated within plus or minus $\frac{1}{16}$ in. between faces and to within plus or minus $\frac{1}{8}$ in. across the corners. Between adjacent sections a maximum $\frac{1}{16}$ in. tolerance in the inside dimension was specified.[15]

In the Kniebrücke Bridge[17] the pylons were fabricated with a camber of 22 in. at the top, which was displaced away from the bridge deck. Individual units of the pylon were fabricated as straight units such that the transverse camber curve of the pylon was a polygonal profile with "kinks' at the joints rather than a smooth curve. It was considered impossible to estimate the relative displacement of one pylon section with respect to an adjacent one during erection. It was impossible to evaluate the temperature deformation precisely even if the temperature variation across the cross section were known. Furthermore, the deformation caused by the erection loading acting upon the previously erected portion of the pylon was unknown. Therefore, it was decided to mill the butting faces in the shop and then have a trial assembly. The trial assembly was to place the inside cell, which was toward the bridge deck, Fig. 9.52, on a transverse face, Fig. 9.86. In this manner the deflection between supports was eliminated and the camber in the pylon was observed and measured as a horizontal curve. Upon being positioned and fitted, the new cell was jacked against the previously positioned cell and the contact of the mating surfaces checked. If the bearing surface had to be corrected it was done on the surface of the new section being fitted.

FIGURE 9.86. Fabrication of pylons, Kniebrücke, Düsseldorf. Ref. 16.)

## REFERENCES

1. Schöttgen, J. and Wintergerst, L., "Die Strassenbrücke über den Rhein bei Maxau," *Der Stahlbau*, Vol. 37, No. 2, February 1968.
2. Kondo, K., Komatsu, S., Inoue, H. and Matsukawa, A., "Design and Construction of Toyosato-Ohhashi Bridge," *Der Stahlbau*, No. 6, June 1972.
3. Fiege, A., "Steel Motorway Bridge Construction in Germany," *Acier-Stahl-Steel* (English version), No. 3, March 1964.
4. Beyer, E. and Ernst, H. J., "Brücke Jülicher Strasse in Düsseldorf," *Der Bauingenieur*, Vol. 39, No. 12, December 1964.
5. Anon., "The Paris-Masséna Bridge: A Cable-Stayed Structure," *Acier-Stahl-Steel* (English version), No. 6, June 1970.
6. Balbachevsky, G. N., "Study Tour of the A.F.P.C.," *Acier-Stahl-Steel* (English version), No. 2, February 1969.
7. Wenk, H., "Die Strömsundbrücke," *Der Stahlbau*, Vol. 23, No. 4, April 1954.
8. Ernst, H. J., "Montage eines seilverspannten Balkens im Gross-Brückenbau," *Der Stahlbau*, Vol. 25, No. 5, May 1956.
9. Taylor, P. R. and Demers, J. G., "Design, Fabrication and Erection of the Papineau-Leblanc Bridge," Canadian Structural Engineering Conference, 1972, Canadian Steel Industries Construction Council, Toronto, Ontario.
10. Demers, J. G. and Simonsen, O. F., "Montreal Boasts Cable-Stayed Bridge," *Civil Engineering*, ASCE, August 1971.
11. Rooke, W. G., "Papineau Bridge Steel Erected in Record Time," *Heavy Construction News*, September 1, 1969.
12. Vogel, G., "Die Montage des Stahlüberbaues der Severinsbrücke Köln," *Der Stahlbau*, Vol. 29, No. 9, September 1960.
13. Fischer, G., "The Severin Bridge at Cologne (Germany)," *Acier-Stahl-Steel* (English version), No. 3, March 1960.
14. Anon., "Opening Batman Bridge 18th May, 1968," Department of Public Works, Tasmania, Australia.
15. Payne, R. J., "The Structural Requirements of the Batman Bridge as They Affect Fabrication of the Steelwork," *Journal of Institution of Engineers*, Australia, Vol. 39, No. 12, December 1967.
16. Tamms and Beyer, "Kniebrücke Düsseldorf," Beton-Verlag GmbH, Düsseldorf, 1969.
17. Schreier, G., "Bridge over the Rhine at Düsseldorf: Design, Calculation, Fabrication and Erection," *Acier-Stahl-Steel* (English version), May 1972.
18. Anon., "The Bridge Spanning Lake Maracaibo in Venezuela," Bauverlag GmbH., Wiesbaden, Berlin, 1963.
19. Gray, N., "Chacos/Corrientes Bridge in Argentina," *Municipal Engineers Journal*, Paper No. 380, Vol. 59, Fourth Quarter, 1973.
20. Rothman, H. B. and Chang, F. K., "Longest Precast-Concrete Box-Girder Bridge in Western Hemisphere," *Civil Engineering*, ASCE, March 1974.
21. Burns, C. A. and Fotheringham, W. D., "Deck Panels for West Gate Bridge (Australia)," *Acier-Stahl-Steel* (English version), June 1974.

# 10

# Structural Behavior of Cables

## 10.1  INTRODUCTION

The purpose of this brief discussion of the structural behavior of cables is to provide a basic understanding of the characteristics of cable action under varying load conditions. Unlike structural steel, structural cables are flexible members and, therefore, do not respond to the usual principles applied to the stiffer components.

Detail technical derivations are not presented because they are available from many other sources on the theory and analysis of structures and/or structural systems. The intent here is to provide an elementary understanding of cables and their behavior as an introduction to the analysis of a complete cable-stayed bridge in Chapter 11. Therefore, only those equations considered to be basic to the understanding of the analysis of cable-stayed bridges are presented.

A fundamental problem encountered in cable structures of all types is the nonlinear behavior of the cable system as a result of the changes in sag and corresponding axial tension. A method for overcoming the nonlinear effect has been proposed; it substitutes an equivalent modulus of elasticity to include the normal modulus together with the effect of change of sag

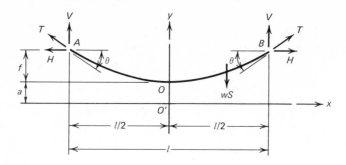

**FIGURE 10.1.   Catenary curve.**

and tension loads. This aspect of cable behavior is discussed in a later section. The usual behavior of cable systems is presented as a prelude to the more meaningful discussion of the several suggestions for the equivalent modulus of elasticity.

## 10.2   CATENARY CURVE

A freely hanging cable supporting its own uniformly distributed weight and connected at the ends to stable anchorages will take the shape of a catenary curve, Fig. 10.1. The equilibrium of the catenary system is maintained by the anchorage forces at the supported ends. These forces are: the cable tension, $T$, with its horizontal component, $H$, and vertical component, $V$. The vertical components, $V$, balance the total weight of the catenary. The horizontal component, $H$, is equal and opposite in direction and, therefore, balanced. Stability is achieved by the general static equilibrium of all the forces.

To determine the forces acting on a catenary curve, it is essential to define the shape of the draped cable. This may be accomplished by considering the equilibrium of a segment of the cable such as $OB$ of Fig. 10.2. The weight of the cable per unit length along the cable axis is denoted by $w$, and the length of the arc $OB$ is denoted by $S$. For a horizontal cable chord, when the support points $A$ and $B$ are at the same elevation, the low point $O$ at the center of the span may be taken as the origin of the coordinate system. However, it is also expedient to assume, $O'$, at a distance of $a$ from the low point $O$, as a convenient origin of the coordinate axes.

The equation for the catenary elastic curve are stated below without derivation.

$$\tan \theta = \frac{V}{H} = \frac{wS}{H}$$

Let

$$a = \frac{H}{w}$$

then

$$\tan \theta = \frac{S}{a}$$

The equation for a catenary can be shown to be:[1,2,3,4]

$$y = a \cosh \frac{x}{a}$$

In order to compare the catenary curve with a corresponding parabolic curve, it is convenient to express the elastic curve in non-dimensional terms.

The sag $f$ may be expressed as

$$f = a \left( \cosh \frac{l}{2a} - 1 \right)$$

When the sag ratio $n = f/l$, and $m = 2a/l$

then

$$n = \frac{m}{2} \left( \cosh \frac{1}{m} - 1 \right) \tag{10.1}$$

which is an expression for the catenary curve in nondimensional terms of $n$ and $m$, which will be referred to in Section 10.4.

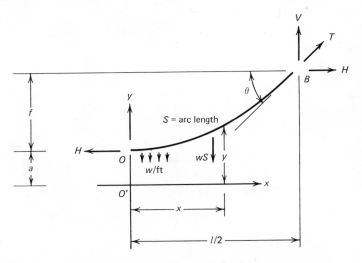

FIGURE 10.2.   Segment of catenary.

## 10.3   PARABOLIC CURVE

The mathematical expression for a parabolic curve is simple when compared with the equation for a catenary curve. Therefore, it is advantageous to compare the two expressions to determine the range for which a parabolic curve may be substituted for a catenary curve with minimal or insignificant error. Therefore, an equation for a parabolic curve in nondimensional terms will be compared to a similar equation for a catenary curve.

The basic equation for a parabolic cable with a horizontal cable chord supporting a uniformly distributed horizontal load and using the general configuration of Fig. 10.3 is stated as follows:[5]

$$y = \frac{wx^2}{2H},$$

with the origin at point $O$.

If the origin is considered to be at $O'$ to agree with the catenary coordinate system, the equation is expressed as:

$$y = a + \frac{x^2}{2a}$$

when cable sag $f = \dfrac{l^2}{8a}$, then at the support point $B$

$$y_b = f + a = a + \frac{l^2}{8a}$$

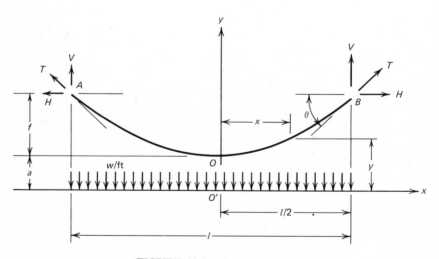

FIGURE 10.3.   Parabolic curve

substituting the terms $n$ and $m$ as previously defined to convert to a non-dimensional equation, the following expression for the parabolic curve is obtained

$$n = \frac{1}{4m} \tag{10.2}$$

## 10.4   CATENARY VERSUS PARABOLA

To compare the parabolic curve with the caternary curve, a full logarithmic plot of equations 10.1 and 10.2 is developed with $n = f/l$ as the abscissa, and $m = 2a/l$ as the ordinate, Fig. 10.4.[6]

The plot indicates that the two curves begin to diverge at a sag ratio $n = f/l$, of approximately 0.15. Therefore, it can be concluded that for sag ratios less than 0.15, the parabolic curve may be substituted for the catenary curve with a reasonably small percentage of error.

The comparison of the parabolic and catenary curves has been computed for a horizontal cable chord with supports at the same elevation and symmetrical about a vertical axis through the low point of the sag of the cables. For an inclined cable chord, Fig. 10.5, which is the usual configuration for cable-stayed bridges, the substitution of a parabolic curve

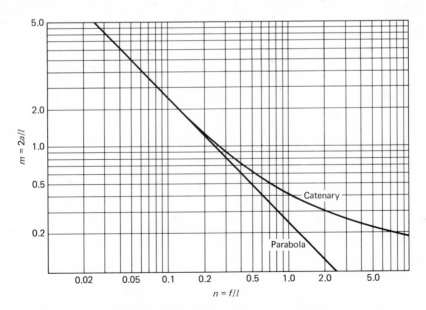

**FIGURE 10.4.   Catenary versus parabola.**

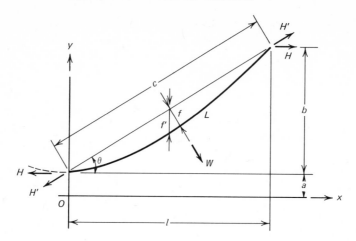

**FIGURE 10 5.** Inclined cable chord. $w$ is the weight per unit cable length, catenary $w$ is the weight per unit horizontal projected length (parabola).

for the catenary curve may also be made. This substitution is based on a comparison of the lengths of the two cables as indicated below.

In a cable-stayed bridge, the inclined cable takes the shape of a catenary because it simply supports its own weight between anchorages with no other intermediate loads, Fig. 10.5. The length of the catenary cable may be expressed as:[7]

$$L^2 = b^2 + 4a^2 \sinh^2 \frac{l}{2a}$$

and letting $a = H/w$, the expression may be written as

$$L^2 = b^2 + 4 \left( \frac{H}{w} \right)^2 \sinh^2 \frac{lw}{2H} \tag{10.3}$$

Similarly the length of a parabolic cable is stated as;[8,9,10] Fig. 10.5:

$$L = c \left[ 1 + \frac{8}{3} \left( \frac{f}{c} \right)^2 \right]$$

$$c = \frac{l}{\cos \theta}$$

$$f = f' \cos \theta$$

$$f' = \frac{wl^2}{8H}$$

and substituting in the above equation:

$$L = \frac{l}{\cos \theta} \left[ 1 + \frac{8}{3} \left( \frac{wl}{8H} \cos^2 \theta \right)^2 \right] \qquad (10.4)$$

In order to compare the two curves, a typical situation for which a comparison of the cable lengths has been investigated is assumed. For convenience of calculation and to represent an inclined cable for a cable-stayed bridge, a typical cable application is assumed with a span of 300 ft and a tower of 125 ft, Fig. 10.6.

The difference in length between a catenary and parabolic curve is calculated using equations 10.3 and 10.4,[3] for various values of the horizontal cable component $H$. The results are tabulated in Table 10.1, which

**TABLE 10.1.   % Error, Catenary vs. Parabola**

| $H$ (kips) | Catenary, $L_c$ (feet) | Parabola, $L_p$ (feet) | $\Delta L = (L_c - L_p)$ (feet) | $\Delta L / L_c$ (%) |
|---|---|---|---|---|
| 50 | 360.1257 | 358.5975 | 1.5282 | 0.4244 |
| 100 | 333.4958 | 333.3993 | 0.0965 | 0.0289 |
| 150 | 328.7727 | 328.7328 | 0.0399 | 0.0121 |
| 200 | 327.1328 | 327.0998 | 0.0330 | 0.0101 |
| 500 | 325.3896 | 325.3359 | 0.0537 | 0.0165 |

indicates that the parabolic curve approaches the catenary curve as the value of $H$ increases.[4] The accompanying percentage error decreases from 0.0289% for a value of $H$ equal to 100 kips to 0.0165% for a value of H equal to 500 kips.

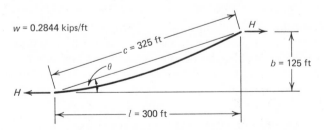

**FIGURE 10.6.   Typical cable application. See FIG. 10.5.**

## 10.5 ASSUMPTIONS FOR ANALYSIS

Although the actual shape of a cable as used in a cable-stayed bridge is a true catenary curve, it has been determined that an assumption of a parabolic curve is within the range of acceptable minimum engineering error for design calculations. Therefore, for analysis and design of cable-stayed bridges with nominal proportions of span to tower height, a parabolic curve will be assumed for the shape of the inclined cable.

Section 10.4 has indicated that for a horizontal cable chord, the parabola may be substituted for the catenary curve below a sag ratio, $n = f/l$, of 0.15 with very little error, Fig. 10.4.

A study of length comparisons tabulated in Table 10.1 indicates that a parabola may be substituted for a catenary with very little error for reasonably large values of the horizontal component.

For the application of a horizontal cable chord, Shaw[5] has indicated that if the angle $\theta$, Fig. 10.1 is small (denoting a sag ratio, $n = f/l$, of approximately 1:6 or less), it is sufficiently accurate to assume the weight of the cable to be distributed as a uniform load per unit length of horizontal projection of the cable.

For the usual application of an inclined cable chord in cable-stayed bridges, Fig. 10.5, Francis[7] has shown that the assumption of a parabola replacing the catenary is valid when $H'/W$ exceeds unity and the angle of inclination of the cable chord to the horizontal does not exceed 70 degrees.

## 10.6 GENERAL CABLE THEOREM

The basic principals of static equilibrium are sufficient to determine the cable forces that develop from the application of gravity loads to the cable system. These cable tension forces may be calculated from the general cable theorem for various configurations of the cable.

The general theory of cables is reviewed in most textbooks on structural analysis and, therefore, only a summary of the theory and a general expression are stated below.[2]

Consider the general case of a cable supported at two points $A$ and $B$, which are at different elevations acted upon by any number of loads, $P_1$, $P_2$, $P_3$, . . . , $P_m$ as indicated in Fig. 10.7. Since it is assumed that the cable is perfectly flexible, the bending moment at any point on the cable must be zero. Because the loads are gravity loads and, therefore, vertical, the horizontal component of the cable tension stress, denoted $H$, must have the same value at any point on the cable and at reaction points $A$ and $B$.

The general cable theorem states that: "At any point, such as $n$, on a cable whch is acted upon by vertical loads (gravity), the product of the horizontal component of the cable stress $H$ and the vertical distance from that point to the cable chord $y_n$ equals the moment that would occur at

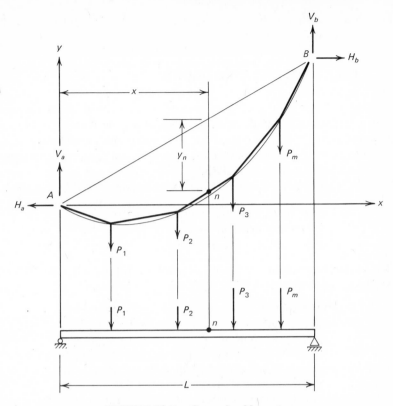

**FIGURE 10.7.   General cable system.**

that section if the same loads were acting on an end-supported beam of the same span as that of the cable."[2]

The theorem is valid and applicable to any set of vertical loads whether the cable chord is horizontal or inclined. Mathematically the theorem may be expressed as

$$Hy_n = \frac{x}{L} \sum M_B - \sum M_n$$

where $H$ = horizontal component of cable stress
    $y_n$ = vertical distance from cable to inclined chord
    $x$ = distance from origin at $A$
    $L$ = span, horizontal distance between supports
$\sum M_B$ = the algebraic sum of the moments about support $B$ of all the loads on the cable
$\sum M_n$ = the algebraic sum of the moments about any point $n$ on the cable of those loads $P_1, P_2, \ldots, P_m$ that act on the cable to the left of point $n$.

The above equation is the basic expression for the determination of the horizontal component of the cable stress.

## 10.7   CABLE WITH INCLINED CHORD

In cable-stayed bridges the cable is always in the inclined chord position, and, as a consequence, the expression to determine the various forces and geometrical changes will be stated for that condition without derivation. For details on the development of the expressions the reader is referred to textbooks on cable structural analysis.[8]

The forces to be determined for analysis and design considerations are:

$$H = \text{the horizontal component of cable stress}$$
$$T_{\max} = \text{the maximum tension stress in the cable}$$
$$V = \text{the vertical component of cable stress}$$

The geometrical quantities to be determined for design and erection considerations are, Fig. 10.8:

$$L = \text{total span length of cable between supports}$$
$$S = \text{cable length}$$
$$\Delta S_s = \text{cable elongation due to cable tension stress}$$
$$\Delta S_t = \text{cable elongation due to temperature change } t \text{ in } °F.$$
$$y = \text{vertical distance from cable to inclined cable chord.}$$

The equations for the determination of the forces and geometrical effects are based on the notation of Fig. 10.8:[2,8,9]

$$H = \frac{wL^2}{8f'}$$

$$T_{\max} = H\left[1 + \left(\frac{h}{L} + 4n\right)^2\right]^{1/2}$$

$$V_r = \frac{Hh}{L} + \frac{wL}{2}$$

$$S \simeq L \sec\theta \left(1 + \frac{8n^2}{3\sec^4\theta}\right)$$

$$\Delta S_s \simeq \frac{HL}{AE}\sec\theta \left(1 + \frac{16n^2}{3\sec^4\theta}\right)$$

$$\Delta S_t = \xi t L \sec\theta \left(1 + \frac{8n^2}{3\sec^4\theta}\right)$$

$$y = \frac{4f'}{L^2}(Lx - x^2)$$

where $w$ = uniform load per unit length of horizontal projection
$f'$ = cable sag
$n = f'/L$, sag ratio
$L$ = span
$S$ = length of cable curve
$\xi$ = thermal coefficient of linear expansion, 0.0000065 in./in./(°F)
$t$ = temperature change in °F

The inclined cable in a cable-stayed bridge is assumed to be a straight line between supports in the analysis of the structure. Although the cable

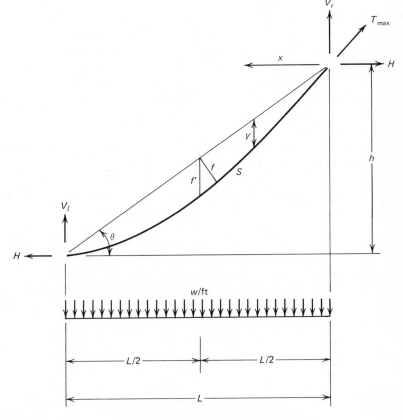

**FIGURE 10.8.   Inclined cable chord.**

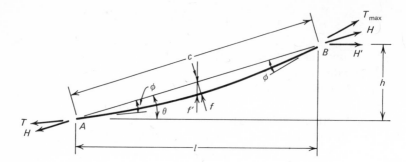

**FIGURE 10.9.    Maximum tension, inclined cable chord.**

is not actually along the chord line, because of the sag produced by its own weight, the tension force thus calculated is assumed to be the tension in the cable. The validity and accuracy of this assumption was investigated by Podolny.[4] He concluded that for design purposes the tension force acting along the inclined chord may be considered as the cable stress.

To arrive at this conclusion, Podolny compared the maximum tension stress in the cable to the tension stress along the chord. His study was made by assuming the cable to have a horizontal chord and then comparing the value for the horizontal component $H$ to the maximum tension force $T_{max}$, Fig. 10.9.

The results are plotted, Fig. 10.10, as percentage error versus the sag ratio $n$ for various angles of inclination of the chord.

It can be seen that for large sag ratios of $1:30$ to $1:100$ and large angles of chord inclination, the percentage error is quite large, but for smaller sag ratios the error is within acceptable limits. If it is assumed, for discussion purposes, that an initial sag ratio of $1:60$ is present and the order of magnitude of tension increases tenfold from initial cable weight only to final tension, then a final sag ratio under load may be of the order of $1:600$ or more, and the percentage error decreases to a value of less than 2%.

Therefore, on the basis of this study, the maximum tension in the cable may be taken as the calculated tension assuming the cable as a straight member between supports.

## 10.8    EQUIVALENT MODULUS OF ELASTICITY

The analysis of a cable-stayed bridge is based on elastic considerations for the materials and, therefore, elastic theories of structural analysis are used to determine the forces acting on each member of the system. As stated

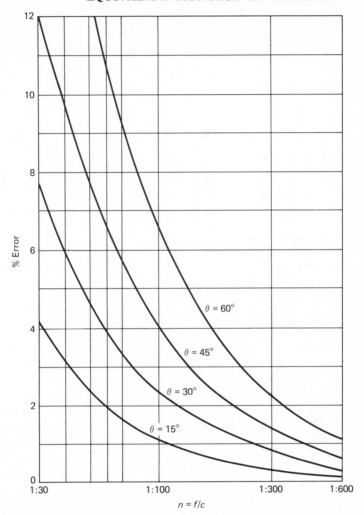

**FIGURE 10.10.** Percentage error of cable tension versus component along inclined chord.

previously, the cable force is considered to act along the inclined chord even though it sags slightly under its own dead weight. As a result of the flexibility of the cable and the changes in length and sag, it is necessary to adopt a corrective technique to account for these nonelastic features. Several methods have been proposed by various authors who suggest the use of an equivalent modulus of elasticity for the cable. This approach is similar to assuming a straight member with a varying modulus of elasticity

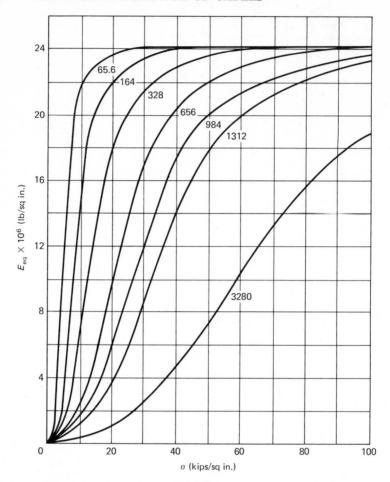

**FIGURE 10.11.    Ernst's equivalent modulus of elasticity. (Ref. 3.)**

that depends on the magnitude of the tension force. The basic principle for analysis is that the behavior of a straight substitute member with an equivalent modulus of elasticity is identical to that of the curved cable.

Although several investigators (Gimsing,[10] Goschy,[11] and Tung and Kudder[12]) have studied the problem of the equivalent modulus of elasticity, each approach results in the solution provided by Ernst,[3] which will be considered herein as the fundamental method.

Ernst developed the following expression for the equivalent modulus of elasticity:

$$E_{eq} = \frac{E}{1 + \frac{(\gamma l)^2}{12\sigma^3} E}$$

where $E_{eq}$ = equivalent modulus of elasticity
$E$ = modulus of elasticity of the cable
$\gamma$ = specific weight of the cable, weight per unit volume
$\sigma$ = unit tension stress in the cable

Ernst presented the results of solving his equation as a plot, Fig. 10.11, of several curves representing different values of the span lengths $L$. The curves are plotted for an $E$ equal to 24,180 kips/sq in. and a specific weight of cable equal to $3.035 \times 10^{-4}$ kips/cu in. The plot indicates the equivalent modulus of elasticity to be used for a particular cable stress and span length. However, it is to be noted that the values of $E_{eq}$ are for a locked-coil strand[3] and for a constant stress in the cable.

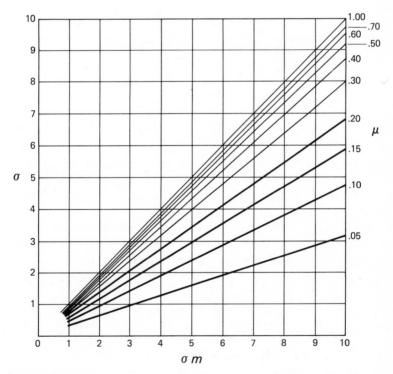

FIGURE 10.12.  Ernst's modified stress for equivalent modulus of elasticity. (Ref. 3.)

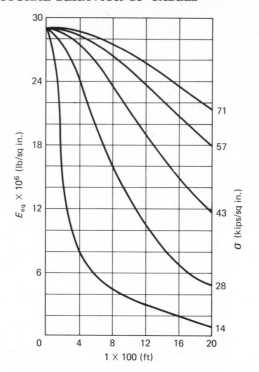

**FIGURE 10.13. Equivalent modulus of elasticity. (Courtesy of Fritz Leonhardt. Ref. 13.)**

The actual action of the cable is such that a change in stress level takes place as the cable forces are increased and the sag decreases. To account for this nonlinear changing stress condition, Ernst modified his basic expression. The modification is based on a mean stress of the cable tension for various values of the parameter $u$, which is the ratio of the final stress to the original stress in the cable, Fig. 10.12. The plot indicates a revised value of $\sigma$ for a mean value of $\sigma_m$ and a specific value of $u$. With the revised value of $\sigma$ from Fig. 10.12 and the plot of Fig. 10.11, an improved value of $E_{eq}$ is determined.

Dr. Leonhardt[13] has presented similar data for parallel wire strands, Fig. 10.13, with the span length as abscissa and the equivalent modulus of elasticity as the ordinate for a family of cable stress levels.

Therefore, in the analysis for the cable tension force, the cable is considered to be a straight member between points of anchorage with an equivalent modulus of elasticity. It is apparent that the analysis becomes an iterative process requiring several determinations of cable stress and the corresponding equivalent modulus of elasticity until a convergence of the values is achieved.

## REFERENCES

1. Seely, F. B. and Ensign, N. E., *Analytical Mechanics for Engineers*, John Wiley and Sons, Inc., New York, 1948, pp. 104–11.
2. Scalzi, J. B., Podolny, W., Jr., and Teng, W. C., "Design Fundamentals of Cable Roof Structures," ADUSS 55-3580-01, United States Steel Corporation, Pittsburgh, 1969.
3. Ernst, J. H., "Der E-Modul von Seilen unter Berücksichtigung des Durchhanges," *Der Bauingenieur*, Vol. 40, No. 2, February 1965, pp. 52/55.
4. Podolny, W., Jr., "Static Analysis of Cable-Stayed Bridges," Ph.D. Thesis, University of Pittsburgh, 1971.
5. Shaw, F. S., "Some Notes on Cable Suspension Roof Structures," *Journal of the Institution of Engineers*, Australia, Vol. 36, April–May 1964.
6. Odenhausen, H., "Statical Principles of the Application of Steel Wire Ropes in Structural Engineering," *Acier-Stahl-Steel* (English version), No. 2, February 1965.
7. Francis, A. J., "Single Cables Subjected to Loads," *Civil Engineering Transactions*, Institution of Engineers, Australia, Vol. CE7, October 1965, pp. 173–180.
8. Maugh, L. C., *Statically Indeterminate Structures*, John Wiley and Sons, Inc., New York, 1964.
9. *Manual for Structural Applications of Steel Cables for Buildings*, American Iron and Steel Institute, Washington, D. C., 1973.
10. Gimsing, N. J., "Anchored and Partially Stayed Bridges," *Proceedings of the International Symposium on Suspension Bridges, Lisbon*, Laboratorio Nacional De Engenharia Civil, 1966, pp. 475–484.
11. Goschy, Bela, "Dynamics of Cable-Stayed Pipe Bridges," *Acier-Stahl-Steel* (English version), No. 6, June 1961, pp. 277/282.
12. Tung, D. H. H., and Kudder, R. J., "Analysis of Cables as Equivalent Two-Force Members, *Engineering Journal*, American Institute of Steel Construction, January 1968, pp. 12–19.
13. Leonhardt, F. and Zellner, W., "Cable Stayed Bridges: Report on Latest Developments," Canadian Structural Engineering Conference, 1970, Canadian Steel Industries Construction Council, Ontario, Canada.

# Design Considerations
and Analysis

## 11.1  INTRODUCTION

Chapter 10 presented basic assumptions, cable theory, and equations. This chapter discusses general design and analysis considerations. Aerodynamic considerations are discussed in Chapter 12.

Once the decision is made to build a bridge structure on a given site, the type of bridge must be selected. Generally, because of the number of considerations to be taken into account (e.g., vertical and horizontal clearances for navigation, terrain conditions at the site, environmental factors, and foundation problems), the geometry of the structure will be dictated by these requirements. Normally, several types of bridge structures will be investigated to meet the various parameters imposed, with each type having its own advantages and disadvantages for the particular site.

We do not intend to discuss the decision-making process in bridge type selection, but will mention a few of the factors that must be considered in the selection of a specific bridge type. Bridge selection is more of an art than a science because there are no universally accepted hard and fast rules. What may be a valid, rational decision in one case may have no validity at all in another case.

We will assume that a cable-stayed bridge is to be built and will proceed to present design and analysis considerations for it. However, at this point the reader should be aware that the cable-stayed bridge is not simply one bridge type, but many different individual types evolving from an extremely versatile concept of bridge design. In selecting a bridge type, the bridge engineer should not limit his thinking to just a cable-stayed bridge; he should consider a number of geometrical variations of cable-stayed bridges.

## 11.2  DESIGN CONSIDERATIONS

A complete design and analysis methodology, criteria, code, or specification is obviously beyond the scope of this book. However, we do present some of the design considerations that are pertinent to this type of structure. Furthermore, wherever possible current design "rules of thumb" and general proportions of existing structures that may prove useful in a trial design are presented. Because the cable-stayed bridge is affected by a wide range of factors, design and analysis considerations will be presented in a general manner. A particular structure would, of course, be subject to the dictates of its design environment. Many of the design considerations that follow have been discussed in previous chapters and we will attempt to avoid duplication by appropriate cross references. However, we do repeat some of these considerations in order to provide the proper synthesis and continuity of presentations.

### 11.2.1  Span Proportions

In any type of bridge structure, one of the first design considerations to be evaluated is the proportioning of the spans. Cable-stayed bridges, for the most part, have been utilized to cross navigable rivers where navigation requirements have dictated the dimensions of the principal spans. Because the girder is supported from above by cable stays, cable-stayed bridges are ideal for spanning natural barriers such as wide rivers. Fig. 1.21, and deep gorges, Fig. 4.12. Similarly, for vehicular or pedestrian bridges, crossing interstate highways or areas of heavy urban development, they can provide long spans unobstructed by piers, Figs. 4.11, 6.6.

Span arrangements are of three basic types: two spans, symmetrical, or asymmetrical, Fig. 1.1; three spans, Fig. 1.2; or multiple spans, Fig. 2.1. A partial tabulation of two-span asymmetrical structures is presented in Table 2.1, which indicates that the longer span is in the range of 60 to 70% of the total length. Exceptions are the Batman and Bratislava bridges, Fig. 2.2, which have ratios of 80%. However, these two structures do not have the back-stays distributed along the short span; they are concentrated into a single back-stay anchored to the abutment. A similar tabulation of three-span structures is presented in Table 2.2, which indicates that the ratio of center span to total length is approximately 55%. In multiple-span structures, the spans are of equal length (although there are exceptions such as the Polcevera Viaduct discussed in Chapter 4), with the exception of flanking spans that connect to the approach spans of abutments. The girder generally has drop-in sections at midspan. The ratio of a drop-in span length to total span length ranges from 20%, when a single stay emanates from each side of the pylon, to 8%, when multiple stays emanate from each side of the pylon.

Experience has indicated that in relatively short-span structures, 400 to 600 ft, a three-span cable-stayed bridge will generally be within 3 to 5% of preliminary estimates of other bridge types. Because this comparison is considered within the accuracy of preliminary estimates, the economic advantage of the cable-stayed bridge is not clear-cut, although it may be said to be competitive. Where feasible, the designer might consider a two-span asymmetrical cable-stayed structure with perhaps the longer span in a range of 800 to 1000 ft. In this type of design, the river piers may be eliminated. If viaduct approach spans are considered then perhaps the approach piers could be used to anchor the back-stays and thus stiffen the longer or center span. The cost of the superstructure may increase, but this may be offset by the decreased substructure cost. To reiterate, the designer must be aware that a cable-stayed bridge is not one bridge type but a number of types.

### 11.2.2 Stay Geometry

The stay geometry of cable-stayed bridge systems varies greatly. The arrangement of stays is subject to numerous considerations, including highway requirements, site conditions, and aesthetic preferences. Chapter 2 presented an extensive discussion of the advantages and disadvantages of the several geometrical variations. Once a general transverse and longitudinal geometry is selected, the decision of the number of stays will be based on the magnitude of force in the stays, how this force will be transferred to the girder and pylon, spacing or distribution of stays at the girder and/on pylon, and the effect of the appropriate component of force on the design of the girder and pylon.

### 11.2.3 Pylons

As in the case of stay geometry, pylons are of various types to suit site conditions, design requirements, aesthetics, and cable geometry. Geometrical variations and their advantages have been discussed in Chapter 2. The height of the pylon in relation to the span will obviously influence the cable forces and thus the amount of cable steel required. The ratio of pylon height to span is discussed in Chapter 3.

### 11.2.4 Girder Types

Girders are constructed of steel or reinforced concrete. The choice of materials are, primarily, a function of availability and economics prevalent at a particular time and specific location of the bridge site. Basically there are two types of girders, the stiffening truss type for steel construction and the solid web type used for steel and concrete. In both steel and concrete construction, girders or box girders or a combination of the two may be used. In addition, concrete can also be cast in place or precast, and segmental construction may also be considered. Variations of roadway structures and depth to span ratios are discussed in Chapter 2. Fabrication and erection methods are presented in Chapter 9.

### 11.2.5    Cable Types and Connections

Cable types are discussed in Chapter 7. In the United States today conventional structural strand and the parallel wire strand are most often used. We predict that in the future the prefabricated parallel wire strand will become the more prevalent type. Cable connections are discussed extensively in Chapter 8.

### 11.2.6    Loads and Forces

Design loadings, their combinations and applications, should be consistent with appropriate specifications, either AASHTO or AREA. These loadings may be modified to suit local conditions and spans in excess of the specification's jurisdiction. It should be noted that the AASHTO specifications are only applicable to spans up to 500 ft and, therefore, do not include the long-span structures. For spans in excess of 500 ft, reductions recommended by Ivy[1] et al (summarized in Table 11.1), are generally accepted criteria.

TABLE 11.1.    Equivalent Lane Loadings

| Loaded Length (ft) | Uniform Live Loads (lb/ft) | Concentrated Live Load | |
|---|---|---|---|
| | | Moment (lb) | Shear (lb) |
| 0–600 | 640 | 18,000 | 26,000 |
| 601–800 | 640 | 9,000 | 13,000 |
| 801–1000 | 640 | 0 | 0 |
| 1001–1200 | 600 | 0 | 0 |
| 1201–over | 560 | 0 | 0 |

There is no universally accepted standard for conditions, loadings, and factors of safety to be used for proportioning the cable sizes. In the past, consultants and cable manufacturers have used $\frac{1}{3}$ the ultimate breaking strength of the cable, as the allowable design value to arrive at a cable size. The design factor is predicated on the assumption of elastic behavior of the structure and the cable. The range of stress in the cable is calculated to be less than the prestretched elastic limit.

Where fatigue effects are likely to occur some designers have used $\frac{1}{5}$ the ultimate breaking strength of the cable for the design allowable load of the cable.

At the present time, the load factor approach for bridge design similar to that recommended by AISI[2] for building design is being considered.

The designer and contractor should bear in mind that when saddles are used, the effective design breaking strength of the cable should be reduced based on the ratio of the saddle radius to the strand diameter. A cable manufacturer should be consulted for specific values for particular applications.

### 11.2.7   Structure Anchorage

The manner in which structure loads are transmitted to the foundation affects the response or stiffness of the structure and the magnitude of axial force in the girder. For long cable-stayed bridges, it may be desirable and practical to consider the effect of expansion joints and their location in the structure. The effect of structure anchorage and expansion joint location has been investigated to some degree by Gimsing.[3]

Consider the structure illustrated in Fig. 11.1a under the action of a uniformly distributed load. Because there is no restraint at the supports to the horizontal component of cable force, the axial force distribution in the girder will be as illustrated, zero force or nearly so in the center of the main span and maximum compression at the pylons. The principal axial forces in the girder are compression loads, thus the system is defined as a self-anchored system.

If, in this system, a horizontal restraint is added at the abutments and expansion joints are added at the pylon, the axial force distribution is altered, Fig. 11.1b. The system now has maximum tension at the center of the main span and zero forces at the pylons. All axial forces in the girder are in tension, consequently, this system is defined as fully anchored externally to abutments or piers.

With expansion joints at the pylon, the center span is only fixed horizontally by the cables and its own lateral flexural stiffness. With live load placed in one-half the center span a horizontal displacement occurs, and because the uniformly distributed dead load has a stabilizing effect, the magnitude of the displacement, aside from stiffness, is a function of the ratio between dead load and live load. If only one expansion joint were used at one of the pylons, the system would act as fully anchored under dead load. Under live load, that portion of the girder containing one side span and the center span would act as a partially self-anchored system. However, symmetry would be destroyed and undesirable bending moments and distortions might result.

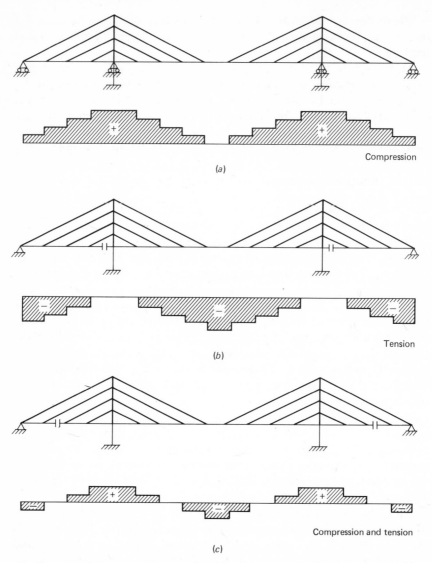

Compression

*(a)*

Tension

*(b)*

Compression and tension

*(c)*

**FIGURE 11.1.** Axial forces in stiffening girder: (*a*) self-anchored; (*b*) fully-anchored; (*c*) partially anchored. (Ref. 3.)

To reduce the magnitude of the axial force, either compression or tension, it may be desirable to combine the self-anchored and fully anchored systems as a partially anchored system. This may be accomplished by providing horizontal restraint at the abutments with no expansion joints, if the induced thermal forces can be accommodated, or with expansion joints located in the end spans, Fig. 11.1c. In this system, the type of girder axial force varies throughout the length, with maximum compression at the pylon and maximum tension in the center of the main span.

Because only the top cable in the end spans are anchored to the abutments, the flexural deformation of the center span will be influenced not only by the deformation of the cables, but also by the flexural deformation in the end spans. The stiffness of the center span can be increased by piers or abutments supporting the end spans at the points of cable attachments. This has been done on the Kniebrücke and the Düsseldorf-Oberkassel structures, and others, Fig. 1.1. In this manner the vertical component of cable force is resisted by the girder. If the pier is not hinged the horizontal force distribution is a function of the relative stiffness between girder and piers with the result of introducing bending in the pier.

The deflection of the girder, in a harp configuration with live load in the center span (illustrated in Fig. 11.2a and Fig. 11.2b), indicates the deflection for live load in one end span and on one-half the main span. As may be assumed, the structure with the supported side spans has the greatest stiffness.

Piers in the end spans are normally acceptable because only the center span width is required for navigation purposes and the piers in the end span will normally be located on land or in relatively shallow water. Economically, the choice should be justified by a comparison of cost savings in the girder stiffness requirements versus the increased pier costs.

## 11.3  METHODS OF ANALYSIS

In order to analyze a cable-stayed bridge an appropriate idealization, or modeling of the structure must be made. The restraints, if any, present at each joint in the structure should be determined in order to mathematically model the structure. The stiffness or flexibility of each member must be known or be determined by the analysis. Connections between the cables, girders, and towers are idealized at their points of intersection. For a single-plane system the structure may be idealized as a two-dimensional plane frame, and torsional forces acting on the girder would have to be superimposed on the girder. A two-plane system may be idealized as a three-dimensional space frame with torsional forces included in the analysis.

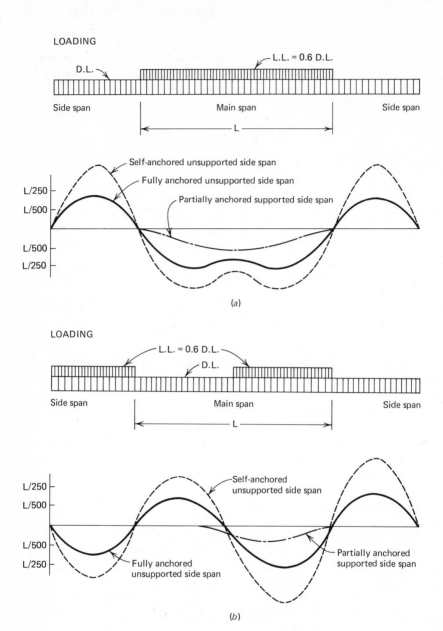

**FIGURE 11.2.** (*a*) and (*b*) Girder deflections. (Ref. 3.)

Several methods have been employed in the analysis of cable-stayed bridges. A mixed method of analysis, where the unknowns in the matrix formulation include displacements and forces, has been developed by Stafford Smith.[4,5] A transfer matrix method has been developed in Germany.[6] Troitsky and Lazar[7] used the flexibility approach while Podolny and Fleming[8,9] used a stiffness approach. Several general computer programs, such as FRAN, STRESS, or STRUDL, are available which use either the stiffness or flexibility approach. However, most of these programs assume linearity and must be modified to accommodate the nonlinear problem inherent in flexible structures. A stiffness approach incorporating an iterative procedure was used by Podolny and Fleming[8,9] to compensate for the nonlinearity of the cables and Tang[10] applied the transfer matrix to the nonlinearity of cable-stayed bridges. The only published material concerning the static behavior analysis of two-plane structures known to us is that of Stafford Smith,[5] Kajita and Cheung,[11] and Baron and Lien,[12,13] the latter also considered dynamic effects in their solution.

## 11.4  STIFFNESS PARAMETERS

The total stiffness of a cable-stayed bridge is subject to the interrelationship of the individual stiffnesses of the girder, the cables and the pylon. To determine the influence of individual stiffnesses with respect to moments in the girder and pylon and to the tension in the cables, a linear study has been conducted[8] using the radiating and harp configuration in a single-plane arrangement, as shown in Fig. 11.3a. The girder is rigidly supported vertically at the pylon, but is independent of the pylon so that no moment is transferred between the girder and the pylon. The girder and the pylon are assumed to have a constant cross section throughout the span and height, respectively. It is further assumed that the cable has an initial prestress capable of resisting a possible compression load by releasing the compression force accordingly.

The STRUDL program was used in the analysis with no correction for the nonlinear behavior of the cables. A modulus of $24 \times 10^6$ psi was used for the cables and a uniform load was applied on all spans of the structure. It is important to note that the type of loading or magnitude of tension in the cables or moment in the girder or pylon is not important for the purpose of this study, because the relative difference in the values due to changes in various stiffness parameters is the prime concern. The parameters investigated for these configurations were as follows: (1) ratio of moment of inertia of the pylon to girder, $\alpha = I_p/I_g$; (2) stiffness ratio of

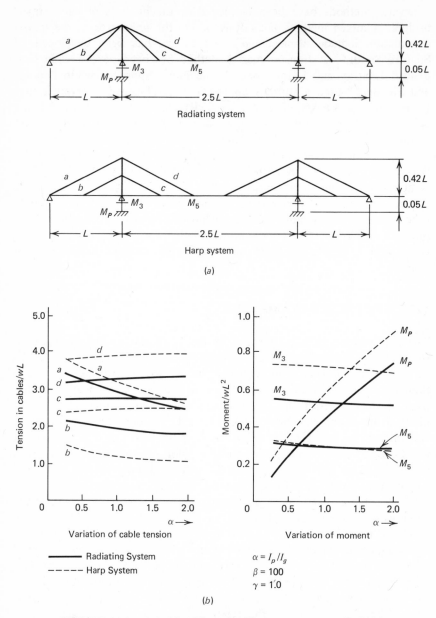

**FIGURE 11.3.** **(a)–(d) Effect of stiffness parameters. (Ref. 8.)**

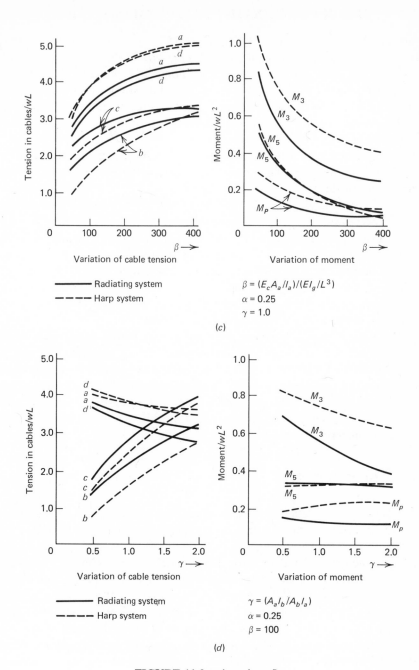

**Variation of cable tension** (upper left)

**Variation of moment** (upper right)

— Radiating system
--- Harp system

$$\beta = (E_c A_a / l_a)/(E I_g / L^3)$$
$$\alpha = 0.25$$
$$\gamma = 1.0$$

(c)

**Variation of cable tension** (lower left)

**Variation of moment** (lower right)

— Radiating system
--- Harp system

$$\gamma = (A_a l_b / A_b l_a)$$
$$\alpha = 0.25$$
$$\beta = 100$$

(d)

**FIGURE 11.3.** (*continued*)

outside cable to girder, $\beta = (E_c A_a / l_a)/(E I_g / L^3)$; and (3) stiffness ratio of inside cable to outside cable, $\gamma = (A_a l_b / A_b l_a)$.

The effect of the parameter $\alpha$ ratio of moment of inertia of the pylon to girder is indicated in Fig. 11.3b. As the bending stiffness in the tower increases the tension in the back-stay cable $a$ decreases and the moment in the pylon increases. There is very little effect on the tension in the other cables or moments in the girder.

The influence of $\beta$ is indicated in Fig. 11.3c. As $\beta$ increases, the tension in the cables increases, and there is a decrease in the moments in the girder and pylon.

As the parameter $\gamma$, Fig. 11.3d, increases, the tension in the outside cables decreases, while that of the inside cables increases. There is very little effect upon the pylon and girder midspan moment. However, there is a marked effect on the girder moment at the tower, but not to the same degree as produced by $\beta$. These conclusions generally confirm those of Okauchi, Yabe, and Ando.[14] The curves are not intended for design purposes, but merely represent a study of the effect of some of the variables involved in the analysis of a cable-stayed bridge.

The significance of the above conclusions may be illustrated as follows. Assuming that in a given trial analysis the overall structure stiffness is inadequate, and that all other design criteria are satisfied by the trial structure, the designer can either increase overall stiffness or adjust the stiffness of the girder, the pylon or the stays. We would submit that from the above discussion, increasing the stiffness of the stay is the most efficient choice to increase the stiffness of the structure as a whole.

## 11.5   NONLINEARITY

Nonlinear considerations in cable-stayed bridges may be classified into three categories: girder, pylon, and cables. Nonlinear behavior in the girder and pylon occurs when they are subjected to compressive loads and bending moments simultaneously. The degree of the nonlinearity depends on the magnitude of the compressive load compared with the Euler load and the magnitude of the deflection produced by the bending action. Normally, it can be assumed that these effects are small. However, for slender girders and pylons this approximation should be verified for extreme loading conditions. Nonlinearity in the cable member occurs when the load increases, and the cable sag decreases, producing an increase in the cable chord length and thus an apparent elongation of the cable. This phenomenon has been discussed in Chapter 10.

## 11.6    MIXED METHOD OF ANALYSIS—SINGLE PLANE

The following procedure and that of the subsequent section have been developed by B. Stafford Smith,[4,5] and are presented here through the courtesy of the Institution of Civil Engineers.

### 11.6.1    Cable-Stayed Bridge Behavior

Before determining a method of analysis for a cable-stayed bridge, it is advantageous to study and understand the behavior of the total structure. A study indicates that the cables support the deck in a vertical direction only, and that torsion due to eccentric loading or wind forces is transmitted to the piers through the deck structure. The overall behavior of the stayed girder is best understood if the various contributing factors are considered separately. Total behavior will then be determined by superposition of all factors influencing the action of the structure.

Consider the single-plane, two-span stayed girder in Fig. 11.4$a$. The girder is simply supported at $A$, $B$, and $C$ with rigid supports and elasti-

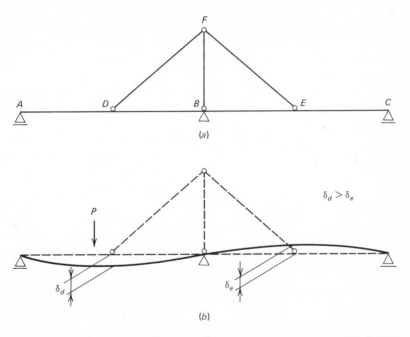

(a)

(b)

**FIGURE 11.4.**   ($a$) and ($b$) Deflection of girder as a continuous beam; ($c$) deflection of girder due to applied load; ($d$) deflection of girder due to elastic elongation of cables; ($e$) deflection of girder due to elastic shortening of tower. (Courtesy of the Institution of Civil Engineers. Ref. 4.)

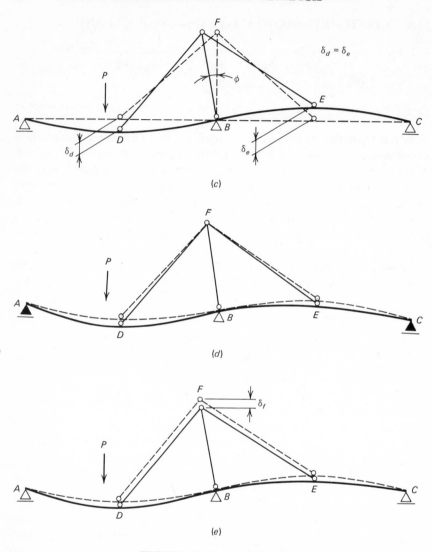

**FIGURE 11.4.** (*continued*)

cally supported at points $D$ and $E$ by the cables. The cables are pinned at the top of the tower, and the tower is hinged at the base. If the cables were omitted and a load were to be applied in the left span of the continuous girder, the point $D$ would drop and point $E$ would lift up, Fig. 11.4b, and the displacement at $D$ becomes greater than that at $E$.

First consider the structure to be supported by rigid cables attached at points $D$ and $E$ on the girder and at $F$ at the top of the rigid tower, Fig.

11.4*a*. Assuming that the cables and the tower are axially rigid, the triangular geometry of *DBF* and *EBF* remains equal. With a load in the left span and a hinge at the base of the tower, the tower will rotate in a counterclockwise direction through a rotation $\phi$, and the displacement at *D* equals that at *E*, Fig. 11.4*c*. Thus, the first action occurs when a hinge is introduced at the base of the tower and the cables and tower remain axially rigid.

A second action occurs as a result of the elastic stretching of the cables due to the tension created by the first action. Because of the elastic stretch of the cables, the tension in the cables is relieved, increasing the deflection in the girder at *D* and decreasing the deflection in the uplift at *E*, Fig. 11.4*d*.

The third action is the axial shortening of the tower due to the compressive load from the stays. This shortening of the tower causes point *F* at the top of the tower to move lower, further relieving the tension in the cables and thus continuing to increase the deflection at D and decrease the uplift at *E*, Fig. 11.4*e*.

### 11.6.2  Fundamental Analysis

A cable-stayed bridge is a statically indeterminate structure in which the girder acts as a continuous beam supported elastically at the points of cable attachments.

The method of consistent displacements or consistent distortions, also known as the general method,[15] may be used in the solution of the indeterminate cable-stayed bridge. The first mathematical operation consists of removing the redundant stresses and/or reaction components, thus reducing the initial indeterminate structure to a determinate and stable structure. Any combination of redundant stresses and /or reactions may be used. Condition equations are then written for the deflection at the point of application of each redundant. On one side of the equation is the summation of all deflection components at the points of application of the redundants (e.g., applied load, redundant load, and temperature), and taken in the direction of the redundant. The other side of the equation is the predetermined sum of all the deflection components. There will be as many equations as there are redundancies.

The general equations for a structure with $n$ redundants are:[16]

$$\delta'_a + X_a\delta_{aa} + X_b\delta_{ab} + \cdots + X_n\delta_{an} = \Delta_a$$

$$\delta'_b + X_a\delta_{ba} + X_b\delta_{bb} + \cdots + X_n\delta_{bn} = \Delta_b \qquad (11.1)$$

$$\delta'_n + X_a\delta_{na} + X_b\delta_{nb} + \cdots + X_n\delta_{nn} = \Delta_n$$

where $X_a, X_b, \ldots, X_n$ = the equivalent redundant forces

$\delta'_a, \delta'_b, \ldots, \delta'_n$ = the displacements due to the applied loads at the points $A, B, \ldots, N$, in the direction of $X_a, X_b, \ldots, X_n$

$\delta_{aa}, \delta_{bb}, \ldots, \delta_{nn}$ = displacements at $A, B, \ldots, N$, due to $X_a, X_b, \ldots, X_n$ equal to unity, no other loads acting

$\delta_{ab}$ = displacement at $A$ (in the direction of the action of the force $X_a$) due to $X_b$ equal to unity acting alone

$\delta_{ba}$ = displacement at $B$ in the direction of $X_b$ due to $X_a$ equal to unity acting alone

$\delta_{ba} = \delta_{ab}$, from Maxwell principle of reciprocal deflections

$\Delta_a, \Delta_b, \ldots, \Delta_n$ = total deflection at A, B, $\ldots$, N

Consider the cable-stayed bridge shown in Fig. 11.5a which is simply supported at $A, B,$ and $C$ and elastically supported by the cables at $D$

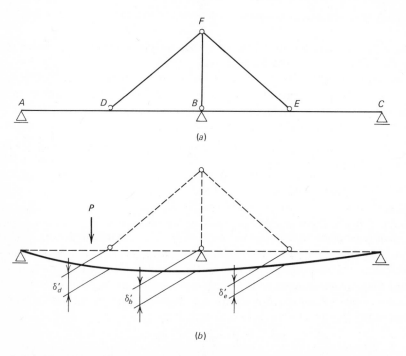

(a)

(b)

FIGURE 11.5.   Deflection of determinate structure with redundants removed. (Courtesy of the Institution of Civil Engineers. Ref. 4.)

and $E$.[4] The cables are pinned at point $F$, the top of the mast, and the mast is pinned at the base. The structure is statically indeterminate to the third degree. The three redundancies may be conveniently taken as the vertical components of tension in the cables and the reaction at the support $B$.

If the indeterminate structure is "cut back" to a determinate one, i.e., the redundancies removed, it reverts to a simple beam spanning from $A$ to $C$ with deflections $\delta'_d$, $\delta'_b$, and $\delta'_e$ due to the applied loading, Fig. 11.5$b$.

The deflections of the girder at points $D$, $B$, and $C$ will be considered separately from the deflections of the cables at points $D$ and $E$. Compatability will require that the girder deflection at $B$ be zero and the girder deflections at $D$ and $E$ be equal to the cable deflection at these points.

Consider the compatability of the girder deflection at $D$ in the vertical direction:

$$\delta'_d - f_{dd}X_d - f_{db}X_b - f_{de}X_e = \Delta_d \qquad (11.2a)$$

where $X_d$ and $X_e$ = vertical components of cable forces at $D$ and $E$

   $X_b$ = support reaction at $B$

   $f_{dd}$ = flexibility coefficient for the deflection at $D$ due to a unit load at $D$ for the simple beam $AC$

   $f_{db}$ = flexibility coefficient for the deflection at $D$ due to a unit load at $B$

   $f_{de}$ = flexibility coefficient for the deflection at $D$ due to a unit load at $E$

   $\delta'_d$ = deflection at $D$ due to the applied loads

   $\Delta_d$ = predetermined total deflection at $D$

In similar manner, girder compatability equations may be written for the girder deflection at $B$ and $E$.

$$\delta'_b - f_{bd}X_d - f_{bb}X_b - f_{be}X_e = \Delta_b \qquad (11.2b)$$

$$\delta'_e - f_{ed}X_d - f_{eb}X_b - f_{ee}X_e = \Delta_e \qquad (11.2c)$$

For a conventional bridge on piers only, the deflections $\Delta_d$, $\Delta_b$, and $\Delta_e$ for the girder would be equal to zero because of the fixed supports. However, in a cable-stayed girder the points $D$ and $E$ are elastically supported and they do not remain at the same elevation as the rigid supports $A$, $B$, and $C$. They deflect as a result of the three previously described actions and must, therefore, be modified as follows:

1. The rotation of the tower causes the position of the cable anchorages at $D$ and $E$ to change by the amount $\phi$ $BD$ and $-\phi$ $BE$, Fig. 11.4$c$. The magnitude of $\phi$ is assumed such that $\phi = \tan \phi$. These additional de-

flection terms must be added to equations 11.2a and 11.2c to maintain compatability.

2. The elastic stretch of the cables will cause further displacement at points $D$ and $E$. For a unit vertical component in the cable tension this displacement becomes,

$$c_d = \frac{l_d}{A_d E_d \sin^2\theta_d} \tag{11.3a}$$

$$c_e = \frac{l_e}{A_e E_e \sin^2\theta_e} \tag{11.3b}$$

the deflection for the total force in the cables is thus,

$$c_d X_d$$
$$c_e X_e$$

and equations 11.2a and 11.2c must be modified accordingly. $A, E, l,$ and $\theta$ are the cross-sectional area, Young's modulus, length and slope to the horizontal, respectively, of the cables.

3. Because of axial shortening of the tower the deflections at $D$ and $E$ must be modified by

$$(X_d + X_e)f_T \tag{11.4}$$

where $f_T$ represents the unit shortening of the tower. This modification must also be included in equations 11.2a and 11.2c. Therefore, to account for the actions of tower rotation, cable stretch, and axial shortening of the tower, the right side of the equations of compatability 11.2a, 11.2b, and 11.2c may be written as:

$$\Delta_d = \phi BD + c_d X_d + (X_d + X_e)f_T \tag{11.5a}$$

$$\Delta_b = 0 \tag{11.5b}$$

$$\Delta_e = -\phi BE + c_e X_e + (X_d + X_e)f_T \tag{11.5c}$$

Because we have interjected an additional variable $\phi$, the rotation of the tower, an additional equation becomes necessary for solution. This condition equation can be readily determined by taking moments for the tower about its hinge.[4]

$$X_d BD - X_e BE = 0 \tag{11.5d}$$

Therefore, by collecting terms and rearranging, the compatability equations may be rewritten in the following form:

$$(f_{dd} + c_d + f_T)X_d + f_{db}X_b + \qquad (f_{de} + f_T)X_e + \phi BD = \delta'_d \tag{11.6a}$$

$$f_{bd}X_d + f_{bb}X_b + \qquad\qquad f_{be}X_e \qquad = \delta_b' \qquad (11.6b)$$

$$(f_{ed} + f_T)X_d + f_{eb}X_b + (f_{ee} + c_e + f_T)X_e - \phi BE = \delta_e' \qquad (11.6c)$$

$$BD\,X_d \qquad\qquad - BE\,X_e \qquad = 0 \qquad (11.6d)$$

Because of the high order of redundancy in most cable-stayed bridges and, therefore, the number of simultaneous equations, the problem is conducive to a computer solution. When the structure is to be subjected to repeated adjustments a computer program is essential to avoid tedious computations.

The simultaneous equations presented in equations 11.6 are conveniently arranged such that they may be expressed in matrix form as shown in equation 11.7.

$$\begin{bmatrix} (f_{dd} + c_d + f_T) & (f_{db}) & (f_{de} + f_T) & BD \\ (f_{bd}) & (f_{bb}) & (f_{be}) & 0 \\ (f_{ed} + f_T) & (f_{eb}) & (f_{ee} + c_e + f_T) & -BE \\ BD & 0 & -BE & 0 \end{bmatrix} \begin{bmatrix} X_d \\ X_b \\ X_e \\ \phi \end{bmatrix} = \begin{bmatrix} \delta_d' \\ \delta_b' \\ \delta_e' \\ 0 \end{bmatrix} \qquad (11.7)$$

The solution of this matrix results in the determination of the vertical components of cable tension $X_d$ and $X_e$, the vertical reaction at the support $X_b$ and rotation of the mast $\phi$. Other loading conditions may be easily determined by changing the $\delta'$ matrix. The upper left-hand part of the coefficient matrix is the flexibility matrix for the structure if it were fixed against rotation of the tower. The right-hand column arises from the deflection of the girder caused by the rotation of the tower. The bottom row results from the consideration of the stability of the tower. Because the unknowns in the X matrix include a displacement along with forces, the method is considered to be of a mixed category.[4]

### 11.6.3   Multicable Structure—Radiating System

A radiating cable system, Fig. 11.6, has actions similar to those described above. The structure is statically indeterminate to the fifth degree. The vertical components of cable forces and the interior support reaction may be taken as the redundants. The rotation of the tower affects the displacement at each cable attachment on the girder in proportion to its distance from the base of the tower. Axial shortening of the tower affects each

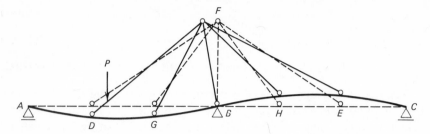

**FIGURE 11.6. Deflection of girder for multicable system. (Courtesy of the Institution of Civil Engineers. Ref. 4.)**

point equally. In the manner described above, a similar set of simultaneous equations may be derived and expressed in matrix form (see equation 11.8).

$$
\begin{bmatrix}
(f_{dd}+c_d+f_T) & (f_{dg}+f_T) & (f_{db}) & (f_{dh}+f_T) & (f_{de}+f_T) & BD \\
(f_{gd}+f_T) & (f_{gg}+c_g+f_T) & (f_{gb}) & (f_{gh}+f_T) & (f_{ge}+f_T) & BG \\
(f_{bd}) & (f_{bg}) & (f_{bb}) & (f_{bh}) & (f_{be}) & 0 \\
(f_{hd}+f_T) & (f_{hg}+f_T) & (f_{hb}) & (f_{hh}+c_h+f_T) & (f_{he}+f_T) & -BH \\
(f_{ed}+f_T) & (f_{eg}+f_T) & (f_{eb}) & (f_{eh}+f_T) & (f_{ee}+c_e+f_T) & -BE \\
BD & BG & 0 & -BH & -BE & 0
\end{bmatrix}
\begin{bmatrix}
X_d \\ X_g \\ X_b \\ X_h \\ X_e \\ \phi
\end{bmatrix}
=
\begin{bmatrix}
\delta_d' \\ \delta_g' \\ \delta_b' \\ \delta_h' \\ \delta_e' \\ 0
\end{bmatrix}
\qquad (11.8)
$$

### 11.6.4 Multicable Structure—Harp System

In a harp structure, Fig. 11.7, at least one pair of cables will be pinned or clamped to the tower and possibly all pairs. The first case will normally

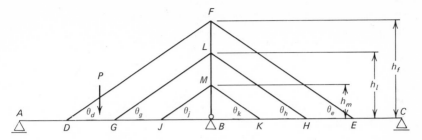

**FIGURE 11.7.** Multicable structure, harp system. (Courtesy of the Institution of Civil Engineers. Ref. 4.)

have the outer pair of cables pinned to the top of the tower so that horizontal movement, with respect to each other, is not permitted. The interior pairs of cables will be supported on saddles in the tower and are free to move horizontally. As a result, the tower is subjected to axial forces only.

Where the rotation of the tower $\phi$ is small, so that the level of the tower supports for the unclamped cables can be assumed not to change, the cables will not contribute to the deflection caused by the rotation of the tower or to the towers rotational stability. Therefore, terms are not required in the bottom row and right-hand column of the flexibility matrix for the tower stability and vertical deflection, respectively, for the unclamped cables. Nonzero terms in the bottom row and right-hand column will only be required for those points relating to clamped cables only.

In the event that a pair of unclamped cables has unequal slopes on either side of the tower, the saddle that is free to move horizontally will shift when there is a change in load. This action will cause a drop in the cable girder connection toward the side of the shift and an uplift on the other side. If in Fig. 11.7 the cable $GLH$ is considered to be supported by a saddle on rollers at $L$ and $\theta_g$ and $\theta_h$ are unequal, then the drop at $G$ and the rise at $H$ can be determined as,*

$$\Delta_g' = \left( X_g \, \frac{\cot^2 \theta_g}{A_g E} \right) \left( \frac{l_h}{\cos^2 \theta_h} - \frac{l_g}{\cos^2 \theta_g} \right) \tag{11.9a}$$

$$\Delta_h' = \left( X_h \, \frac{\cot^2 \theta_h}{A_h E} \right) \left( \frac{l_g}{\cos^2 \theta_g} - \frac{l_h}{\cos^2 \theta_h} \right) \tag{11.9b}$$

* *Author's note:* The original equations presented in reference 4 were incorrect, they have been corrected here by private correspondence with B. Stafford Smith.

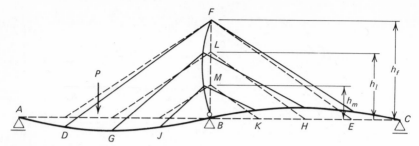

**FIGURE 11.8.** Bending of tower, multicable system. (Courtesy of the Institution of Civil Engineers. Ref. 4.)

If $\theta_g$ equals $\theta_h$ and $l_g$ equals $l_h$ the vertical drop at $G$ and the rise at $H$ can be determined as

$$\Delta'_g = \frac{X_g l_g}{A_g E \sin^2 \theta_g} \tag{11.9c}$$

$$\Delta'_h = \frac{X_h l_h}{A_h E \sin^2 \theta_h} \tag{11.9d}$$

Therefore, the right side of the appropriate compatability equation for $G$ and $H$ would have to be modified (refer to equations 11.2 and 11.5) by the addition of equations 11.9. The appropriate flexibility matrix coefficients would also be modified.

When all the cable pairs are clamped to the tower, Fig. 11.7, there is the added action of tower bending as a simple beam, between $F$ and $B$, as well as tower rotation. The resulting lateral deflections at $L$ and $M$ will influence the tension in the cables as well as the moments in the girder.

The isolated action of bending in the tower is illustrated in Fig. 11.8. The lateral deflection of $M$ caused by the bending in the tower is determined by:

$$\Delta'_m = (X_j \cot \theta_j - X_k \cot \theta_k) f_{mm} + (X_g \cot \theta_g - X_h \cot \theta_h) f_{ml} \tag{11.10}$$

where

$$f_{mm} = \frac{h_m^2 (h_f - h_m)^2}{3E \, I_T h_f}$$

and

$$f_{ml} = \frac{h_m (h_f - h_l)[h_f^2 - (h_f - h_l)^2 - h_m^2]}{6E \, I_T h_f}$$

Thus the deflection at $J$ due to bending of the tower may be determined as:

$$\Delta'_j = X_j \frac{(h_f - h_m)^2 BJ}{3E\,I_T h_f} BJ - X_k \frac{(h_f - h_m)^2 BJ}{3E\,I_T h_f} BK$$

$$+ X_g \frac{(h_f - h_l)[h_f^2 - (h_f - h_l)^2 - h_m^2]BJ}{6E\,I_T h_f h_l} BG$$

$$- X_h \frac{(h_f - h_l)[h_f^2 - (h_f - h_l)^2 - h_m^2]BJ}{6E\,I_T h_f h_l} BH \qquad (11.11)$$

Therefore, to account for the bending in the tower, equation 11.11 would have to be added to the compatibility equation for $J$ and similar modifications would have to be made for the compatibility equations of $G$, $K$, and $H$. It is also to be noted that the value of the unit shortening of the tower, $f_T$, will vary for each pair of cables because of the different heights of connection to the tower.

### 11.6.5   Axial Force in the Girder

In long spans where the angle of the cable inclination to the horizontal girder is shallow, the cable tension will be relatively large when compared with the cable tension for a short span where the slope of the cable is greater. Consequently, when the cable has a shallow slope, the horizontal component of force along the axis of the girder will also be high. If in addition the cross section of the girder is relatively light, the axial shortening of the girder should be taken into account in the design calculations.

If the girder is free to move at $A$, Fig. 11.4a, the axial force in $DB$ is $X_d \cot \theta_d$ and the girder shortening is $X_d \cot \theta_d \, l_{db}/A_G E_G$. This girder shortening produces an increase in the slope of the cable $FD$, which results in a deflection at the cable anchorage $D$ amounting to:

$$\Delta'_d = X_d \frac{\cot^2 \theta_d l_{db}}{A_G E_G} \qquad (11.12)$$

This additional deflection must be taken into account in equation 11.5a and therefore the coefficient of $X_d$ in equation 11.6a and also the flexibility matrix equation 11.7 must be modified by adding

$$f^G_{bd} = \frac{l_{db} \cot^2 \theta_d}{A_G E_G} \qquad (11.13)$$

When the support at $A$ is fixed in position and resists the axial force in the girder, the horizontal component of tensile force in the cable must be distributed as a tensile force component in $AD$ and a compressive force component in $DB$. The magnitude of the axial girder force is proportional to the inverse ratio of the lengths over which they act. Therefore, the axial shortening of $DB$ would be:

$$\frac{l_{ad}}{l_{ab}} \frac{X_d \cot \theta_d l_{db}}{A_G E_G}$$

and the deflection at $D$ then becomes

$$\Delta_d' = \frac{l_{ad}}{l_{ab}} \frac{X_d \cot^2 \theta_d l_{db}}{A_G E_G} \tag{11.14}$$

and the flexibility equation is modified by

$$f_{bd}^G = \frac{l_{ad}}{l_{ab}} \frac{l_{db} \cot^2 \theta_d}{A_G E_G} \tag{11.15}$$

In the case of a multicable system as in Fig. 11.6 the deflection of $D$ and $G$ due to axial shortening of the girder is

$$\Delta_d' = (X_g \cot \theta_g + X_d \cot \theta_d) \frac{\cot \theta_g l_{gb}}{A_G E_G} \tag{11.16}$$

$$\Delta_g' = (X_g \cot \theta_g l_{gb} + X_d \cot \theta_d l_{db}) \frac{\cot \theta_d}{A_G E_G} \tag{11.17}$$

and the flexibility coefficient would require modification in the compatibility equations of $G$ and $D$.

### 11.6.6    Fixed Base Tower

In a bridge with a tower fixed at the base, the rotation of the tower is prevented. Therefore, the right-hand column and the bottom row of the flexibility matrix, which accounts for the girder deflection and column stability normally caused by tower rotation, may be omitted.

The horizontal movement of point $L$, caused by forces in the cables on either side of point $L$, Fig. 11.9, may be determined as:

$$\Delta_l' = (X_g \cot \theta_g - X_h \cot \theta_h) f_{ll} + (X_d \cot \theta_d - X_e \cot \theta_e) f_{lf} \tag{11.18}$$

and for a cantilever of uniform cross section

$$f_{ll} = \frac{h^3}{3E I_T} \tag{11.19}$$

and

$$f_{lf} = \frac{2h_f^3 - 3h_f^2(h_f - h_l) + (h_f - h_l)^3}{6E I_T} \tag{11.20}$$

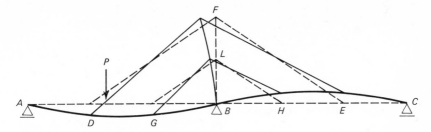

**FIGURE 11.9. Deflection of tower due to applied load, $P$. (Courtesy of the Institution of Civil Engineers. Ref. 4.)**

Therefore, the vertical deflection at $C$ due to the bending of the tower is:

$$\Delta'_g = X_g \frac{h_l BG}{3E\,I_T} BG - X_h \frac{h_l BH}{3E\,I_T} BH$$

$$+ X_d \frac{[2h_f^3 - 3h_f^2(h_f - h_l) + (h_f - h_l)^3]GB}{6E\,I_T h_f h_l} BD$$

$$- X_e \frac{[2h_f^3 - 3h_f^2(h_f - h_l) + (h_f - h_l)^3]GB}{6E\,I_T h_f h_l} BE \qquad (11.21)$$

and the flexibility coefficients in the row representing compatibility at $G$ must be modified accordingly. Similarly the rows representing compatibility of points $D$, $H$, and $E$ must also be modified.

### 11.6.7 Multitower Continuous Girder Cable-Stayed Bridge

Where large crossings are required, the normal procedure is to require two towers at either end of a large central span, Fig. 11.10. The analysis of a continuous girder system involving several towers and cables may be performed by initially releasing all girder connections between the end supports $A$ and $D$. As a result, a single coefficient matrix may be constructed and modified as previously outlined. The size of the coefficient matrix will be equal to the total sum of the number of cables, the interior rigid supports, and the number of hinged towers.

### 11.6.8 Cables Attached to Rigid Supports

When a cable is attached to a fixed support as at $A$ and $D$, Fig. 11.10, the terms $f_{aa}$ and $f_{dd}$ of the coefficients of $X_a$ and $X_d$ in the coefficient matrix are zero. The coefficients of these two forces will consist only of appropriate modifications. The forces $X_a$ and $X_d$ are the vertical components of

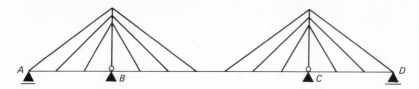

**FIGURE 11.10.** Long span multitower continuous girder bridge. (Courtesy of the Institution of Civil Engineers. Ref. 4.)

the force in the cables at $A$ and $D$; they are not the abutment reactions at $A$ and $D$. The abutment reactions may be evaluated after the cable forces and interior support reactions are determined by considering the equilibrium of the girder as a whole.[4]

Consider the structure, illustrated in Fig. 11.11, wherein the back-stay cable is attached to the girder at $A$ and the fore-stay cables are attached to the girder at $C$ and $D$. The continuous girder is supported at $A$, $B$, and $E$. The tower is pinned at its base at $B$, and all cables are pin connected to the tower at $F$.

As previously indicated, the number of equations required for a solution is equal to the sum of the number of cables, the interior supports, and the number of hinged towers. Therefore, for this structure, five equations are required.

Four deflection compatibility equations may be written as:

$$\delta_a' - f_{aa}X_a - f_{ab}X_b - f_{ac}X_c - f_{ad}X_d = \Delta_a$$
$$\delta_b' - f_{ba}X_a - f_{bb}X_b - f_{bc}X_c - f_{bd}X_d = \Delta_b \qquad (11.22)$$
$$\delta_c' - f_{ca}X_a - f_{cb}X_b - f_{cc}\dot{X}_c - f_{cd}X_d = \Delta_c$$
$$\delta_d' - f_{da}X_a - f_{db}X_b - f_{dc}X_c - f_{dd}X_d = \Delta_d$$

It may be seen that terms $\delta_a$, $f_{an}$, and $f_{na}$ are equal to zero, and the term $X_a$ used here denotes the vertical component of tension in the cable at $A$. The term $\Delta_a$ will reflect appropriate modifications as they affect $\mathrm{X}_a$.

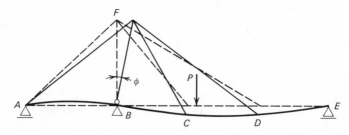

**FIGURE 11.11.** Single tower, cable attached to rigid support. (Courtesy of the Institution of Civil Engineers. Ref. 4.)

Because of a tower rotation of $\phi$ there will be a reduction of tension in cables $C$ and $D$, with corresponding deflections at $C$ and $D$. There will also be an increase in the tension of cable $AF$. These changes may be represented as

$$
\begin{aligned}
\Delta_a' &= -\phi BA \\
\Delta_c' &= \phi BC \\
\Delta_d' &= \phi BD
\end{aligned}
\tag{11.23}
$$

Similarly the cable elongation may be represented by

$$
\begin{aligned}
\Delta_a' &= c_a X_a \\
\Delta_c' &= c_c X_c \\
\Delta_d' &= c_d X_d
\end{aligned}
\tag{11.24}
$$

Elastic shortening of the tower is given by

$$
(X_a + X_c + X_d) f_T
\tag{11.25}
$$

Because the unknown term $\phi$ representing tower rotation has been introduced, a stability equation for the tower must be formulated by summing moments about the base,

$$
- X_a\, BA + X_c\, BC + X_d\, BD = 0
\tag{11.26}
$$

By rewriting the deflection compatibility equations and taking into account the tower stability equation, the simultaneous equations may be represented in matrix form as follows:

$$
\begin{bmatrix}
(c_a + f_T) & 0 & f_T & f_T & -BA \\
0 & f_{bb} & f_{bc} & f_{bd} & 0 \\
f_T & f_{cb} & (f_{cc} + c_c + f_T) & (f_{cd} + f_T) & BC \\
f_T & f_{db} & (f_{dc} + f_T) & (f_{dd} + c_d + f_T) & BD \\
-BA & 0 & BC & BD & 0
\end{bmatrix}
\begin{bmatrix}
X_a \\ X_b \\ X_c \\ X_d \\ \phi
\end{bmatrix}
=
\begin{bmatrix}
0 \\ \delta_b' \\ \delta_c \\ \delta_d' \\ 0
\end{bmatrix}
\tag{11.27}
$$

The redundants are the vertical components of cable force and the tower rotation.

## 11.7   MIXED METHOD OF ANALYSIS—DOUBLE PLANE

### 11.7.1   Structural Behavior

The primary difference in single-plane and double-plane cable-stayed bridge structures is that in the former only vertical actions of the continuous girder are restrained by the cables, and torsional forces are transmitted through the deck section into the piers. In the latter, the cables assist in restraining both vertical and torsional forces. Double-plane structures require a three-dimensional analysis whereby the deflections of the two cable planes are considered in conjunction with the bending and rotation of the girder.

In the basic structure, illustrated in Fig. 11.12, the continuous girder is supported by pairs of rigid supports at $AA'$, $BB'$, and $CC'$ and by pairs of elastic cable supports at $DD'$ and $EE'$. The cables are pinned or clamped at the top of the masts at $F$ and $F'$. The cables are attached such that the ends are restrained from a relative displacement to the mast, which are hinged at their bases at $B$ and $B'$. The girder is assumed to be completely restrained against torsion at each of the rigid supports.

If a load is placed at any point other than on the longitudinal center line of the bridge as $P_2$, Fig. 11.12, the two systems of the towers and cables undergo different displacements, causing different values of tension in the adjacent parallel cables of each system. Therefore, each adjacent pair of parallel cables must assist in restraining the torsion and vertical displacement of the girders.

The amount of restraint provided by the tower cable system is a function of several factors, such as, the tower and cable stiffness, type of connection at the base of the tower, and whether structural connection exists between the towers to resist relative rotation of the towers.

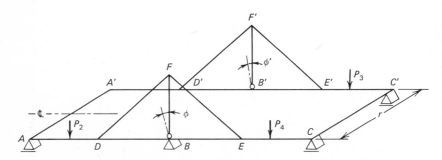

**FIGURE 11.12.   Double-plane system, eccentric load, $P_2$. (Courtesy of the Institution of Civil Engineers. Ref. 5.)**

As an illustration of the last factor, consider the structure in Fig. 11.12 in which the towers have no structural connection between them, i.e., no portal frame. If a single eccentric load, $P_2$, is applied to one of the girders, both towers will rotate in a counterclockwise direction. However, because of the eccentricity of the load, $P_2$, there will be a differential rotation $\phi$ and $\phi'$ of towers $BF$ and $B'F'$, respectively. As a consequence of the differential rotation, the deflection of $D$ will be larger than that of $D'$, causing a torque in the girder at section $DD'$. If the towers were connected to each other to restrain a relative rotation between them, there would be less torque action and the system would be torsionally stiffer.

In the extreme case, when the structure is loaded with equal and antisymmetrically placed loads, $P_2$ and $P_3$, tower $BF$ rotates in a counterclockwise direction and tower $B'F'$ rotates in a clockwise direction. Therefore, the torque loads due to applied loads $P_2$ and $P_3$, are cumulative for each position of load. If the structure is loaded with $P_2$ and $P_4$ symmetrically placed with respect to the transverse center line of the bridge, the towers will be parallel and remain in their original position, thus providing maximum torsional restraint to the girder.

### 11.7.2  Basic Analysis

The analysis of a double-plane, cable-stayed structure is similar to that developed for single-plane, cable-stayed structures. Like single-plane structures, all interior rigid and elastic supports of the girder are released and a set of deflection compatibility equations is formulated for each support and for the stability of the tower. The double-plane structure, in addition, requires that the torque at each interior support be released and additional equations of compatibility be formulated for the torsion at each interior support.

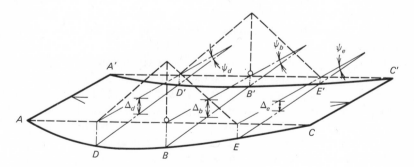

FIGURE 11.13.  Deflections and rotations of primary structure. (Courtesy of the Institution of Civil Engineers. Ref. 5.)

If all redundant releases are applied to the structure, Fig. 11.13, and the external loads are applied to the simple span, $AA'CC'$, the center line deflections and twists, $\delta'_d$, $\delta'_b$, $\delta'_e$, $\psi'_d$, $\psi'_b$, and $\psi'_e$, may be determined by conventional methods. Assuming the restraints provided by the cables at points $D$, $D'$, $E$, and $E'$ are completely rigid against deflection and torque, the compatibility equations are

$$\delta'_d - (X_d + X'_d)f_{dd} - (X_b + X'_b)f_{db} - (X_e + X'_e)f_{dc} = \Delta_d$$

$$\delta'_b - (X_d + X'_d)f_{bd} - (X_b + X'_b)f_{bb} - (X_e + X'_e)f_{bc} = \Delta_b$$

$$\delta'_e - (X_d + X'_d)f_{ed} - (X_b + X'_b)f_{eb} - (X_e + X'_e)f_{ee} = \Delta_e$$

$$\psi'_d - \frac{(X_d - X'_d)r\psi_{dd}}{2} - \frac{(X_b - X'_b)r\psi_{db}}{2} - \frac{(X_e - X'_e)r\psi_{de}}{2} = \psi_d \quad (11.28)$$

$$\psi'_b - \frac{(X_d - X'_d)r\psi_{bd}}{2} - \frac{(X_b - X'_b)r\psi_{bb}}{2} - \frac{(X_e - X_e)r\psi_{be}}{2} = \psi_b$$

$$\psi'_e - \frac{(X_d - X'_d)r\psi_{ed}}{2} - \frac{(X_b - X'_b)r\psi_{eb}}{2} - \frac{(X_e - X'_e)r\psi_{ee}}{2} = \psi_e$$

where $X_d$ = the vertical component of cable tension at $D$
$\quad\quad f_{dd}$ = the center line deflection at $D$ due to a unit load at $D$
$\quad\quad f_{db}$ = the center line deflection at $D$ due to a unit load at $B$
$\quad\quad \delta'_d$ = the center line deflection at $D$ due to the applied loads
$\quad\quad r$ = distance between sets of cables
$\quad\quad \psi_{dd}$ = twist of the girder at $D$ due to a unit torque applied at $D$
$\quad\quad \psi_{db}$ = twist of the girder at $D$ due to a unit torque applied at $B$
$\quad\quad \psi'_d$ = twist of the girder at $D$ due to the applied loads
$\quad\quad \Delta_d$ = predetermined total center line deflection at $D$
$\quad\quad \psi_d$ = predetermined total twist at $D$

However, the restraints at $D$ and $E$ are not rigid because of the rotation of the tower, cable elongation, and elastic shortening of the tower. As before, the compatibility equations 11.28, must be modified to account for these actions.

Assuming that the towers $BF$ and $B'F'$ rotate counterclockwise $\phi$ and $\phi'$, respectively, with the application of load, the center line deflection at $D$ will be $(\phi + \phi')l_{bd}/2$ and the uplift at E will be $-(\phi + \phi')l_{be}/2$, when the magnitude of $\phi$ is such that the tan $\phi = \phi$. Corresponding twist at $D$ and $E$ will be $(\phi - \phi')l_{bd}/r$ and $-(\phi - \phi')l_{be}/r$, assuming a positive twist sign convention such that an upward movement normal to the deck moves out of the paper toward the reader.

The elongation of the cables will cause the center line of the girder at $D$ and $E$ to deflect by $(X_d + X'_d)c_d/2$ and $(X_e + X'_e)c_c/2$ respectively, where the vertical flexibility of the cable is formulated as before for the single-plane, cable-stayed girder. The twist then becomes $(X_d - X_d')c_d/r$ and $(X_e - X_e')c_e/r$ at $D$ and $E$ respectively.

Similarly the elastic shortening of the tower will produce a center line deflection of

$$\frac{[(X_d + X_e) + (X_d' + X_e')]f_T}{2} \tag{11.29}$$

at $D$ and $E$, where $f_T$ is the flexibility coefficient of the tower, and a twist

$$\frac{[(X_d + X_e) - (X_d' + X_e')]f_T}{r} \tag{11.30}$$

at $D$ and $E$.

The right-hand side of equations 11.28 may, therefore, be expressed as

$$\Delta_d = \frac{(\phi + \phi')l_{bd}}{2} + \frac{(X_d + X_d')c_d}{2} + \frac{[(X_d + X_e) + (X_d' + X_e')]f_T}{2}$$

$$\Delta_b = 0$$

$$\Delta_e = \frac{-(\phi + \phi')l_{be}}{2} + \frac{(X_e + X_e')c_e}{2} + \frac{[(X_d + X_e) + (X_d' + X_e')]f_T}{2}$$

$$\psi_d = \frac{(\phi - \phi')l_{bd}}{r} + \frac{(X_d - X_d')c_d}{r} + \frac{[(X_d + X_e) - (X_d' + X_e')]f_T}{r} \tag{11.31}$$

$$\psi_b = 0$$

$$\psi_e = \frac{-(\phi - \phi')l_{be}}{r} + \frac{(X_e - X_e')c_e}{r} + \frac{[(X_d + X_e) - (X_d' + X_e')]f_T}{r}$$

Similar to the single-plane, cable-stayed structures the additional, unknown rotations of the towers $\phi$ and $\phi'$ have been introduced, which require two more equations for a solution. These equations are formulated from the equilibrium of the tower by summing moments about the base. Thus

$$X_d l_{bd} - X_e l_{be} = 0 \tag{11.32}$$

$$X_d' l_{bd} - X_e' l_{be} = 0 \tag{11.33}$$

By rewriting the deflection compatibility equations and using the tower stability equations the matrix form can be written as stated in equation 11.34.

The solution of equation 11.34 produces the vertical reactions at each interior support and cable connection, and the rotation of the towers. With these values, the other reactions, moments, shears, and torques can be determined.

$$
\begin{bmatrix}
\left(f_{dd}+\dfrac{c_d}{2}+\dfrac{f_T}{2}\right) & \left(f_{dd}+\dfrac{c_d}{2}+\dfrac{f_T}{2}\right) & (f_{db}) & (f_{db}) & \left(f_{de}+\dfrac{f_T}{2}\right) & \left(f_{de}+\dfrac{f_T}{2}\right) & \left(\dfrac{l_{bd}}{2}\right) & \left(\dfrac{l_{bd}}{2}\right) \\[8pt]
(f_{bd}) & (f_{bd}) & (f_{bb}) & (f_{bb}) & (f_{be}) & (f_{be}) & 0 & 0 \\[8pt]
\left(f_{ed}+\dfrac{f_T}{2}\right) & \left(f_{ed}+\dfrac{f_T}{2}\right) & (f_{eb}) & (f_{eb}) & \left(f_{ee}+\dfrac{c_e}{2}+\dfrac{f_T}{2}\right) & \left(f_{ee}+\dfrac{c_e}{2}+\dfrac{f_T}{2}\right) & -\left(\dfrac{l_{be}}{2}\right) & -\left(\dfrac{l_{be}}{2}\right) \\[8pt]
\left(\dfrac{r\psi_{dd}}{2}+\dfrac{c_d}{r}+\dfrac{f_T}{r}\right) & -\left(\dfrac{r\psi_{dd}}{2}+\dfrac{c_d}{r}+\dfrac{f_T}{r}\right) & \left(\dfrac{r\psi_{db}}{2}\right) & -\left(\dfrac{r\psi_{db}}{2}\right) & \left(\dfrac{r\psi_{de}}{2}+\dfrac{f_T}{r}\right) & -\left(\dfrac{r\psi_{de}}{2}+\dfrac{f_T}{r}\right) & -\left(\dfrac{l_{bd}}{r}\right) & -\left(\dfrac{l_{bd}}{r}\right) \\[8pt]
\left(\dfrac{r\psi_{bd}}{2}\right) & -\left(\dfrac{r\psi_{bd}}{2}\right) & \left(\dfrac{r\psi_{bb}}{2}\right) & -\left(\dfrac{r\psi_{bb}}{2}\right) & \left(\dfrac{r\psi_{be}}{2}\right) & -\left(\dfrac{r\psi_{be}}{2}\right) & 0 & 0 \\[8pt]
\left(\dfrac{r\psi_{ed}}{2}+\dfrac{f_T}{r}\right) & -\left(\dfrac{r\psi_{ed}}{2}+\dfrac{f_T}{r}\right) & \left(\dfrac{r\psi_{eb}}{2}\right) & -\left(\dfrac{r\psi_{eb}}{2}\right) & \left(\dfrac{r\psi_{ee}}{2}+\dfrac{c_e}{r}+\dfrac{f_T}{r}\right) & -\left(\dfrac{r\psi_{ee}}{2}+\dfrac{c_e}{r}+\dfrac{f_T}{r}\right) & -\left(\dfrac{l_{be}}{r}\right) & \left(\dfrac{l_{be}}{r}\right) \\[8pt]
(l_{bd}) & 0 & 0 & 0 & -(l_{be}) & 0 & 0 & 0 \\[8pt]
0 & (l_{bd}) & 0 & 0 & 0 & -(l_{be}) & 0 & 0
\end{bmatrix}
\begin{bmatrix}
X_d \\ X'_d \\ X_b \\ X'_b \\ X_e \\ X'_e \\ \phi \\ \phi'
\end{bmatrix}
=
\begin{bmatrix}
\delta'_d \\ \delta'_b \\ \delta'_e \\ \psi'_d \\ \psi'_b \\ \psi'_e \\ 0 \\ 0
\end{bmatrix}
$$

Eq. (11.34)

### 11.7.3   Effects of Other Actions

The horizontal components of the cable forces will induce an axial compressive force in the girder. This force will cause an elastic shortening of the girder, the extent of which depends on the magnitude of the axial force and area of the cross section. As a result, a modification of the flexibility matrix may be required.

Depending on the geometrical configuration of a particular structure, other actions that may modify the matrix are the bending of the tower when the base is fixed or when the clamped cables are attached to the tower at different heights. A pair of cables that are carried by saddles on rollers at the tower will also affect the flexibility matrix. All of these factors may be accommodated by appropriate adjustment of the flexibility matrix as previously outlined for single-plane cable structures, with the exception that the induced torque forces must be taken into account.

### 11.7.4   Double-Plane Structure with an A-Frame Tower

When an A-frame tower is used, Fig. 11.14, and the cables are concentrically connected, the previous separate tower rotations $\phi$ and $\phi'$ will become one single value of $\phi$. Rotational equilibrium for the combined tower is now dependent on the horizontal components of all cables in the total system, requiring only one equation instead of two and equation 11.34 reduces to equation 11.35. Because of the elimination of the relative rotation of the towers, the structure becomes torsionally stiffer.

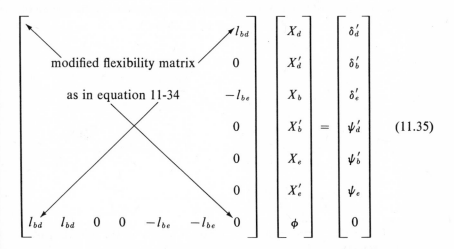

$$(11.35)$$

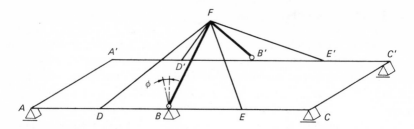

FIGURE 11.14.  Bridge with A-frame tower. (Courtesy of the Institution of Civil Engineers. Ref. 5.)

### 11.7.5   Double-Plane Structure with a Portal Tower

An example of a portal tower arrangement, Fig. 11.15, has the advantage of limiting the relative rotation of the tower and thus increasing the torsional resistance of the system. However, it is not as effective in limiting the relative rotation as the previously discussed A-frame arrangement.

If complete rigidity of the portal frame within its plane were practical it would be as effective as the A-frame. Because complete rigidity is not possible, it becomes necessary to include the effect of out-of-plane warping of the tower in the analysis.

If a load $P$ is eccentrically applied to one of the girders as indicated in Fig. 11.15, the resulting differential rotation of the towers will produce a twist in the portal beam, as well as the usual counterclockwise rotation of the towers. The resistance provided by the portal beam to the differential rotation of the towers is partly a function of the bending and torsional stiffnesses of the portal columns and beam, and partly due to the degree of fixity against twist provided at the base of the towers.

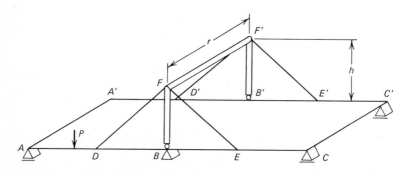

FIGURE 11.15.  Bridge with portal tower. (Courtesy of the Institution of Civil Engineers. Ref. 5.)

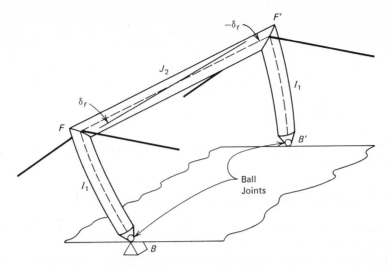

**FIGURE 11.16. Portal columns in bending only; portal beam in torsion only. (Courtesy of the Institution of Civil Engineers. Ref. 5.)**

If the base of the towers are free to twist, the deformation will be as illustrated in Fig. 11.16, in which the columns are subjected only to bending and the beam only to torsion. For this condition it can be shown that the out-of-plane deflection at the top of the tower is

$$\delta_f = (X_d \cot \theta_d - X_e \cot \theta_e)\overline{X} \tag{11.36}$$

where

$$\overline{X} = \left[ \left( \frac{h^2 r}{2GJ_2} \right) + \left( \frac{h^3}{3EI_1} \right) \right]$$

These warping deflections $\pm \delta_f$ at $F$ and $F'$, result in downward deflection at $D$ and an uplift at $D'$ equal to $\delta_f l_{bd}/h$ and producing a twist to the girder at $D$ given by:

$$\psi_{df} = \frac{2(X_d \cot \theta_d - X_e \cot \theta_e)l_{bd}\overline{X}}{hr} \tag{11.37}$$

Similarly at $E$ the twist is:

$$\psi_{ef} = \frac{-2(X_d \cot \theta_d - X_e \cot \theta_e)l_{be}\overline{X}}{hr} \tag{11.38}$$

The behavior of this structure is the same as that for the double-plane, single tower structure except for the portal warping effect. Therefore, the

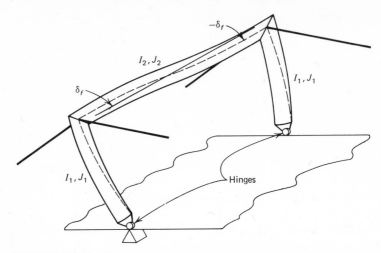

**FIGURE 11.17.** Portal columns in bending and torsion; portal beam in bending and torsion. (Courtesy of the Institution of Civil Engineers. Ref. 5.)

matrix formulation for this structure is the same as that stated in equation 11.35, provided the coefficients for $X_d$ and $X_e$ in the fourth row of the matrix are modified by the addition of $2l_{bd} \cot \theta_d \, \overline{X}/hr$ and $-2l_{bd} \cot \theta_e$ $\overline{X}/hr$, and modified in the sixth row by $-2l_{be} \cot \theta_d \, \overline{X}/hr$ and $2l_{be} \cot \theta_e$ $\overline{X}/hr$, respectively. Thus, the torsional compatibility at $D$ and $E$ are maintained.

When the base of the towers are restrained against rotation in the horizontal plane, the resulting deformed portal will be as illustrated in Fig. 11.17. In this condition the portal columns and the portal beam are subjected to bending and torsional moments. By an energy analysis, the resulting twist at $D$ and $E$, respectively, may be represented by

$$\psi_{df} = \frac{2(X_d \cot \theta_d - X_e \cot \theta_e)l_{bd}}{hr} \frac{\overline{X}\,\overline{Y}}{(\overline{X} + \overline{Y})} \qquad (11.39)$$

$$\psi_{ef} = \frac{-2(X_d \cot \theta_d - X_e \cot \theta_e)l_{be}}{hr} \frac{\overline{X}\,\overline{Y}}{(\overline{X} + \overline{Y})} \qquad (11.40)$$

where $\overline{X}$ is as previously defined and

$$\overline{Y} = \left[ \left( \frac{r^2 h}{4GJ_1} \right) + \left( \frac{r^3}{24EI_2} \right) \right]$$

and the coefficient of $X_d$ and $X_e$ in the fourth and sixth rows of equation 11.35 must again be modified for the torsional compatibility at $D$ and $E$.

### 11.7.6    Multitower Continuous Girder—Double-Plane Configuration

In a manner similar to single-plane structures, multitower structures may be accommodated by the analysis outlined above. The procedure is to release all interior supports and thereby produce a simple beam for which the "free" deflections and twists may be calculated. Appropriately modified compatibility and stability equations may be formulated to produce a single "mixed" matrix for the entire structure.

## 11.8    SUMMARY OF THE MIXED METHOD

The overall behavior of a cable-stayed bridge is determined by superimposing the actions of tower rotation, stay elongation, and tower shortening. Consideration of these actions evolves into a mixed force-displacement analysis in which modifications are made to the coefficients to account for each action. Additional modifications are introduced to accommodate bending of the towers, fixity of the tower at its base, shortening of the girder, twist or torsion of the girder, etc. The compatibility equations are appropriately modified and the equilibrium equations formulated to produce a single mixed matrix for the total structure.

The mixed method of analysis as presented above is essentially linear. The method assumes that the deflection is proportional to the load at all portions of the structure and for the structure as a whole. The stiffness of the cable can be taken into account by the utilization of an effective modulus for the stays as in Chapter 10. This method does not consider nonlinearity and in that sense is restricted to elastic behavior of the structure. The reader should bear in mind that this methodology was developed when only first-generation computers were available and there were no standard structural framework programs. Today the problem can be solved by standard two-dimensional plane frame or three-dimensional space frame programs that are appropriately modified.

## 11.9    NONLINEAR BEHAVIOR OF CABLE-STAYS

The nonlinear behavior of suspension bridges is well known and is considered in design. It has been assumed by some designers, however, that a cable-stayed bridge is a linear structure since the cables act as direct tension members. This is not the case. The cables exhibit nonlinear behavior because of the change in sag, which occurs with changing tension in the cable. A parametric study conducted at the University of Pittsburgh[8] is presented here to illustrate the nonlinear behavior of the cable stays.

In order to develop a rational design procedure for cable-stayed bridges, it is necessary to understand how various factors and parameters affect the behavior of the structure. There are many types of geometry that may be selected in this type of structure. This discussion will confine itself to the static analysis of a single-plane transverse cable geometry. Further, it will be limited to the radiating and harp type of longitudinal cable geometry configuration.

The behavior of any structure is subject to the stiffness of its members and their relation to each other. Therefore, the behavior of a cable-stayed bridge is influenced by the relative stiffness of the girder to the tower, the girder to the cables, and the individual cables to each other. The stiffness of the cables, and, therefore, of the structure, depends on the amount of initial tension in the cable as measured by an initial sag ratio. The type of live load distribution on the structure will influence the individual members in the system.

The purpose of the investigation was to determine the effect several of the parameters enumerated above may have on the bending moments of the main girder and tower and the tension in the cables. The curves plotted in Figs. 11.19 to 11.30 inclusive are not intended for design purposes, but are merely to illustrate the effect of some of the variables involved.

### 11.9.1    Effect of Initial Sag Ratio

The initial sag ratio, defined as the ratio of the initial cable sag to the chord length, is a convenient measure of pretension in the cables of a specified structure. As the cable tension increases, the sag ratio decreases. Therefore, the initial sag ratio is considered to be a measure of the tension that exists in the cable prior to the application of design load.

To determine the influence of initial sag ratio on the stiffness of the cables and thus on the stiffness of the structure, a nonlinear analysis was conducted using both the radiating and harp configuration in a single-plane arrangement, Fig. 11.18. Cable attachments are at the end and third points of the end span and at the seventh point of the center span. The girder is rigidly supported vertically at the tower, but is independent of the tower and there is no moment transfer between girder and tower. The tower is fixed at its base. The girder and tower are assumed to have a constant cross section throughout their span and height. It is assumed that the cable has an initial prestress capable of resisting a possible compression load (negative tension).

For purposes of this mathematical model, a single load bearing plane of the Nordbrücke Bridge at Düsseldorff was modified. For convenience, the dimensions were rounded off when converting from metric to English units.

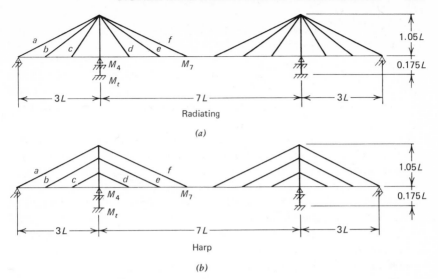

**FIGURE 11.18.** Single-plane, longitudinal cable configurations for parametric studies: (*a*) radiating; (*b*) harp.

Cable areas were similarly rounded off to suit dimensions of cables manufactured in the United States. Whereas the actual structure had a tapered tower, the mathematical model has a constant area and moment of inertia equal to that at the base of the actual tower. Although the actual girder has a varying cross section, the mathematical model uses the average cross-sectional area and moment of inertia. In comparing the harp and radiating configuration, the same area for the cables was used for the two study models. On the basis of this assumption, the cable stiffnesses were not constant between the two models as a result of the difference in length of the interior cables.

The nonlinear analysis was performed using an iterative procedure to account for the change in cable sag when the load was applied. A linear elastic analysis using STRUDL was also performed for comparative purposes.

### 11.9.2    Cable Tension Comparison

The variation of final cable tension with the initial sag ratio, for a uniformly loaded bridge, is presented in Fig. 11.19 for the harp configuration and in Fig. 11.20 for the radiating configuration. A plot of percentage error between the linear and nonlinear analyses for the harp and radiating configurations are presented in Figs. 11.21 and 11.22, respectively. These

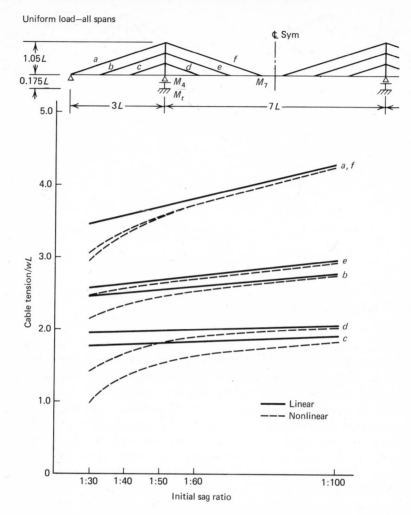

**FIGURE 11.19.    Cable tension comparison, harp.**

plots indicate that the linear analysis overestimates the final tension in all cables in the harp configuration, while in the radiating configuration the linear analysis overestimates the final tension in the two outside cables of the end span, and the outside cable of the center span. The linear analysis underestimates the final tension in the inside cable of the end span and the two inside cables of the center span.

In analyzing the plots presented in Fig.s 11.19 through 11.22, it becomes apparent that at an initial sag ratio of approximately 1:60, the non-

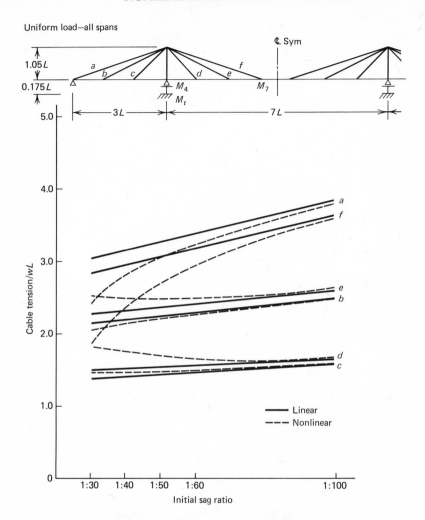

**FIGURE 11.20.   Cable tension comparison, radiating.**

linear curves begin to approach linearity in character and in some cases very closely parallel the linear curves.

### 11.9.3   Moment Comparison

Curves are plotted for bending moments at selected points in the harp and radiating bridge structures, Figs. 11.23 and 11.24 and the curves plotted in Figs. 11.25 and 11.26 indicate percentage error versus initial sag ratio.

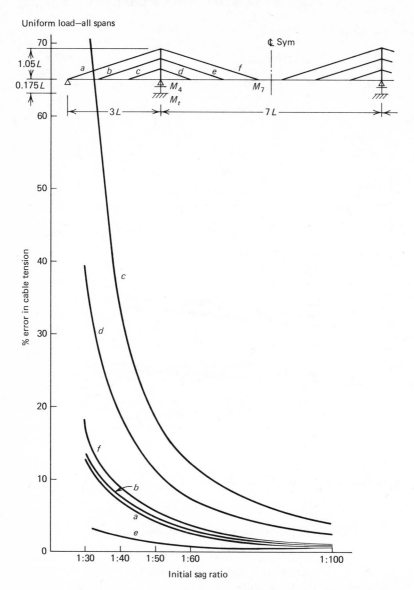

**FIGURE 11.21.** Cable tension comparison, harp—percentage error.

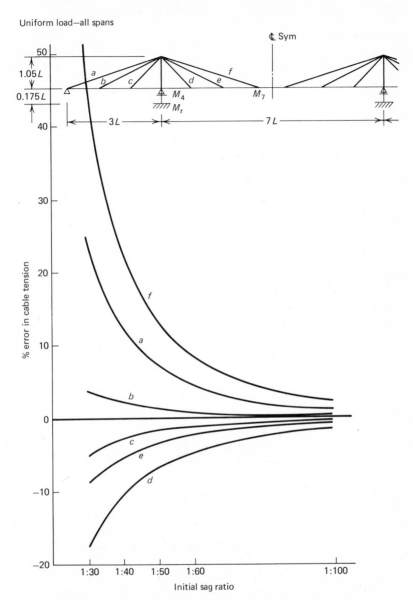

**FIGURE 11.22.** Cable tension comparison, radiating—percentage error.

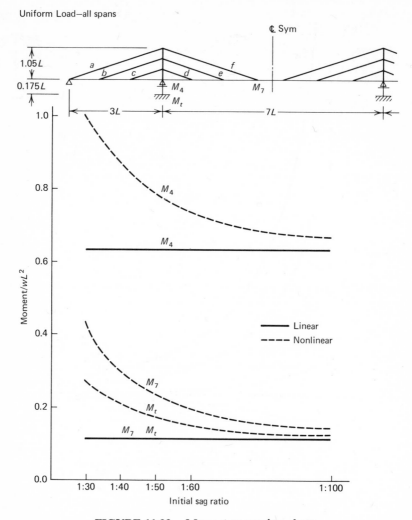

**FIGURE 11.23.   Moment comparison, harp.**

The curves indicate the bending moments at the base of the tower, in the girder where it is supported at the tower, and in the girder at the point where the outside cable is connected to it in the center span. These moments do not include the effect of the initial tension in the cables. The foremost observation is that the linear analysis underestimates all values of moments at these locations.

The tower bending moment does not approach a linear character until an initial sag ratio of approximately 1:80 for the radiating system and

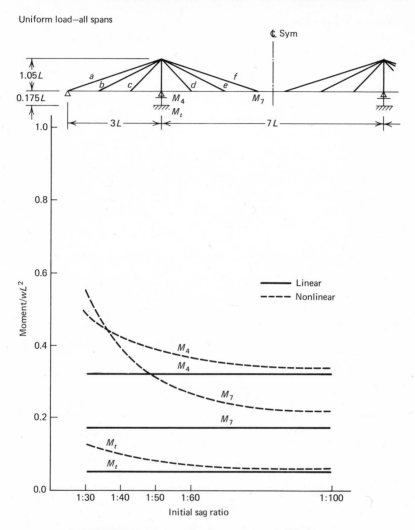

**FIGURE 11.24. Moment comparison, radiating.**

approximately 1:90 for the harp system have been reached. The moment at midspan for both configurations does not achieve a linear character until an initial sag ratio of 1:100 or beyond is reached.

### 11.9.4 Effect of Load Application

To determine the effect of the application of load on the cable tension and moments in the structure, data was obtained for various fractions of load

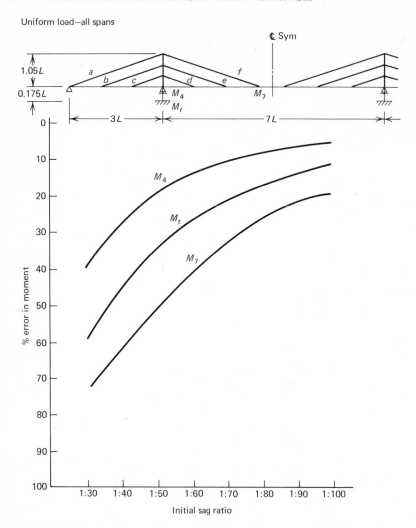

**FIGURE 11.25.   Moment comparison, harp—percentage error.**

for the radiating configuration. Cable tension curves are presented for initial sag ratios of 1:60 and 1:00, respectively, Figs. 11.27 and 11.28. It is noted that the cable tension curves for an initial sag ratio of 1:100 are essentially linear in character, although a slight nonlinearity exists during the dead load application for a sag ratio of 1:60 which become linear when the full dead load is applied.

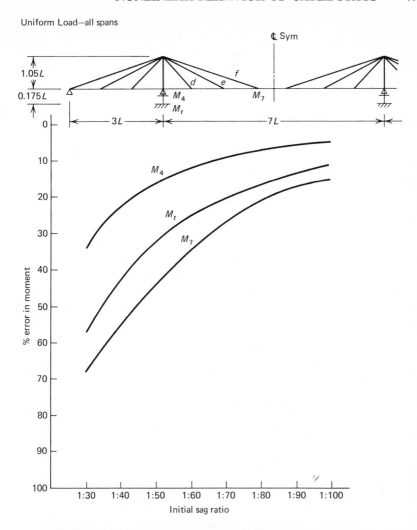

**FIGURE 11.26.   Moment comparison, radiating—percentage error.**

Curves for the moments in the structure are presented in Figs. 11.29 and 11.30. It is noted that the greatest nonlinearity occurs in the moment near the midspan. The degree of nonlinearity is diminished as the initial sag ratio decreases until linearity is achieved at approximately the application of full dead load for the 1:60 sag ratio. For the sag ratio of 1:100, linearity is achieved at an earlier stage, approximately $\frac{2}{3}$ dead load. It

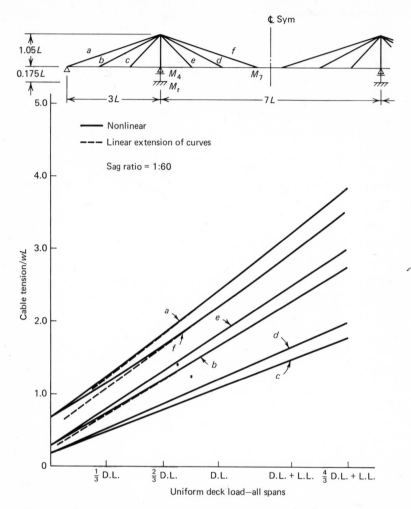

**FIGURE 11.27.    Cable tension comparison, radiating—sag ratio, 1:60.**

may, therefore, be concluded that the nonlinear action is confined to the dead load or erection stage. Thus, under full dead load the sag ratio has decreased to a degree of tautness that the response of the structure may be considered to be linear during the application of live load. This conclusion is in agreement with that of Seim, et al.[17,18]

It has been shown that a significant indicator of the stiffness of a cable-stayed bridge is in the initial stiffness of the cables. It is concluded from

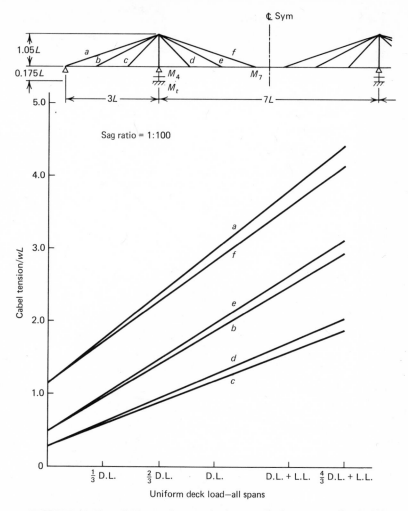

**FIGURE 11.28.    Cable tension comparison, radiating—sag ratio, 1:100.**

the data presented here for the two cable configurations investigated that the nonlinear cable behavior will be confined to the application of dead load and that for application of live load the structure will behave in a linear manner.

The conclusion of linearity in the live load range is important because it permits the construction of influence lines. The application of the Müller-Breslau principle and Maxwell's theorem requires that a linear relationship

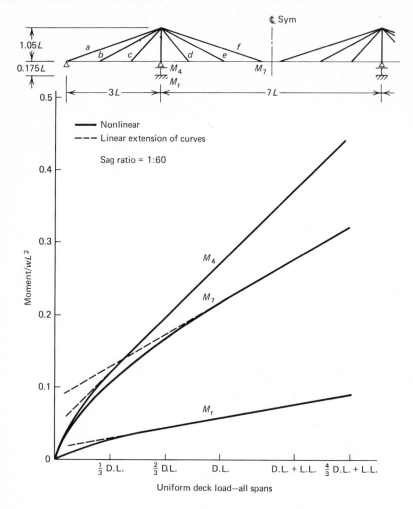

**FIGURE 11.29.**    Moment comparison, radiating—sag ratio, 1:60.

exist or be assumed to exist, for loads, stresses, and deflection. As demonstrated above, linearity may be assumed in the live load range because of the increase in cable stiffness due to the reduction in the sag with increased tension.

## 11.10    INFLUENCE LINES

Shortly after the construction of the Strömsund Bridge, Homberg[19] published the analytical results of several cable-stayed systems proposed for

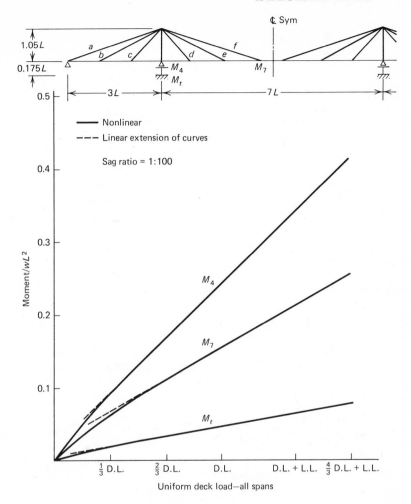

**FIGURE 11.30.    Moment comparison, radiating—sag ratio, 1:100.**

the bidding of the North Bridge at Düsseldorf, and the Rhine Bridge at Speyer. The result of the studies are extracted from the published reports and reproduced below.[19] Some of the cable-stayed systems studied are illustrated schematically in Fig. 11.31. Influence lines for System 1 are presented in Fig. 11.32. The dimensions are those proposed for the Rhine Bridge at Speyer. The corresponding influence lines for a self-anchored suspension bridge under similar conditions is indicated in Fig. 11.33. Comparing Figs. 11.32 and 11.33, the influence lines for the cable-stayed system are seen to be of a different character than those for the suspension

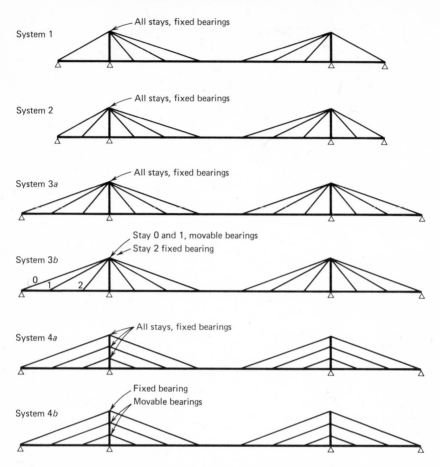

**FIGURE 11.31.** Schematic representation of cable-stayed bridges. (Courtesy of *Der Stahlbau*. Ref. 20.)

bridge. The influence line for bending moment in the girder at the tower indicates large positive and negative areas for the suspension bridge and the cable-stayed system indicates primarily negative moments. It is also seen that the influence lines for moments in the center span of the cable-stayed bridge vary in magnitude and direction as do those of the suspension bridge.

Influence lines for cable forces and bending moments in the girder for system 3*a* are depicted in Fig. 11.34. The system indicates truss-like characteristics as a result of the fixed saddle cable supports at the towers. If the two interior sets of cables were supported at the tower by movable

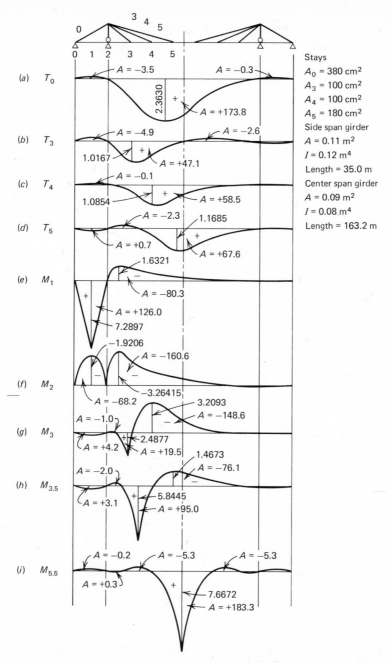

**FIGURE 11.32. Influence line for cable-stayed system 1. (Courtesy of** *Der Stahlbau.* **Ref. 20.)**

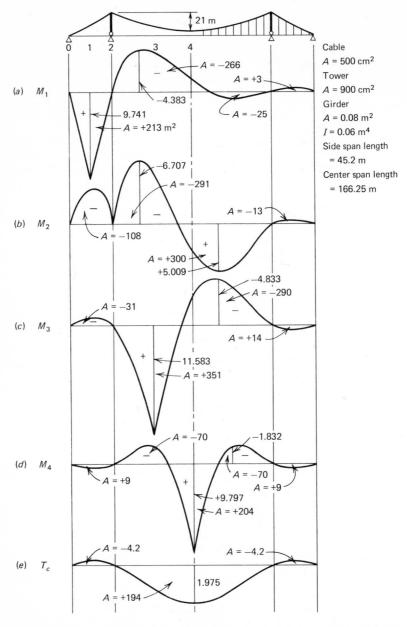

**FIGURE 11.33.** Influence line for self-anchored suspension bridge. (Courtesy of *Der Stahlbau.* Ref. 20.)

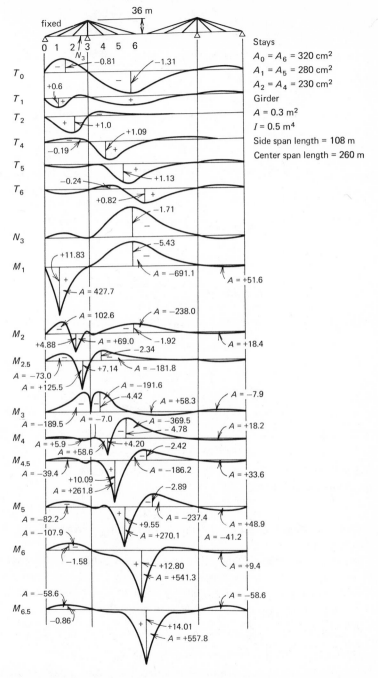

**FIGURE 11.34.** Influence line for cable-stayed system 3*a*. (Courtesy of *Der Stahlbau*. Ref. 20.)

**409**

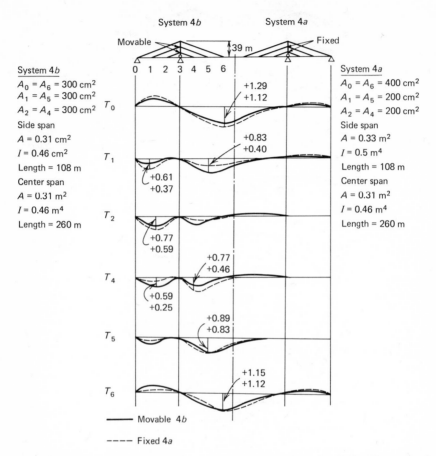

**FIGURE 11.35. Influence lines for cable-stayed systems 4a and 4b. (Courtesy of Der Stahlbau. Ref. 20.)**

saddles the bending moments in the stiffening girder would increase considerably.

The cable tension influence lines for the two cable-stay systems 4a and 4b, of Fig. 11.31, as proposed for the North Bridge at Düsseldorf are illustrated in Fig. 11.35. From a stress point of view, system 4a has a better balance in the girder, which is attributed to the fixed saddle supports at the tower for all three sets of stays. However, this condition produces considerably higher bending moments in the towers and differentials of horizontal stresses which must be resisted by the saddle bearings.

O'Connor[20] produced similar influence lines and made comparisons, which are duplicated in the following discussion. Influence lines were constructed for the radiating and harp configurations illustrated in Fig. 11.36.

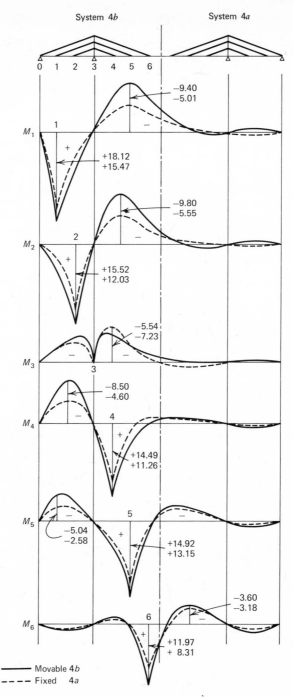

**FIGURE 11.35.** (*continued*)

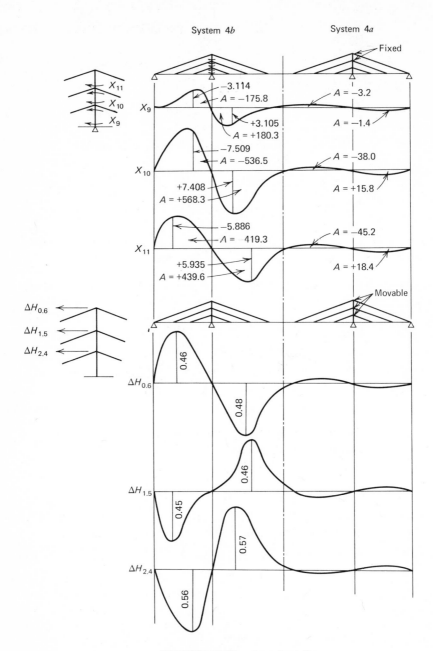

**FIGURE 11.35.** (*continued*)

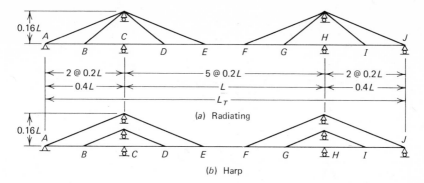

FIGURE 11.36.  Bridge geometries used for analysis of radiating and harp types. (Ref. 21.)

The influence lines, shown in Figs. 11.37 through 11.40 do not account for axial strains in the tower and girder and assume that the stay connection at the tower is such that no bending moment is produced in the tower. Cable areas are assumed constant for all stays and the moment of inertia of the girder is constant along the length of the girder.

Cable-stay area is presented as a dimensionless coefficient

$$\frac{E_c A_c L_T^2}{E_G I_G}$$

where $E_c$ = modulus of elasticity for the cable stay
$A_c$ = cross-sectional area of the cable stay
$E_G$ = modulus of elasticity for the girder
$I_G$ = moment of inertia for the girder
$L_T$ = overall length of the structure

Tabulated values for $E_c A_c L_T^2 / E_G I_G$ range from 250 to 16,000, with practical values ranging from 2,000 to 10,000.

O'Conner's method of analysis for these influence lines is as follows:[20]

1. Form a statically determinate base structure by removing the supports at $C$ and $H$, and cutting the cables at $B$, $D$, $E$, $F$, $G$, and $I$ in the case of the radiating, and at $D$, $E$, $F$, and $G$ in the case of the harp, Fig. 11.36. The vertical components of force at these points will be considered as the redundant forces. There are eight such forces in the case of the radiating system and six for the harp system.

2. Compute deflections corresponding to each of the unit redundant forces. This flexibility matrix is made up of terms from the girder deflections plus terms from the cable extensions.

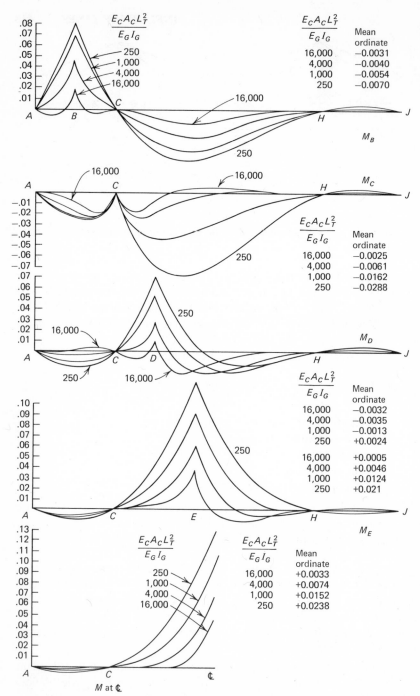

**FIGURE 11.37. Influence lines for deck bending moments, radiating type bridge; bending moment ordinates $\times L$. (Ref. 21.)**

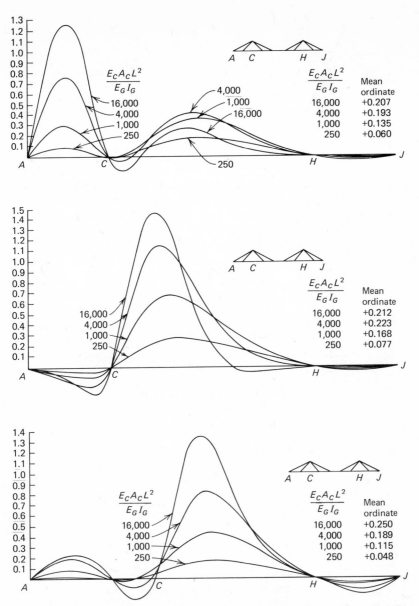

**FIGURE 11.38.** Influence lines for cable forces, radiating type bridge. (Ref. 21.)

415

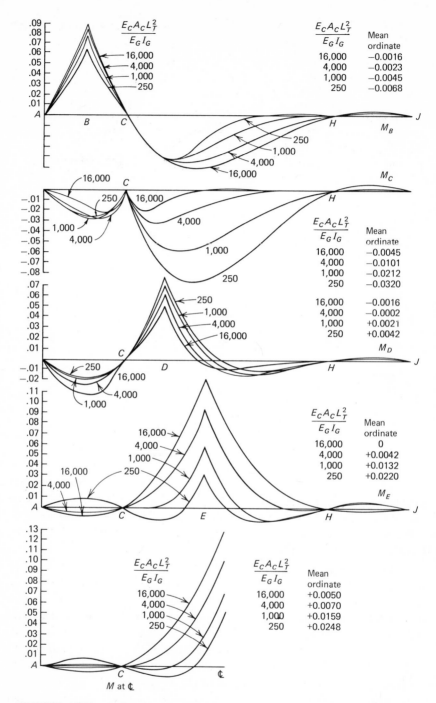

**FIGURE 11.39.** Influence lines for deck bending moments, harp type bridge; bending moment ordinate × L. (Ref. 21.)

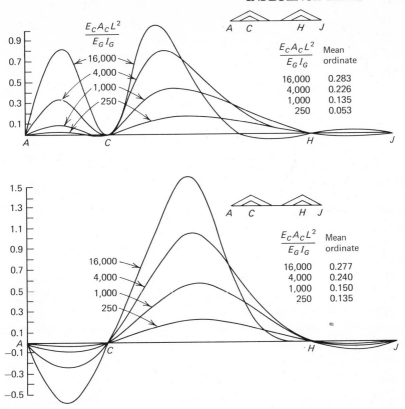

**FIGURE 11.40.  Influence lines for cable forces, harp type. (Ref. 21.)**

The matrix of girder deflections can be obtained by a simple standard program, loading the girder with the $M/EI$ diagram for the particular load and obtaining deflections using moment area methods. The matrix of cable deflections can be computed manually, read as data, and added algebraically to the girder deflections.

3. From the principle of Müller-Breslau, an influence line for bending moment is identical with the deflected shape corresponding to a unit angular rotation across an element at the point in question.

4. For the base structure, compute the initial deflections of the base structure using the angle-load analogy. Insert these deflections as the constant terms in a set of simultaneous equations with the redundant forces, and the coefficients of the unknowns being the flexibility matrix computed previously. Solve for these redundant forces.

5. Apply the redundant forces to the girder alone. Compute its deflected shape. Add this new shape to the initial deflected shape to determine the required influence line.

6. The influence line for the vertical component of a cable force corresponds to a unit relative vertical displacement across the cut in the cable. This unit displacement can be introduced to form a set of simultaneous equations for the redundant forces. The redundant forces may be applied to the girder alone, and the resulting deflected shape is the influence line for the redundant. The influence line for cable force may be obtained by multiplying each ordinate by $1/\sin \theta$, where $\theta$ is the cable slope from the horizontal.

Based on these influence lines, O'Conner arrived at the following generalized conclusions:[20]

1. The harp arrangement of cables has larger bending moment ordinates and smaller cable force ordinates than the radiating type. This results from the less direct load transfer to the abutments and piers. Consequently, the radiating type configuration is generally to be preferred. However, the difference between the two forms is not as marked as would be expected.

2. For the harp arrangement, the influence lines for bending moments in the girder at the cable connections nearest the towers are hardly affected by the cable stiffness. These bending moments are hardly benefited by the presence of the cables.

3. With this exception, the influence lines for the girders are greatly dependent on the parameter $E_c A_c L_T^2 / E_G I_G$.

4. Plots for: (a) the maximum positive bending moment at the center of the main span; (b) the maximum positive bending moment at the center of the side span; and (c) the maximum negative bending moment at the tower for a moving unit load are illustrated in Fig. 11.41. Observe that the cable stiffness has a small effect on item b for the harp but a relatively large effect on the other bending moments. This effect of the cable stiffness extends well beyond the normal practical range. Therefore, it would be inaccurate to attempt an analysis that ignored elongations in the cables.

5. The bending moment at the center of the side spans for the harp may become the absolute maximum bending moment as observed in Fig. 11.41. This same conclusion is also evident in the work performed by Homberg.[19]

6. The maximum bending moment at the center of the main span is virtually the same value for the radiating or the harp system.

7. Curves of maximum moment versus location on the girder for a moving unit load are plotted in Fig. 11.42. In general, the negative bending moments are smaller than the positive bending moments in the deck structure. For the harp system, Fig. 11.42 indicates the relatively high bending moments in the side span and near the towers.

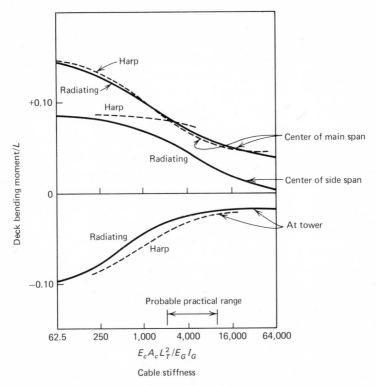

**FIGURE 11.41. Variation of deck bending moment with cable stiffness for a unit moving load. (Ref. 21.)**

8. The influence lines for cable force have negative ordinates for the radiating and harp arrangements. However, only moderate dead load tensions are necessary to avoid resultant compressive forces, and will prevent the cables from becoming slack. There is little difference between the radiating and harp systems in this respect and, therefore, both are equally acceptable.

9. For the radiating configuration of cables a better distribution of design bending moments in the girder could be achieved with larger side spans and a smaller gap between the central cables.

10. The above comparisons between the harp and radiating arrangements have assumed equal cable areas. In actual fact, this assumption is not valid. The harp would be expected to have smaller cable forces than the radiating system and, therefore, smaller cable areas. As a result, design bending moments for the harp arrangement should be greater than those used in the previous discussion.

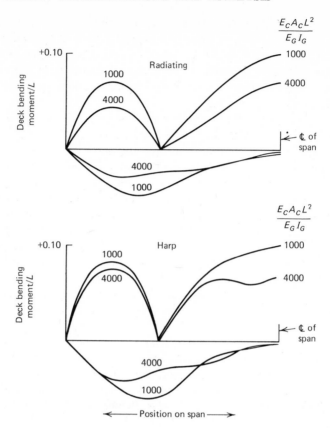

**FIGURE 11.42.   Curves of maximum deck moment due to a unit moving load. Ref. 21.)**

## 11.11   LIVE LOAD STRESSES

In the design of girder bridges an influence line for bending moment is normally sufficient to determine the stress in the girder. The stress at a point $r$ in section $i$ is denoted simply by $f_{i,r} = M_i/S_{i,r}$, where $M_i$ is the moment at section $i$ and $S_{i,r}$ is the section modulus at point $r$ in section $i$. Thus, $f_{i,r \, max} = M_{i \, max}/S_{i,r}$. However, in cable-stayed bridges, in addition to the bending moment in the girder, there is the effect of axial load and stress concentrations produced by the cable anchorages in the girder and the tower.

Tang[10,21] has suggested the following formulation to obtain the stress at a point $r$ in section $i$:

$$f_{i,r} = a_1 M_i + a_2 N_i + a_3 T_k \tag{11.41}$$

where $f_{i,r}$, $M_i$, $S_{i,r}$ are as previously designated, $N_i$ is the axial force at section $i$, and $T_k$ is the cable force of a nearby cable $k$. The coefficients are $a_1 = 1/S_{i,r}$, $a_2 = 1/A_i$; $a_3$ is the effect of nonuniform stress distribution due to the anchorage of cable $k$. Note that $S_{i,r}$ and $A_i$ are the effective section modulus for point $r$ at section $i$ and the effective cross-sectional area at section $i$, respectively.

The nonuniform bending stress distribution across the bridge deck must be considered when calculating an effective section modulus $S_{i,r}$. This is especially true when the girder is slender and wide.

Additionally, at the anchorage point of the cable to the girder or tower a local stress concentration will occur, which is accounted for by the coefficient $a_3$. This can be calculated[21] by assuming that a uniform stress distribution is not attained until a distance from the anchorage point equal to the width of the bridge is reached. Between the anchorage point and the point of uniform distribution, the stress may be determined by an approximate analysis, by finite element computer methods, or by some approximate assumption based on sound engineering judgment.

Because the live load that produces a maximum $M_{i,r}$ or maximum $N_i$ does not necessarily produce a maximum $f_{i,r}$, influence lines for the three elements on the right-hand side of equation 11.41 must be evaluated simultaneously, or, as suggested by Tang,[10,21] an influence line in the form of equation 11.41 may be evaluated to obtain a maximum and minimum $f_{i,r}$. This can be accomplished by superposition. Instead of the unit deformation $\Delta\phi_i$ for rotation, $\Delta u_i$ for horizontal displacement of the girder and vertical displacement of the tower, and $\Delta l_c$ for change in cable chord length, a combined distortion equal to

$$\Delta\phi_i = a_1, \qquad \Delta u_i = a_2, \qquad \text{and } \Delta l_c = a_3 \qquad (11.42)$$

is applied. The resulting deflection curve of the bridge becomes the influence line of $f_{i,r}$.[10,21]

## 11.12   OTHER METHODS OF ANALYSIS

Nonlinearity in the bridge structure, other than cable nonlinearity discussed above, involves the girder and tower legs and results from deflections of these components under the combined effect of bending moments and axial forces. Generally, these effects of nonlinearity are negligible unless the girder and/or tower are very slender. In any event, after all details of the design of a cable-stayed bridge are completed, a nonlinear analysis, subject to overload conditions, should be performed to assure the adequacy of the total design. The nonlinear analysis may be necessary to determine stress considerations during erection.

Standard plane frame and space frame computer programs are available, which, then properly modified to account for nonlinearity, can be used for the analyses. Troitsky and Lazar[7] have used a flexibility technique for comparison with models to study the behavior of cable-stayed bridges. Lazar, Troitsky, and Douglass[22] have further proposed a load balancing analysis to partially reduce the effects of loads acting on a structure by applying a prestressing force. Lazar[23] also employed the stiffness method of analysis. The transfer matrix method which was developed in Germany,[6] has been extended by Tang[10,21] to accommodate nonlinearity by using fictitious loads.

Kajita and Cheung[11] employed a finite element method for a linear method of analysis to consider the torsion in the deck for a two-plane, three-dimensional structure. This method was further extended to consider dynamic analysis. Baron and Lien[12] presented static and dynamic analyses of the proposed Southern Bay Crossing Cable-Stayed Bridge, in San Francisco.

References 7, 10, 11, 12, 21, 22 and 23 provide detailed explanations of the various methodologies used for the analysis of cable-stayed bridges.

## REFERENCES

1. Ivy, R. J., Lin, T. Y., Mitchell, S., Raab, N. C., Richey, V. J., and Scheffey, C. F., "Live Loading for Long-Span Highway Bridges," *ASCE Transactions*, Vol. 119, Paper No. 2708, 1954.
2. *Manual for Structural Applications of Steel Cables for Buildings*, American Iron and Steel Institute, Washington, D. C., 1973.
3. Gimsing, N. J., "Anchored and Partially Anchored Stayed Bridges," *Proceedings of the International Symposium on Suspension Bridges, Lisbon*, Laboratorio Nacional de Engenharia Civil, 1966.
4. Smith, B. Stafford, "The Single Plane Cable-Stayed Girder Bridge: A Method of Analysis Suitable for Computer Use," *Proceedings of the Institution of Civil Engineers*, May 1967.
5. Smith, B. S., "A Linear Method of Analysis for Double-Plane Cable-Stayed Girder Bridges," *Proceedings of the Institution of Civil Engineers*, January 1968.
6. Protte, W. and Tross, W., "Simulation als Vorgehensweise bei der Berechnung von Schrägseilbrüchen," *Der Stahlbau*, No. 7, July 1966.
7. Troitsky, M. S. and Lazar, B. E., "Model Analysis and Design of Cable-Stayed Bridges," *Proceedings of the Institution of Civil Engineers*, March 1971.
8. Podolny, W., Jr., "Static Analysis of Cable-Stayed Bridges," Ph.D. thesis, University of Pittsburgh, 1971.
9. Podolny, W., Jr. and Fleming, J. F., "Cable-Stayed Bridges—Single Plane Static Analysis," *Highway Focus*, Vol. 5, No. 2, August 1973, U.S. Dept. of Transportation, Federal Highway Administration, Washington, D. C.
10. Tang, M. C., "Analysis of Cable-Stayed Girder Bridges," *Journal of the Structural Division*, ASCE, Vol. 97, No. ST 5, May 1971, Proc. Paper 8116.

11. Kajita, T. and Cheung, Y. K., "Finite Element Analysis of Cable-Stayed Bridges," IABSE, Pub. 33-II, 1973.
12. Baron, F. and Lien, S. Y., "Analytical Studies of a Cable Stayed Girder Bridge," *Computers & Structures*, Vol. 3, Pergamon Press, New York, 1973.
13. Baron, F. and Lien, S. Y., "Analytical Studies of the Southern Crossing Cable Stayed Girder Bridge," Report No. UC SESM 71-10, Vol. I & II, Department of Civil Engineering, University of California, Berkeley, California, June 1971.
14. Okauchi, I., Yabe, A. and Ando, K., "Studies on the Characteristics of a Cable-Stayed Bridge," *Bull. of the Faculty of Science and Engineering*, Chuo University, Vol. 10, 1967 (in Japanese).
15. Kinney, J. S., *Indeterminate Structural Analysis*, Addison-Wesley Publishing Co., Inc., 1957.
16. Scalzi, J. B., Podolny, W., Jr., and Teng, W. C., "Design Fundamentals of Cable Roof Structures," United States Steel Corporation, ADUSS 55-3580-01, Pittsburgh, Pennsylvania, October 1969.
17. Seim, C., Larsen, S. and Dang, A., "Design of the Southern Crossing Cable Stayed Girder," Preprint Paper 1352, ASCE National Water Resources Engineering Meeting, Phoenix, Arizona, January 11–15, 1971.
18. Seim, C., Larsen, S. and Dang, A., "Analysis of the Southern Crossing Cable Stayed Girder," Preprint Paper 1402, ASCE National Structural Engineering Meeting, Baltimore, Maryland, April 19–23, 1971.
19. Homberg, H., "Einflusslinien von Schrägseilbrücken," *Der Stahlbau*, No. 2, February 1955.
20. O'Conner, C., *Design of Bridge Superstructures*, New York, John Wiley & Sons, Inc., 1971.
21. Tang, M. C., "Design of Cable-Stayed Girder Bridges," *Journal of the Structural Division*, ASCE, Vol. 98, No. ST 8, August 1972, Proc. Paper 9151.
22. Lazar, B. E., Troitsky, M. S. and Douglass, M. McC., "Load Balancing Analysis of Cable-Stayed Bridges," *Journal of the Structural Division*, ASCE, Vol. 98, No. ST 8, August 1972, Proc. Paper 9122.
23. Lazar, B. E., "Stiffness Analysis of Cable-Stayed Bridges," *Journal of the Structural Division*, ASCE, Vol. 98, No. ST 7, July 1972, Proc. Paper 9036.

# 12

# Design Considerations— Wind Effects

## 12.1  INTRODUCTION

Because of their flexibility, cable-supported systems may be subject to potentially large dynamic motions induced by wind forces. In the past 16 decades wind oscillations have severely damaged at least eleven suspension structures, including the Tacoma Narrows Bridge in Washington. A popular misconception is that wind forces are only significant for long-span bridges. This is obviously incorrect, as seen from Table 12.1,[1] which lists

**TABLE 12.1.  Bridges Severely Damaged or Destroyed by Wind**

| Bridge | Location | Designer | Span (ft) | Failure Date |
|--------|----------|----------|-----------|--------------|
| Dryburgh Abbey | Scotland | John and William Smith | 260 | 1818 |
| Union | England | Sir Samuel Brown | 449 | 1821 |
| Nassau | Germany | Lossen and Wolf | 245 | 1834 |
| Brighton Chain Pier | England | Sir Samuel Brown | 255 | 1836 |
| Montrose | Scotland | Sir Samuel Brown | 432 | 1838 |
| Menai Straits | Wales | Thomas Telford | 580 | 1839 |
| Roche-Bernard | France | Le Blanc | 641 | 1852 |
| Wheeling | U.S.A. | Charles Ellet | 1010 | 1854 |
| Niagara-Lewiston | U.S.A. | Edward Serrell | 1041 | 1864 |
| Niagara-Clifton | U.S.A. | Samuel Keefer | 1260 | 1889 |
| Tacoma Narrows | U.S.A. | Leon Moisseiff | 2800 | 1940 |

bridges in a span range from 245 to 2800 ft that have been damaged or destroyed. It has been known for 150 years that suspension bridges are susceptible to serious vibration problems induced by wind forces. Yet on November 7, 1940 the Tacoma Narrows Bridge began to oscilliate in a mild gale (Fig. 12.1), and its oscillations increased to a destructive ampli-

**FIGURE 12.1.** Torsional motion of the Tacoma Narrows Bridge just before failure, November 7, 1940.

tude until, in a final agonizing spasm, the main span broke loose from its supporting cables and crashed into the water 208 ft below, Fig. 12.2. This catastrophic failure shocked the engineering profession, and many professionals were surprised to find that such a wind-induced failure of a suspension bridge was not without precedent.

### 12.1.1 Descriptions of Bridge Failures

In 1817 a 4 ft wide, 260 ft span footbridge was constructed across the Tweed River at Dryburgh Abbey in Berwick County, Scotland. This structure was distinguished by a side parapet, which served as a stiffening member, and by supporting inclined (stay) chains. In 1818, 6 months after completion, it collapsed when the chain stays broke at the joints as a result of wind-induced oscillations.[1,2,3,4,5]

Sir Samuel Brown built the first vehicular suspension bridge (Union) in England in 1820 across the Tweed at Nordham near Berwick. This structure had a 12 ft roadway, with 3 ft walks on each side, and a span of 449 ft. It was the first eyebar suspension bridge completed in England and had 12 chains of eyebars, each 2 in. in diameter. Failure occurred in a violent wind 6 months after completion.[1,2,4,5]

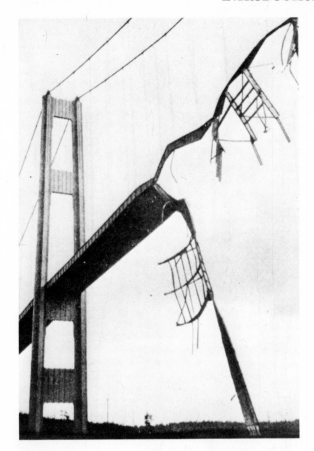

**FIGURE 12.2.   The failure of the Tacoma Narrows Bridge.**

The Nassau Bridge was constructed with eyebar chains and without stiffening members across the Lahn River in Germany. Twelve chains were broken in a wind storm 4 years after it had been opened to traffic.[1,4]

The four-span Brighton Chain Pier structure was built in 1823 and partially wrecked by a wind storm in 1833. This structure was rebuilt and was destroyed 3 years later in 1836. It consisted of four 255 ft spans supported by four chains of 2 in. diameter eyebars on each side of a 12 ft 8 in. roadway. This collapse was witnessed and documented by Lt. Col. William Reid of the British Army.[6] His sketches (Fig. 12.3) and a few paragraphs of his report are reproduced below.[1,2]

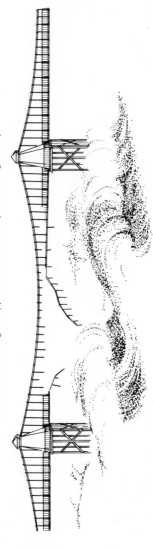

SKETCH Showing the manner in which the 3rd Span of the CHAIN PIER at BRIGHTON undulated just before it gave way in a storm on the 29th of November 1836.

255 Ft

a

b

The part (a) represents the under surface of the road way and (b) the upper surface which were both visible at the same time.

SKETCH Showing the appearance of the 3rd span after it gave way.

**FIGURE 12.3.  Damage to the Brighton Chain Pier. (Lt. Col. Reid's sketches. Ref. 1.)**

The same span of the Brighton chain-pier (the third from the shore) has now twice given way in a storm. The first time it happened in a dark night, and the storm was accompanied by much thunder and lightning: the general opinion of those who do not inquire into the cause of such matters was, that it was destroyed by lightning; but the persons employed about the pier, and whose business it was to repair it, were satisfied that the first fracture was neither caused by lightning nor by the waters, but by the wind.

The fracture this year was similar to the former, and the cause evidently the same. This time, it gave way half an hour after midday, on the 30th of November 1836, and a great number of persons were therefore enabled to see it.

The upper one of the two sketches annexed, shows the greatest degree of undulation it arrived at before the roadway broke; and the under one shows its state after it broke; but the great chains from which the road is suspended remained entire.

When this span became relieved from a portion of its load by the roadway falling into the sea, its two piers went a little to one side, and the curve of the chain became less, as in the sketch. The second and fourth spans in these sketches are drawn straight, merely to show better the degree of undulation of the third span.

Those also undulated greatly during the storm, but not in the same degree as the third span. A movement of the same kind in the roadway has always been sensibly felt by persons walking on it in high winds; but on the 29th of November 1836, the wind had almost the same violence as in a tropical hurricane, since it unroofed houses and threw down trees. To those who were at Brighton at the time, the effect of such a storm on the chain-pier was matter of interest and great curiosity. For a considerable time the undulations of all the spans seemed nearly equal. The gale became a storm about eleven o'clock in the forenoon, and by noon it blew very hard. Up to this period many persons, from curiosity, went across the first span, and a few were seen at the further end; but soon after mid-day the lateral oscillations of the third span increased to a degree to make it doubtful whether the work could withstand the storm; and soon afterwards the oscillating motion across the roadway seemed to the eye to be lost in the undulating one, which in the third span was much greater than in the other three; the undulatory motion which was along the length of the road is that which is shown in the first sketch; but there was also an oscillating motion of the great chains across the work, though the one seemed to destroy the other, as they did not both (at least as far as could be seen) take place in a marked manner at the same time.

At last the railing on the east side was seen to be breaking away, falling into the sea; and immediately the undulations increased; and when the railing on this side was nearly all gone, the undulations were quite as great as represented in the drawing.

Lt. Col. Reid's sketches indicate the characteristic sine-curve oscillations that eventually lead to the classical flutter mode of failure.

The Montrose Bridge over the South Esk River in Forfarshire, Scotland, had a 432 ft span and was built in 1829. Shortly after construction, in 1830, approximately 700 people congregated on one side of the structure to witness a boat race. As the boats passed under the bridge the crowd rushed to the other side. The sudden impact snapped the chains on one side resulting in great loss of life. The bridge was repaired but was subsequently destroyed 8 years later in 1838 by a violent gale. Eyewitness reports indicate the span undulated in two segments.[1,2,4,5]

The world-famous Menai Straits Bridge in Wales was built in 1826 by the renowned British engineer, Thomas Telford, and had, for its day, a record span—580 ft. It was damaged by wind in 1826, 1836, and 1839. The following excerpt is from a paper by Mr. W. A. Provis[7] and relates the motion that was observed in the 1826 storm.[2]

It was observed, that the character of the motion of the platform was not that of simple undulation, as had been anticipated, but the movement of the undulatory wave was oblique, both with respect to the lines of the bearers, and to the general direction of the bridge. It appeared, that when the summit of the wave was at a given point on the windward side, it was not collateral with it on the leeward side, but, in relation to the flow of the wave, consistently behind it, and forming a diagonal line of wave across the platform.

This description indicates that the structure was undergoing a torsional oscillation. After necessary repairs were made the structure survived 10 years without any distress until 1836 when an unusually heavy gale caused violent undulations of the deck and fractured several suspenders. A gatekeeper reported waves or undulations of 16 ft double amplitude. The damage was repaired, but the structure was again subjected to winds of hurricane intensity in 1839. This storm so severely damaged the structure that, with the exception of the pylons, it had to be entirely rebuilt.[1,2]

The Roche-Bernard Bridge over the Vilaine River had, for its day, a record span—641 ft—and is notable in that it was among the earliest suspension bridges to use wire cable. It was damaged in a hurricane in 1852 and the suspended roadway dropped into the river.[1,2,5]

In 1848, Charles Ellet, Jr., built the record-breaking 1010 ft span suspension bridge over the Ohio River at Wheeling, West Virginia. Six years later, in 1864, the structure was destroyed in what was described as a hurricane. The bridge deck was completely destroyed and 10 of its 12 cables were pulled from their anchorages. The following account of the failure appeared in the *Wheeling Intelligencer*.

For a few moments we watched it with breathless anxiety, lunging, like a ship in the storm; at one time it rose to nearly the height of the towers, then fell, and twisted and writhed, and was dashed almost bottom upward. At last there seemed to be a determined twist along the entire span, about one-half of the flooring being nearly reversed, and down went the immense structure from its dizzy height to the stream below, with an appalling crash and roar.

For a mechanical solution of the unexpected fall of this stupendous structure, we must await further developments. We witnessed the terrific scene and saw that it was brought about by the violence of the gale. The great body of the flooring and the suspenders, forming something like a basket swung between the towers, was swayed to and fro, like the motion of a pendulum. The cables on the south side were finally blown off the apex of the eastern tower, retaining their position on the opposite side of the river. This destroyed the equilibrium of the swinging body; and each vibration giving it increased momentum, the cables, which sustained the whole structure, were unable to resist a force operating on them in so many different directions, and were literally twisted and wrenched from their fastenings.

Steinman commented on the Wheeling newspaper account as follows:

The newspaperman who wrote the foregoing dramatic account unknowingly summarized the crux of the aerodynamic phenomenon he had observed when he used the significant phrase: "Each vibration giving it increased momentum." And when he stated that the mechanical solution of the failure "must await further developments," he wrote better than he knew.

A structure with a span of 1043 ft. was constructed in 1850 over the Niagara River between Lewiston, New York, and Queenston, Ontario. To partially stabilize the bridge against motion, it was built with guy cables extending from the roadway deck to the sides of the gorge. Because these guys were jeopardized by an ice jam which had formed upstream during the winter of 1863–64, it was decided that they would be temporarily removed in the spring when the ice jam would break up. While the guys were detached, a heavy wind arose and the entire suspended system was destroyed, leaving only the cables which stood for another 34 years but were never used.[2]

The 1268 ft. Niagara-Clifton Bridge at Niagara Falls was built in 1868 by Samuel Keefer and rebuilt by G. W. McNulty in 1888. Seven months later on the night of January 9, 1889 it was completely destroyed by wind action. An eyewitness described the bridge as rocking "like a boat in a heavy sea," tipping up "almost on its very edge." By morning the structure was destroyed. It lay a crumpled mass of steel and wire bottom side up beneath the waters of the river below.[5]

At the time of its construction the Tacoma Narrows Bridge had the third longest center span of any type bridge constructed. It had a main span of 2800 ft. It was stiffened by a shallow plate girder and had a narrow roadway width such that the ratio of its length to width was much larger than any similar contemporary structure. Early in the morning of November 7, 1940 a fairly strong wind arose which caused some movement in the structure. By 7:30 or 8:00 A.M. the wind velocity increased to 38 miles per hour (verified by an anemometer reading taken at the time). The main span was oscillating in a vertical mode of moderate amplitude at a frequency of approximately 36 cycles per minute. The vibration continued in this manner until about 10:00 A.M. when the velocity increased to 42 miles per hour. At this point there was a drastic change in the motion. The frequency of motion changed from 36 cycles per minute to 12 cycles per minute. The changed motion was in a violent torsional mode with angular distortion reaching an amplitude of approximately 45 degrees from the static position. The vertical acceleration was rapidly approaching that of gravity. The torsional motion of the structure is illustrated in Fig. 12.1. The structure withstood this torsional deformation for about an hour when it finally yielded and a section of deck and laterals near the center of the span dropped out. The motion decreased momentarily and then built up again. In a few minutes the rest of the deck tore away from the cables and fell into the water. As a result of the loss of dead weight in the center span, the center cables rose up and the side spans sagged deeply, causing the towers to deflect shoreward and the motion to die out.[2]

The Tacoma Narrows Bridge was observed to have undesirable motions during and immediately after construction. It was recognized that some correction was necessary, and there were suggestions for stop-gap provisions. However, no one recognized the catastrophic potential. It was assumed there was plenty of time to make laboratory studies and arrive at corrective measures in due course.

The authors wish to emphasize that the designer of the Tacoma Narrows Bridge, Leon Moisseiff, was one of the most highly regarded and forward thinking bridge engineers of his day. At the time, aerodynamic considerations for bridge designs were unheard of. Mr. Moisseiff was a pioneer in a bridge design concept which was not yet in the structural research laboratories and certainly not in the average bridge designer's offices.

### 12.1.2    Adverse Vibrations in Other Bridges

Obviously not all suspended bridge structures have been destroyed or badly damaged as a result of wind forces. However, a few structures have been observed to have undesirable oscillations as a result of wind-induced vibration. They are listed in Table 12.2.[1]

The Fyksesund Bridge in Norway is a rolled I beam stiffened bridge located at a site where high winds are frequent. This structure was re-

TABLE 12.2. Modern Bridges which Have Oscillated in Wind

| Bridge | Year Built | Span (ft) | Type of Stiffening |
|---|---|---|---|
| Fyksesund (Norway) | 1937 | 750 | Rolled I-beam |
| Golden Gate (U.S.A.) | 1937 | 4200 | Truss |
| Thousand Island (U.S.A.) | 1938 | 800 | Plate girder |
| Deer Isle (U.S.A.) | 1939 | 1080 | Plate girder |
| Bronx-Whitestone (U.S.A.) | 1939 | 2300 | Plate girder |
| Long's Creek (Canada) | 1967 | 713 | Plate girder |

ported to have oscillations of 1 to 3 nodes in the center span and amplitudes of plus or minus 2 ft $7\frac{1}{2}$ in. In 1945 stays were added below the structure and have apparently been effective in controlling the motion.[1]

The Golden Gate Bridge is the only structure listed in Table 12.2 that has a stiffening truss. Motions were observed twice. Vibration was observed on February 9, 1938, but the wind velocity was not recorded. The second occurred during a storm on February 11, 1941 with a recorded wind velocity of 62 mph. On both occasions the amplitude of vibration was estimated at 2 ft.[1,8]

The Thousand Island Bridge project connecting the United States and Canada consists of two suspension bridges. The span on the American side is 800 ft while that on the Canadian side is 750 ft. Both structures are stiffened by plate girders. Frequent mild motion has been reported but apparently has been effectively corrected by the addition of guys.[1,9]

Before the Deer Isle Bridge was completed in 1939 it had to be stabilized against wind-induced vibrations by the addition of diagonal stays, which joined the stiffening girder at the pylon to the main cables in the center and side spans. However, serious damage, including breaking of some of the stays, was incurred during storms occurring during the winter of 1942–43. A more extensive system of longitudinal cable-to-girder diagonals and transverse stays has since been installed and no excessive motions have been reported.[1]

Ever since the new floor system was installed on the Bronx-Whitestone Bridge crossing the East River in New York it has had a tendency for mild vertical motions. The vibrations have not been serious but have been observed by those crossing the structure. A single-stay system was installed to reduce the vibrations, but the structure was not effectively stiffened until truss members in the plane above the existing plate girder were installed in 1946. The plate girder then became the bottom chord of the truss.[1,10,11]

Shortly after its erection in 1967, the Long's Creek Bridge, located 20 miles west of Fredericton on the Trans-Canada Highway, was subjected to wind-induced vibrations. This structure is a cable-stayed bridge with a main span of 713 ft. The deck structure consists of an orthotropic deck forming the top flange for two 8 ft deep inverted T-plate girders, 33 ft on center. The vibration frequency was observed to be 0.6 cycles per second with the amplitude of vibration reaching as much as 8 in. when the handrail was blocked with snow. The vibration was in the shape of a half-sine wave. When wind tunnel tests were undertaken to study the damping of the vibration, ten boxes, 6 × 6 × 1 ft. 6 in. deep, filled with rocks were hung from the bottom flange of the structure and submerged in the water. The addition of a soffit plate on the bottom changed the cross section into a box of increased torsional stability and the vibration was diminished by 40%. The addition of triangular edge fairing on the outside of the structure virtually eliminated the vibration. An interesting fact is that an essentially twin structure, 10 miles upstream from the first structure, had no observed motion because the structure is sheltered by a high embankment.[12,13]

### 12.1.3 Lessons from History

The failure of the Tacoma Narrows in 1940 startled and shocked the engineering profession. As indicated above, the phenomenon was not without precedent. However, the failures of the past had been discounted; a later generation of engineers, not recognizing the lessons of history, espoused the virture of flexibility without careful consideration of its hazards.

In the recent past there have been a number of innovations in structural analysis, materials, fabrication and erection procedures. These include the use of high-strength steels, welding, orthotropic decks, box girders, computer technology, cable stays, monocable suspension, transverse as well as inclined hangers, cantilever construction, prestressed concrete, parallel wire strands and many other technological advances. As a result, recent bridge structures have larger dimensions and increased flexibility and decreased dead weight and damping characteristics. Reduction of dead weight produces a magnification of wind effects relative to the inertia of the structure; increased flexibility decreases the natural frequency of vibration. Thus, modern fabrication techniques have decreased the structure's ability to absorb energy by sliding friction between component parts and thus less energy is required to initiate and maintain vibration.[14]

Therefore. recently constructed structures are more sensitive not only to static wind effects but to dynamic effects as well. Some existing and relatively recent structures have been so affected by wind oscillations that they required additional reinforcement. As a consequence, wind forces have taken on an increased significance and can be a major problem in cable-supported bridge systems. Serious consideration of these forces is required of the designer. If we, as a new generation of engineers, do not take cognizance of these forces then history will repeat its cycle and we will suffer another series of disasters culminating in future disasters as shocking and as dramatic as the Tacoma Narrows failure.

## 12.2 WIND ENVIRONMENT

Before wind instability studies are conducted for a particular bridge, it is important to estimate the wind environment at the particular site. It is desirable, in the determination of wind action on a suspended bridge structure, to obtain information of strong wind activity at the site over a period of years. Required are the wind velocity, direction, and frequency. This type of data is generally obtainable from meteorological records of the U.S. Weather Bureau and similar local weather records. However, these data are generally recorded at an airport or federal building at a nearby city which may be some distance from the bridge site. These records should be carefully used because the effects of the terrain at the instrument location may be somewhat different from those at the bridge site.

From meteorological data it is possible to plot high wind speeds and probability of occurrence. With this information, statisticians can estimate the highest expected winds, their expected direction, and their recurrence interval.

### 12.2.1 The Natural Wind

The composition of the wind in the atmosphere is a complex subject which is beyond the scope of this text. However, it is important for the bridge engineer to have a basic knowledge of the nature of wind. Wind may be defined in an elementary manner as air movement caused by atmospheric pressure differentials which occur over the earth's surface. Pressure differentials are produced by complex atmospheric processes from temperature differentials resulting from solar radiation. Solar radiation is strongest at the equator and weakest at the poles along with radiation away from the earth. Pressure differences are indicated on weather maps by isobars or contour lines of equal barometric pressure. Differences in pressure produce acceleration of air particles. Additional accelerators are geostrophic acceleration, resulting from the curvature and rotation of the earth, and centripetal acceleration. The sum of these accelerations is a movement in the free air unaffected by friction at the ground surface. The velocity of this free air is referred to as a gradient velocity, $\overline{V}_G$. The height above the earth's surface at which the gradient velocity is attained, usually between 1000 and 2000 ft, is referred to as the gradient height, $z_G$. At heights above this value the wind velocity is regarded as constant with height.[15,16,17]

The concern of structural engineers is with wind velocity at or near the earth's surface. At this level the wind velocity is affected by the surface friction, which is a function of surface roughness or terrain conditions. Davenport[16,17] has suggested that the mean wind speed $\overline{V}_z$ at a height $z$, less than $z_G$, may be related to the gradient velocity $\overline{V}_G$ by a power law relationship of the form

$$\frac{\overline{V}_z}{\overline{V}_G} = \left(\frac{z}{z_G}\right)^a \tag{12.1}$$

where the gradient height $z_G$ and the exponent $a$ are a function of the surface roughness. Suggested values are presented in Table 12.3 with a graph-

**TABLE 12.3.  Surface Roughness Influence on Gradient Height and Power Law Exponent**

| Surface Condition | Gradient Height $z_G$ (ft) | Power Law Exponent $a$ |
|---|---|---|
| Open water surface | 900 | 0.12 |
| Open flat land (open grass, prairie or farm land with few obstacles; shores, desert) | 900 | 0.16 |
| Forest and suburban (uniformly covered with obstacles 30–50 ft. in height) | 1300 | 0.28 |
| Urban areas (large and irregular objects, centers of cities) | 1700 | 0.40 |

From Davenport.[16,17]

ical representation shown in Fig. 12.4 for a uniform wind gradient of 100. The term mean wind speed is generally assumed to be an average of the wind speed over a given time period which may vary from one hour to ten mintues duration.

The above discussion is applicable to relatively flat terrain. The wind velocity can be significantly modified by hilly or mountainous terrain or by

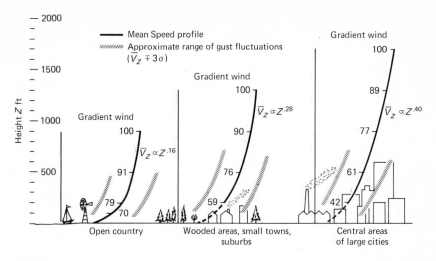

**FIGURE 12.4.   Effect of surface roughness on mean wind velocity. (Courtesy of A. G. Davenport. Ref. 16.)**

abrupt changes in terrain roughness. These effects are discussed in detail by Davenport.[16] The terrain roughness is also significant when trying to assess the wind environment at a bridge site from records made some distance away.

The interaction of the natural wind with the surface roughness or friction of the earth's surface produces a wind character that is gusty or turbulent as opposed to being smooth and uniform. Turbulence or gusts produce velocity fluctuations that are spatial and temporal in character. That is, the wind force acting on a structure will vary in direction as well as magnitude in a vertical as well as horizontal direction at any point in time and will also vary with time.

Terrain conditions may also significantly effect the design wind considerations for a particular site, depending on the structure's exposure. A bridge spanning a large river may be exposed to strong steady winds in a direction perpendicular to its longitudinal axis and turbulent winds parallel to the structure.

### 12.2.2   Design Wind Velocity

In the structural design for wind resistance of any structure a fundamental problem is the estimation of the expected wind velocity. It has been previously mentioned that from a statistical interpretation of meteorological records it is possible to determine wind velocity, direction, and frequency.

From this meteorological information, a local mean velocity or fundamental wind velocity, $V_f$, is obtained, with reference to some arbitrary datum or height, $z_f$, which is often at 33 ft (10 m). The structure may then be designed for this fundamental wind velocity with proper modification to account for spatial distributions as functions of the dimensions of the structure.

### 12.2.3    Fundamental Wind Velocity

The definition of a fundamental wind velocity may be stated as the expected value of a maximum mean wind velocity of 10 minute duration, at a height above the water or ground of 33 ft (10 m.), recurring at a specified time interval. The value of the time interval is the expected life span of the structure. For bridge structures it is recommended that the fundamental wind velocity be determined for a minimum time interval of 100 years.

### 12.2.4    Design Wind Velocity at Structure Altitude

Design conditions at a particular site will require that the altitude of the structure (bridge deck elevation) be set at some predetermined value, for example, the height above water for navigation clearance. The design wind velocity at the specified height may be determined by the power law relationship given by equation 12.1, where the fundamental wind velocity and height, $V_f$ and $z_f$, are substituted for the gradient velocity and height, $\overline{V}_G$ and $z_G$, respectively.

$$\frac{V_z}{V_f} = \left( \frac{z}{z_f} \right)^a \tag{12.2}$$

The power law exponent $a$ should be determined from observations at the bridge site; however, where this is impossible, the values presented in Table 12.3 may be used.

### 12.2.5    Effect of Structure Length on Design Wind Velocity

In the previous section, a design wind velocity was determined for some datum elevation or height of the structure, such as the elevation of the bridge deck. If it is assumed that the direction of the wind is perpendicular to the longitudinal axis of the bridge, then the wind velocity will vary from point to point along the length of the bridge and will further vary with time.

The wind speed $V_z$ is the mean or average speed over some time period, say 10 minutes. As a result of surface roughness, a turbulence or gust effect is produced in the natural wind such that for some interval of time $S$ within this 10 minutes, the actual wind velocity $V_s$ will be larger than the mean velocity $V_z$. The mean wind velocity may be considered as a constant wind velocity uniformly distributed across the length of the bridge. Because of the gust effect of the wind, an increased velocity is randomly dispersed and superimposed on the mean velocity along the structure length.

Hirai and Okubo[18] have reported the following empirical formula of Ishizaki and Mitsuda,[19] based upon observations of Sherlock[20] and Deacon[21]

$$\frac{V_s}{V_z} = \left(\frac{S}{600}\right)^{-r}; \qquad r = 0.09 \left(\frac{z}{12.2}\right)^{-0.42} \qquad (12.3)$$

where $V_s$ and $V_z$ are wind velocities at a height $z$ (in meters) for the time duration $S$ in seconds. If the value of $z$ is expressed in feet in the equation for $r$ above, then the term $z$ should be divided by the constant 3.72 instead of 12.2.

The width of the turbulence $L$ in the direction of the wind is stated as the product of $V_s$ and $S$. If the horizontal width of the turbulence along the length of the bridge and perpendicular to the wind direction is taken as $B$, then $L$ may be related to $B$ by some factor $k$ such that

$$L = kB = V_s S \qquad (12.4)$$

then from equations 12.3 and 12.4

$$M_1 = \frac{V_s}{V_z} = \left(\frac{kB}{600 \, V_z}\right)^{-r/(1-r)} \qquad (12.5)$$

Assuming that the duration $S$ is determined by equating $B$ to the horizontal length of the structure, the modification factor $M_1$ for the design wind velocity can be evaluated. Thus the design wind velocity along the length of the bridge can be determined as the product of the modification factor and the $V_z$ term determined from equation 12.2.

Hirai and Okubo[18] assumed a value of 7.0 for $k$ in equation 12.5 and a value of 40 m/sec for $V_f$ and 0.16 for $a$ in equation 12.2. The value of $z_f$ was taken as 10 m. Based on these values, corresponding values for $M_1$ are shown in Table 12.4. From this table it can be seen that the effect of the spatial distribution of gusts decreases as the length and elevation or height of the structure increases.

### 12.2.6  Effect of Structure Height on Design Wind Velocity

In the same manner that the wind velocity varies along the length of the deck of a bridge structure, it will also vary along the height of the pylon.

However, the determination of a modification factor for height, $M_h$, is considerably more complex than that for the length of the structure. Empirically, Hirai and Okubo[18] have suggested that a height modification factor may be determined in the same manner as for length by substituting the height $H$ of the structure for the length $L$ (equating $H$ to $B$ in equation 12.5) and $H/2$ for the structure altitude $z$. In this manner the modification factor for height can be determined similarly to that determined for length. Values for the height modification factor are given in Table 12.5, for the same conditions as those evaluated for length in Table 12.4. From this tabulation it can be seen that the effect of spatial distribution of gusts decreases as height increases.

### 12.2.7   Wind Force and Angle of Attack

The wind force assumed on an object is normally not in the same line as the direction of the actual wind. In conventional aerodynamic analysis (e.g., airfoil design), the wind force is divided into components: drag and lift parallel and perpendicular to the wind direction (Fig. 12.5). This same convention may be applied to a bridge deck (Fig. 12.6), wherein the resultant wind is oriented to the structure by the angle of attack $\alpha$ assumed to be positive when striking the section from the underside. It is convenient in considering wind effects on bridge structures to consider lift acting perpendicular to the normal position of the bridge deck and drag to be acting parallel to the normal position of the bridge deck (Fig. 12.7). The application of the orientation criteria requires the conversion of the wind tunnel test data to the changed orientation of the lift and drag forces.

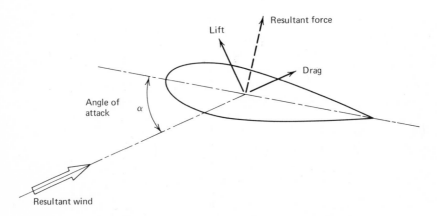

**FIGURE 12.5.   Force components on an airfoil section.**

**TABLE 12.4.  Modification Factors $M_1$ for Horizontal Lengths of Structures**

| $z(m)$ | B(m) | | | | | | | | | |
|---|---|---|---|---|---|---|---|---|---|---|
| | 150 | 300 | 450 | 600 | 750 | 900 | 1050 | 1200 | 1350 | 1500 |
| 10 | 1.400 | 1.300 | 1.244 | 1.206 | 1.177 | 1.155 | 1.136 | 1.119 | 1.105 | 1.093 |
| 30 | 1.240 | 1.186 | 1.155 | 1.133 | 1.117 | 1.104 | 1.093 | 1.083 | 1.075 | 1.068 |
| 50 | 1.196 | 1.153 | 1.129 | 1.112 | 1.099 | 1.088 | 1.080 | 1.072 | 1.065 | 1.059 |
| 70 | 1.168 | 1.132 | 1.112 | 1.097 | 1.086 | 1.078 | 1.070 | 1.064 | 1.058 | 1.053 |
| 100 | 1.144 | 1.114 | 1.097 | 1.085 | 1.076 | 1.069 | 1.062 | 1.057 | 1.052 | 1.048 |

**TABLE 12.5.  Modification Factors $M_h$ for Vertical Heights of Structures**

| $H(m)$ | 30 | 40 | 60 | 80 | 100 | 120 | 140 | 160 | 180 | 200 | 220 | 240 |
|---|---|---|---|---|---|---|---|---|---|---|---|---|
| $M_h$ | 1.500 | 1.432 | 1.316 | 1.256 | 1.221 | 1.192 | 1.171 | 1.157 | 1.144 | 1.132 | 1.121 | 1.114 |

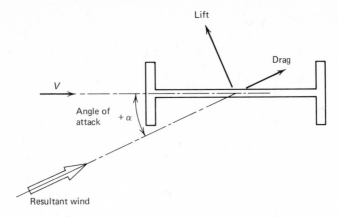

FIGURE 12.6.    Force components on a bridge section.

The wind also produces an angular motion or torsional moment force acting on the cross section.

When evaluating wind forces on a structure, the possible direction of critical wind velocity should be determined. In plan it is generally assumed that the critical wind direction is perpendicular to the longitudinal axis of the bridge. An obvious question arises as to the maximum value of the angle of attack $\alpha$ that should be considered on the deck cross section. The angle of attack is a function of wind velocity and site conditions. Data for the relationship between maximum observed angle of attack and wind velocity were made at the site of the Severn Bridge[22] and are plotted in Fig. 12.8. Preliminary data obtained by the Federal Highway Administration on the Newport Suspension Bridge during hurricane Doria is also

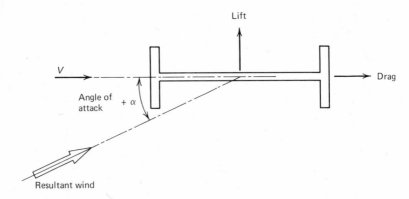

FIGURE 12.7.    Reoriented force components on a bridge section.

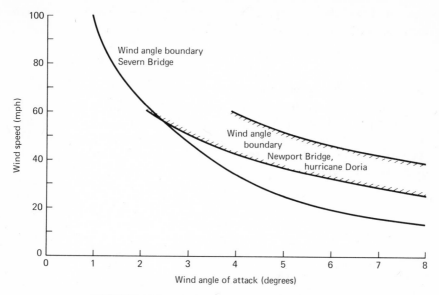

**FIGURE 12.8.    Angle of attack versus wind speed.**

plotted for maximum and minimum values in Fig. 12.8. From these curves it can be seen that the angle of attack decreases with increasing wind speed. Therefore, at lower wind speeds the structure must be made stable for larger values of the angle of attack. The curves shown in Fig. 12.8 may serve as a guide, but are not necessarily applicable to other sites and possibly may impose unnecessary constraints to the analysis.

Wind forces are dynamic considerations because they represent the effect of a moving fluid around a cross section. A common design assumption is to separate wind effects into two major classifications, static and dynamic effects. Under an idealized condition which never occurs in practice, an object subjected to a wind stream of constant velocity and direction that does not vary with time may be thought of as being subjected to static effects. Dynamic effects of flow around an object arise from turbulence in the natural wind, vortex separation, and changes in the flow pattern as a result of the movement of the object. Thus, these mechanisms cause time-dependent variations of the wind force.

## 12.3   WIND EFFECTS—STATIC

The bridge structure should be designed for static as well as dynamic wind effects. Static wind loads are derived from an assumption of a steady, uniform wind, and the lift, drag, and moment forces may be determined from the following basic equation:

$$F = \frac{1}{2} \rho V^2 C A \tag{12.6}$$

where $F$ is the lift or drag force, $C$ is a dimensionless lift or drag coefficient as a function of the angle of attack $\alpha$, $A$ is the projected (frontal) area exposed to the wind, $V$ is the wind speed normal to the bridge, and $\rho$ is the density of air (mass per unit volume assumed to be 0.002378 slugs/ft$^3$ or 0.0766 lb/ft$^3$ at sea level and at 15°C). Thus, the lift and drag forces are stated as:

$$L = \frac{1}{2} \rho V^2 C_L A \tag{12.7}$$

$$D = \frac{1}{2} \rho V^2 C_D A \tag{12.8}$$

and for a unit length of span, equations 12.7 and 12.8 become

$$L = \frac{\rho V^2 C_L A}{2S} \tag{12.9}$$

$$D = \frac{\rho V^2 C_D A}{2S} \tag{12.10}$$

where $S$ is the span length.

Under steady conditions, the lift force will generally be displaced from the axis of rotation of the section and cause a moment or torque about the axis of rotation. This moment can be expressed as

$$M = \frac{1}{2} \rho V^2 C_M A B \tag{12.11}$$

and for a unit span length

$$M = \frac{\rho V^2 C_M A B}{2S} \tag{12.12}$$

where $B$ is the deck width.

The magnitude of these forces will vary with changes in the angle of attack and with the cross-sectional shape. Because the effect of shape generally can be determined only by wind tunnel tests, the actual effects are usually not subject to control at the design stage. Empirical values, based on previous tests for similar cross sections, are normally used in design and then verified in wind tunnel tests. By testing properly scaled models in a wind tunnel, it is possible to obtain scaled forces of lift, drag, and tor-

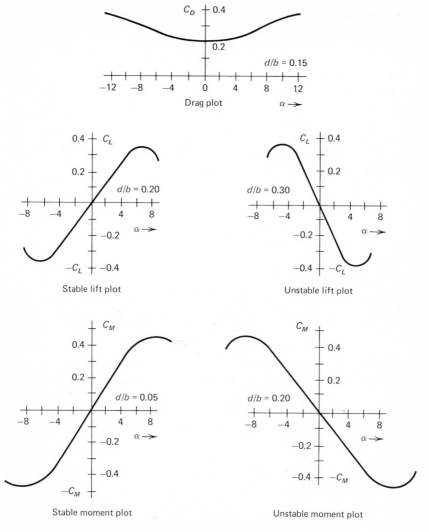

**FIGURE 12.9.   Lift, drag, and moment plots. (Ref. 23.)**

sional moment. Typical nondimensional plots of lift, drag, and moment coefficients for varying angles of attack are shown in Fig. 12.9.[23,24]

## 12.3.1   Lateral Buckling

The lateral out-of-plane buckling of a bridge deck may be visualized as a simple beam loaded by the wind force. The wind force acts approximately along the beam's centroidal axis in the direction of the drag force;

it has a lateral bending stiffness $EI_l$ and torsional stiffness $GJ$.[25] A solution for the critical lateral uniform buckling load based on the above assumptions is attributed to Prandtl and is given as:

$$q_{cr} = \frac{28.3[EI_lGJ]^{1/2}}{S^3} \tag{12.13}$$

where $S$ is the span. Equation 12.13 presents a first estimate of the uniform load $q_{cr}$, that will produce a lateral buckling of the bridge deck. From equation 12.10 the corresponding critical wind velocity, normal to the section, can be determined as:

$$V_{cr} = \left[\frac{2q_{cr}}{\rho C_D(A/S)}\right]^{1/2} \tag{12.14}$$

### 12.3.2   Torsional Divergence

The cross section of a bridge deck may twist under the wind action as a result of excessive lift and/or drag loads, which increase the angle of attack causing an increased twisting moment in the deck.[25] On a truss bridge, this phenomenon is analogous to an overturning moment such as that which caused the failure of the Chester, Illinois, bridge over the Mississippi River (July 29, 1944).[5]

A simplified analysis of torsional divergence may be derived by considering an element of the bridge deck at midspan.[25] The twisting moment per unit length of span due to wind is determined by equation 12.12 as:

$$M = \frac{\rho V^2 C_M AB}{2S}$$

The value of $C_M$ as a function of the angle of attack may be approximated from a moment plot (Fig. 12.9) as

$$C_M = a\alpha + b \tag{12.15}$$

where $a$ is the slope of the moment curve and $b$ is the intercept of $C_m$ at $\alpha$ equal to zero.

$$a = \frac{dC_M}{d\alpha} \qquad b = C_{Mo}$$

substituting equation 12.15 into 12.12 results in

$$M = \frac{1}{2}\rho V^2 \left(\frac{dC_M}{d\alpha}\alpha + C_{Mo}\right)\frac{AB}{S} \tag{12.16}$$

and represents the moment force acting on the section. The resisting moment per unit span is related to the twist angle $\alpha$ by

$$M_r = k\alpha \tag{12.17}$$

where $k$ is the structural stiffness coefficient obtained from the torsional stiffness properties of the deck structure.

Equating equations 12.16 and 12.17 results in a linear relationship in $\alpha$ given by

$$k\alpha = \frac{1}{2}\rho V^2 \left(\frac{dC_M}{d\alpha}\alpha + C_{Mo}\right)\frac{AB}{S} \tag{12.18}$$

which can be rewritten in the form

$$\alpha = \frac{1/2\rho V^2 C_{Mo}AB/S}{k - 1/2\rho V^2 (dC_M/d\alpha)(AB/S)} \tag{12.19}$$

In equation 12.19, as the denominator approaches zero, the value of $\alpha$ approaches infinity. The torsional divergence velocity can thus be defined as

$$\frac{1}{2}\rho V^2 \frac{dC_M}{d\alpha}\frac{AB}{S} = k$$

$$V = \left(\frac{2k}{(dC_M/d\alpha)(AB/S)}\right)^{1/2} \tag{12.20}$$

The above derivation considered only an element at the midspan of the ·structure. For an actual structure the solution must consider each element along the span simultaneously. Scanlan[25] has presented the following solution.

When assuming a uniform wind velocity across the deck section and along the span, the twisting moment force applied by the wind to all elements of the deck is given by equation 12.16. The torsional resisting moment (equation 12.17) for the total span is given by

$$\{M_r\} = [K]\{\alpha\} \tag{12.21}$$

where $[K]$ is a symmetrical torsional stiffness matrix and $\{\alpha\}$ is a column matrix of torsional deformation (angle of attack) for $N$ positions along the span.

By equating equations 12.16 and 12.21 and letting

$$\lambda = \frac{1}{2}\rho V^2 \left(\frac{dC_M}{d\alpha}\right)\frac{AB}{S} \tag{12.22}$$

the following matrix relationship is obtained

$$[K]\{\alpha\} = \lambda\{\alpha\} + \left(\frac{1}{2}\rho V^2 C_{Mo}\frac{AB}{S}\right)\{1\} \tag{12.23}$$

where $\{1\}$ is a column matrix of ones

Equation 12.23 may be rewritten as

$$([K] - \lambda I)\{\alpha\} = \lambda\{\alpha_0\} \tag{12.24}$$

where

$$\alpha_0 = \frac{C_{Mo}}{dC_M/d\alpha}.$$

By substituting $\beta = \alpha + \alpha_0$, equation 12.24 may be rewritten as

$$([K] - \lambda I)\beta = [K]\{\alpha_0\} \tag{12.25}$$

To solve for $\beta$ the evaluation of $([K] - \lambda I)^{-1}$ is required. However, as before, the value of $\beta$ grows without bound as the determinate $([K] - \lambda I)^{-1}$ approaches zero. By setting this determinate to zero the critical values of $\lambda$ can be determined. Thus, the smallest root $\lambda_r$ which is a solution to

$$([K] - \lambda I)^{-1} = 0 \tag{12.26}$$

results in the determination of critical wind velocity for torsional divergence

$$V_{cr} = \left[\frac{2\lambda_r}{\rho(dC_M/d\alpha)(AB/S)}\right]^{1/2} \tag{12.27}$$

Normally, torsional divergence is not a problem because the deck structure usually has adequate torsional stiffness as a result of other structural design considerations. However, for conventional suspension or cable-stayed structures that are very slender and have low torsional stiffness, the torsional divergence may manifest itself.

### 12.3.3   Turbulence Effects

In the previous two sections the presentation was based on the assumption that the wind was uniform in intensity along the length of the structure. In fact, as previously pointed out (Section 12.2.1), natural wind is not steady but turbulent in character. Consequently, the wind pressure at all points along the span will not be constant at any given moment. As a result, some authorities assume that the average pressure is less than that of the mean. Thus, as discussed above, the static phenomenon wherein the wind pressure is considered uniform is assumed to be a conservative approach. However, as discussed in Sections 12.2.5 and 12.2.6, tentative specifications require an increase in the mean velocity (Tables 12.4 and 12.5).

## 12.4 VIBRATION

A vibration is a reciprocating or oscillating motion that repeats itself after an interval of time, Fig. 12.10. This interval of time is referred to as the *period* of vibration and is measured as a unit of time per cycle, such as seconds per cycle. The *frequency* of vibration is numerically equal to the reciprocal of the period and is measured as the number of cycles per unit of time, cycles per second. The maximum ordinate of the curve of Fig. 12.10 represents the maximum displacement of the vibrating body from its position of equilibrium and is referred to as the *amplitude* of vibration.

All structures have a *natural frequency* such that when an external dynamic force acting on the structure comes within the natural frequency range, a state of vibration may be reached whereby the driving force frequency and the body's natural frequency are in tune, a condition referred to as *resonance*. At resonance the structure undergoes violent vibration, often resulting in major structural damage. A familiar example of resonance is the shattering of a glass by a particular musical note or sonic vibration.

When an external dynamic force (wind) is applied to a mass, such as a cable or bridge deck, the mass will be set into a vibratory motion. This vibratory motion can be represented by an infinite number of superim-

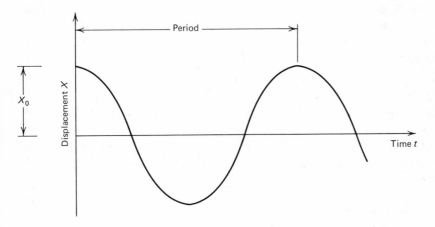

**FIGURE 12.10.  Undamped free vibration starting from an initial displacement.**

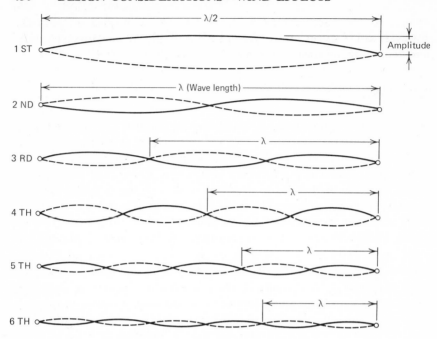

**FIGURE 12.11.    Modes of vibration.**

posed modes of vibration, Fig. 12.11. If one of these modes coincides with the natural frequency of one of the lower modes, and if the imposed force has a periodic component of the same frequency, then resonance will occur. The amplitude or displacement will be increased and the effect of the pulsating force may also increase. This is why marching troops break step when crossing a bridge. The two dynamic response phenomena associated with flexible systems are forced vibrations and self-induced vibrations, which are discussed in subsequent sections.

### 12.4.1   Free Vibrations

When an externally applied disturbance acting on a structure is removed, the structure will respond to the removal of the excitation by vibrating. When a mass is displaced and then released, for example, as a result of an initial velocity produced by an impulse or impact such as a moving load, motion will occur as the result of the initial displacement.

### 12.4.2   Forced Vibrations

Forced vibrations are those produced by a time-dependent externally applied force that is impressed on the structure; examples are wind, or displacements caused by earthquakes. The magnitudes of these forces or externally applied displacements are independent of the motion of the structure. The response of the structural system diminishes with time as a result of damping. While the excitation force is active, there is a forced vibration. When the excitation is removed the response is a free vibration.

### 12.4.3  Self-Excited Vibrations

Self-excited vibrations are those caused by forces produced by the displacements or deformations of the system. Forces causing these vibrations cease when the motion stops, whereas in forced vibrations the external force is independent of the motion. In this type of vibration, the structure's own movement is exerting an additional energy to that of the exciting force, Fig. 12.12.

### 12.4.4  Damping

The vibratory motion indicated in Fig. 12.10 is classified as undamped vibration and theoretically would continue indefinitely. Vibrations can be diminished if a damping force can be provided that will act in a direction opposite to the movement of the structure. In aerodynamic vibration, the wind is exerting energy into the structure. This energy is then dissipated by the internal frictional resistance of the molecules of the material, the drag effect of the medium surrounding the structure, or dry frictional resistance resulting from the slippage of structural connections between members or between the structure supports. When the cause of vibration is removed

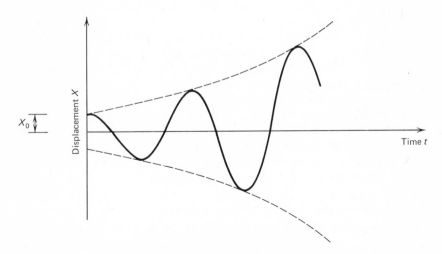

**FIGURE 12.12.    Self-excited vibration.**

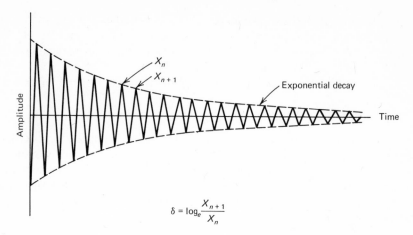

$$\delta = \log_e \frac{X_{n+1}}{X_n}$$

**FIGURE 12.13.   Exponential decay of a free vibration with viscous damping.**

from a vibrating structure, the vibration will decay or damp as a result of the dissipation of energy within the structure, Fig. 12.13. A piano string that has been set to vibrating by the striking of a hammer will gradually stop due to air resistance and internal friction.

The degree of structural damping is expressed as the logarithmic decrement $\delta$. It can be defined as the logarithm of the ratio of two consecutive peaks (amplitudes).[13,36]

$$\delta = \log\left(\frac{X_{n-1}}{X_n}\right) \tag{12.28}$$

see Fig. 12.13.

Lack of available data prevents estimation of damping in a bridge. The impracticability of exciting a large bridge prohibits the measurement of damping of existing structures. However, any slipping motion between component parts of a structure will increase the damping value because it helps to dissipate the energy. Thus, bolted connections which may slip a little will have better damping characteristics than welded connections. The friction developed by stringers sliding on transverse floor beams will develop a higher damping value than an orthotropic plate deck. From observations of movements on the shop welded, field bolted cable stayed Long's Creek Bridge, the logarithmic decrement $\delta$ was estimated to be 0.065.[12] Model tests of the Golden Gate Bridge were compatible with the prototype if a value $\delta = 0.031$ were assumed.[27] The logarithmic decrement was assumed as zero for the deck of the all-welded Severn Bridge;

however, the diagonal hangers were estimated as having a value of $\delta =$ 0.052. The actual damping value for the deck was used as "insurance" or taken as a bonus.[22] Damping devices are not normally built into bridges. However, as a temporary measure, weighted boxes were suspended from the deck of the Long's Creek Bridge and immersed in the water below as a means of damping the oscillations until permanent corrective measures could be applied (Section 12.1.2).

## 12.5  WIND EFFECTS—AERODYNAMIC

There are several mechanisms of interaction between wind and structure that produce a vibration in the structure, but only three of the mechanisms are important to bridge design. They are vortex shedding, flutter, and turbulence.

### 12.5.1  Vortex Shedding

If a steady wind blows against a cylinder or other obstruction, the wake consists of a special turbulence termed the Von Karman vortex street or, for brevity vortex shedding, Fig. 12.14. Vortexes are formed as a result of air flowing around a cross section and separating from the boundary of that section. The shedding of these vortexes on the leeward side is representative of forces acting on the cross section at right angles to the direction of the wind in an alternating and periodic fashion.

It is important to note, that the oscillations of the section are not produced by the vortexes. They are merely a convenient physical and mathematical indication from which the air flow around the cross section, and the resulting forces that are acting, may be inferred, formulated, and computed.[23,24,28]

Any bluff object will shed vortexes when placed in a wind stream. If a bridge deck represents a solid section, such as a plate girder or box girder, the entire deck, as a unit, will shed vortexes. Bridge towers, cables, and hangers will also produce a vortex trail.

The unsteadiness of forces produced by the wind flow around an object can be separated into three components: the time-dependent fluctuation of force resulting from the separation of the wind flow around the section, even in a steady flow; the unsteadiness due to the structure's own movement; and the unsteadiness or turbulence of the wind itself. These components are not necessarily independent and they frequently occur simultaneously.

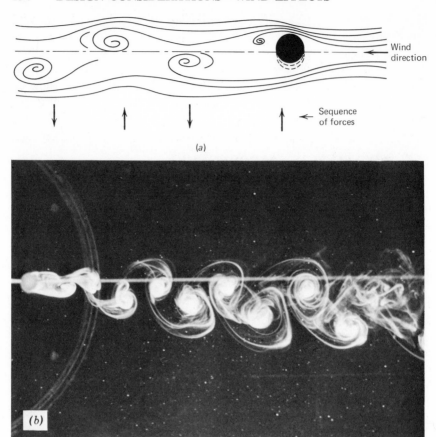

(a)

(b)

**FIGURE 12.14.** (*a*) Schematic vortex street; (*b*) vortex shedding behind a cylinder in water using fluorescein dye at a very low Reynolds number (between 2000 and 3000). (Courtesy of R. L. Wardlaw.)

Although the vortex shedding is a complicated phenomenon in fluid mechanics, the frequency of shedding can be expressed by the simple equation

$$S = \frac{fD}{V} \tag{12.29}$$

where $f$ is the frequency of vortex formation on one side of the wake, $D$ is the dimension of the body normal to the wind flow, $V$ is the wind velocity, and $S$ is the dimensionless Strouhal number. The value of the

Strouhal number is a function of the geometry of the section and a non-dimensional Reynolds number defined as:

$$R_e = \frac{\rho V D}{\mu} \tag{12.30}$$

where $\rho$ is the mass density of the fluid (air), $V$ is the wind velocity, $D$ is the dimension normal to the flow, and $\mu$ is the viscosity of the air. The ratio $\mu/\rho$ is defined as the kinematic viscosity of air $\nu$; thus the Reynolds number becomes

$$R_e = \frac{V D}{\nu} \tag{12.31}$$

The value of the Strouhal number for a given shape is reasonably constant over large ranges of the Reynolds number. The Strouhal number for circular cylinders has been experimentally established for a Reynolds number range of $10^2$ to $10^5$ as approximate values of 0.2 in a smooth flow and 0.25 in a turbulent flow. Experimental work at the Fairbank Highway Research Station wind tunnel by Robert A. Komenda has established a value of 0.22 for the Strouhal number of a helical strand cross section (thesis is unpublished at this writing). For square cross sections in natural wind, $S \approx 0.11$.

The frequency of the vortex trail is representative of the oscillatory force on the section. When the wind velocity is such that the frequency of the vortex trail corresponds to the natural frequency of the structure or member, it is possible for large amplitudes of vibration to occur. The magnitude of the amplitude will depend on the structural damping and geometry of the section. Slenderness and streamlining of a deck section reduces the tendency of the air flow to separate from the section and will, in effect, narrow the vortex wake, reducing the intensity of the vortexes and thus the magnitude of the oscillatory forces.[13]

The motion response of the structure will occur over a narrow band of wind velocity near the critical resonant value. After passing through resonance, increasing velocity will cause the exciting or vortex frequency to be larger than the natural frequency and the amplitude will decrease.

Wind tunnel tests on a model of the Long's Creek Bridge were conducted by Wardlaw,[12,26] in Canada, Fig. 12.15, and show that vortex shedding is limited in amplitude and is confined to narrow velocity ranges. It should be noted, however, that for flexible structures in a natural wind environment, the periodic vortex shedding frequency may be slightly altered by turbulence in the air stream and the structure's own motion.

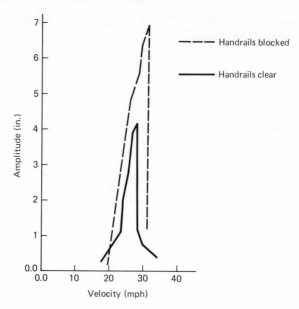

**FIGURE 12.15. Long's Creek Bridge, wind tunnel observation of vertical motion. Ref. 13.)**

Therefore, there is a tendency to arrive at the structure's natural frequency over a wide range of wind speeds.[14]

A contributing factor to the collapse of the original Tacoma Narrows Bridge was an approximately 39 ft long Karman vortex which occurred during a steady 42 mph wind. Oscillations had occurred for two hours during a 38 mph wind velocity. When the wind increased to 42 mph, the structure entered a torsional vibration mode, which had an angular rotation of about 45 degrees with the deck in each direction. This continued for another hour until failure finally occurred.

Analytical solutions for the elimination of vortex excitations are not available. Wind tunnel tests can indicate the type of cross sections which cause minimal excitations. Desirable cross sections are those that allow a laminar flow pattern around them.

### 12.5.2 Flutter

A self-induced vibration is produced by a change in the wind force as a result of the structures' own motion. If this vibration opposes the motion, then it is said to have a damping effect. If it adds to the motion, the oscillations can build up to dangerous amplitudes. Only a few sectional shapes are sensitive to this condition. However, all shapes develop a simultaneous aerodynamic coupling of torsional and vertical motion known as flutter, which lies between the natural frequencies of the structure for vertical flexure and torsion. As wind velocity increases, a critical velocity will be reached whereby flutter is incipient. As described for the Tacoma Bridge, flutter is characterized by a rapid buildup in amplitude with little or no increase in wind speed, and there is a distinct possibility that a catastrophic amplitude may be produced in a few cycles of motion.

Mass and the ratio of torsional to vertical bending natural frequencies are the factors that govern the wind speed that will cause flutter. No analytical methods are available to predict critical velocity of bridge deck structures. Bleich[29] has presented a method whereby critical velocities can be achieved for a flat plate, which might be considered an idealization of a bridge deck. Experience has indicated that bridge decks will have a lower critical velocity than a flat plate.

German[30] wind tunnel tests on various types of cross sections have produced sufficient data to experimentally determine shape factors in some instances and obtain a bridge critical velocity. This velocity is a function or percentage of the value for a flat plate with an equivalent inertia and elastic properties. Taylor[14] has published flutter speeds for various bridge cross sections, Fig. 12.16. They indicate that the percentage variation approaches 100 for streamline shapes and decreases to 20 for bluff, open shapes. However, some care must be exercised when generalizing these results. Where there is doubt, model tests should be used.[31]

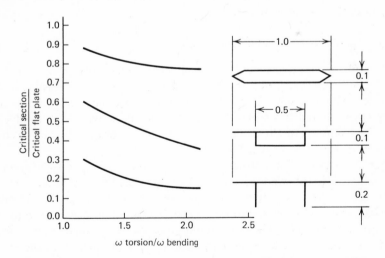

**FIGURE 12.16. Flutter speeds for various bridge cross sections. (Taylor, *Engineering Journal* (Canada), November, 1969, Ref. 14.)**

### 12.5.3   Turbulence

The natural wind is turbulent or gusty rather than smooth and uniform in character. This turbulence results in velocity fluctuations in both vertical and horizontal directions. These fluctuations are random in nature because they do not occur at a particular frequency but are distributed over a band of frequencies. The structure will respond to these random fluctuations when they occur at or near frequencies of the bridge. Wind acting on a structure will also have spatial variations; that is, distribution along the height and length of the structure will not be uniform. At any instant in time, the velocity will not be constant in direction or magnitude.

As a consequence of turbulence, vortex shedding excitation becomes more difficult to determine than that in smooth flow. Critical flutter velocities for vortex shedding are higher than those in smooth flow. The combined effects of vortex shedding and higher critical flutter velocities have a greater influence on the longer spans.

### 12.6   WIND TUNNEL TESTING

Because of nonexistent or cumbersome analytical procedures, wind tunnel tests are a convenient and economical method to establish aerodynamic characteristics and stability of a structure. There are two types of wind tunnel tests, the full model test simulating the atmospheric boundary layer, and the so-called static section model test. These will be discussed separately in the following sections.

### 12.6.1   Boundary Layer Full Model Test

A full model is a scaled down, 1:200 or higher, reproduction of a structure with suitably scaled dimensions, moments of inetria, and elastic characteristics. There is some doubt by engineers concerning the validity of testing at greatly reduced scales. There are scale effects within limits which cannot be ignored. Usually in this type of testing the terrain, as well as the structure, is usually modeled, Fig. 12.17.

Often the exposure of a structure to aerodynamic effects is significantly different for different wind directions. A bridge crossing a wide river may be exposed to strong steady winds perpendicular to the bridge, but turbulent winds may occur parallel to the bridge. The question of terrain roughness becomes important when attempting to asses the wind strength at the site of a structure from weather data recorded at a distance from the site. Davenport[32] has indicated in a recent investigation of the Narrows Bridge in Halifax that sectional model tests are not a reliable representation of

FIGURE 12.17.   The Narrows Bridge, Halifax, Canada, wind tunnel surfaces used in aeroelastic study. (Courtesy of A. G. Davenport.)

the total stability of the structure. However, tests of this type may be useful to determine general stability under conditions of partial erection. Full model tests of the Narrows Bridge will be discussed in Section 12.10.6.

### 12.6.2   Sectional Model Test

In a sectional model test only a representative portion of the bridge suspended structure is tested. The scale ratio may be in the range of 1:30 to 1:50. The scale test specimens are larger and modeling costs are lower than for the full model method. The model is mounted on springs to have the appropriate scaled mass, moment of inertia, and frequencies.

In a static wind tunnel test, the sectional model of the deck is subjected to various wind velocities at various angles of attack. The reaction forces of lift, drag, and moment are carefully measured. Static drag, lift, and moment curves are obtained by plotting the dimensionless coefficients $C_D$, $C_L$, and $C_M$, for various angles of attack, Fig. 12.9.

These plots are significant because the slope of the lift and moment curves indicate stability or instability of the section. The steeper the positive slope in the central range, the greater the stability, and, conversely, the steeper the negative slope, the greater the instability.[23]

In full model testing it is possible to test with a properly simulated turbulent flow. Techniques have not been developed as yet to properly simulate turbulent flow for the larger sectional model tests. It is normal practice to test sectional models in a steady flow on the assumption that the results are conservative for turbulent flow. The assumption of conservatism is based on the fact that turbulent flow reduces the susceptibility to vortex excitation and raises the critical flutter speed. However, this assumption is questioned by the writers of the tentative Japanese specifications for wind[18] (Sections 12.2.5 and 12.2.6).

### 12.6.3   Dynamic Similarity

So that the prototype structure and the model agree, there must be an equality of the several nondimensional parameters between prototype and model[12,13,33] indicated below by category, such as:

$$\frac{V}{N_y D}, \qquad \frac{V}{N_\theta D} \tag{12.32}$$

where $N_y$ and $N_\theta$ are the natural frequencies in vertical flexure and torsion respectively;

$$\frac{m}{\rho D^2}, \qquad \frac{J}{\rho D^4} \tag{12.33}$$

where $m$ and $J$ are the mass per foot of span and mass moment of inertia per foot of span, respectively, and $\rho$ is the air density, and

$$\delta_y, \qquad \delta_\theta \tag{12.34}$$

are the logarithmic decrement in vertical flexure and torsion, respectively.

In addition to the above parameters for agreement, there must be agreement of the center of gravity and the axis of torsional movement between prototype and model. The wind velocity scale is established from equation 12.32 such that

$$\frac{V_p}{V_m} = \frac{D_p}{D_m} \frac{N_p}{N_m} \tag{12.35}$$

where the subscripts $p$ and $m$ refer to the prototype and model, respectively.

### 12.6.4   Aerodynamic Similarity

The parameter that establishes the equivalency between prototype and model is the Reynolds number, $VD/\nu$. It is impractical, in wind model tests, for the Reynolds number of the model to achieve similarity to that of the prototype. However, it has been shown[34] in model tests conducted for the Severn Bridge that, for commonly used cross sections of sharp-edged bluff bodies, the flow similarity can be practically achieved even with large variations between model and prototype parameter values. In the Severn Bridge investigations it was found that the forces on the model are sensitive to Reynolds number below $R_e = 2 \times 10^6$.

However, at larger values it was found that the forces were virtually independent of the Reynolds number. The above investigations were conducted on sectional models. Therefore, the validity of aerodynamic testing on full models at greatly reduced scales is questioned by some engineers because there are some limits for which the scaling effects can no longer be safely ignored.[13]

## 12.7    STABILITY OF STAYED-GIRDER BRIDGES

Having established some degree of understanding of wind forces and their effects, let us now investigate and compare the aerodynamic capabilities of a cable-stayed system with that of a conventional suspension system. The modes of vibration of a suspension system are the symmetric and antisymmetric modes, Fig. 12.18.[26] In the symmetric mode, the towers are deflecting toward each other, causing the center span to deflect while the end spans are cambering. In the antisymmetric mode, the towers are deflecting in the same direction, resulting in the antisymmetric deformation pattern with respect to the midspan.[26]

In a suspension bridge, the most dangerous mode of oscillation is the antisymmetric flutter mode, Fig. 12.19a. This type of oscillation caused the destruction of the Tacoma Narrows Bridge. This specific mode is easily developed by the pitching moment of wind forces; the two cables move in opposite directions in half the span length—one goes up while the other goes down. Since the cable system provides no resistance to the

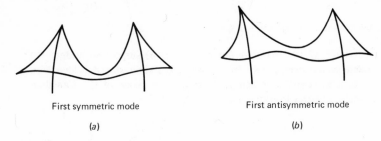

First symmetric mode

(a)

First antisymmetric mode

(b)

FIGURE 12.18.    Longitudinal modes of vibration: (a) first symmetric mode; (b) first antisymmetric mode. (Ref. 26.)

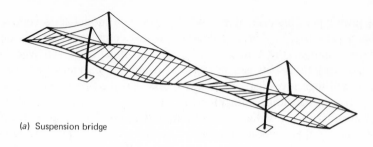

(*a*) Suspension bridge

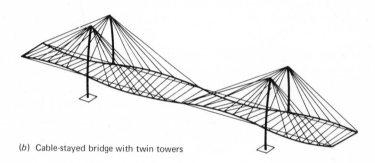

(*b*) Cable-stayed bridge with twin towers

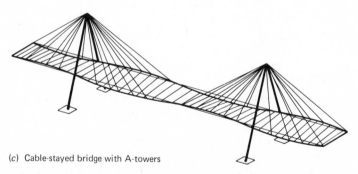

(*c*) Cable-stayed bridge with A-towers

**FIGURE 12.19.   Relative deformations: (*a*) suspension bridge; (*b*) cable-stayed bridge with twin towers; (*c*) cable-stayed bridge with A-towers. (Leonhardt and Zellner. By permission of the Canadian Steel Industries Construction Council. Ref. 3.)**

induced torsional deformation of the deck, a relatively high torsional and bending restraining deck system is required.[3,35] This mode of oscillation can be restricted if the cables are attached to the girder or truss at the center of the structure, because the cable at this point is attempting to move in opposite directions simultaneously. This method was used in an effort to stabilize the Tacoma Narrows Bridge, but the connection broke and the structure reverted into the first antisymmetric mode and was consequently demolished within an hour.[26]

It is relatively more difficult to provide and maintain a resonant oscillation with attendent large amplitudes in a multicable stayed structure. The cables of different lengths and different frequencies tend to disturb the formation of the first or second mode of oscillation by interfering with smaller wave lengths of higher order. Thus, the inherent system damping of the cable-stayed structure produces relatively smaller amplitudes compared with the suspension system. The difference in deflection of the girders in the two cable planes of a cable-stayed system results primarily from the different deflections of the pylons in each plane, Fig. 12.19b. For an A-frame tower, the differential deflection of the towers in each cable plane is negated, and the resistance of the cable-stay system to torsional oscillations of the roadway deck is further enhanced, Fig. 12.19c.[3,35]

As a result of the inherent system stiffness and damping, the cable-stayed bridge is not as sensitive to wind oscillations as the suspension bridge. Therefore, the cable-stayed bridge requires less torsional stiffness in its suspended deck system. However, this conclusion is only qualitative. The difference in response of the two systems has yet to be demonstrated by wind tunnel tests of sectional models.

## 12.8   DECK STABILITY

Numerous wind tunnel tests by investigators in several countries have indicated that bluff cross sections have characteristics that produce intense Karman vortex shedding and large fluctuating vertical forces, which result in vertical bending coupled with a torsional response. These tests have led to the development of a cross-sectional shape that is considered to have favorable aerodynamic characteristics. Aerodynamic stability of suspension and cable-stay bridges can be achieved by shaping the cross section such that:[3,35]

1. The wind eddies that produce the Karman vortex shedding effect will be diminished or eliminated.

2. A minimum of lift and pitching moments will be produced to minimize the bending and torsional oscillation.

**FIGURE 12.20.    Verrazano Narrows Bridge, New York.**

Additional studies and tests have tended to validate this conclusion. Therefore, it can be stated that the conventional stiffening truss of the Verazzano Narrows suspension bridge (Fig. 12.20) is designed for increased flexural and torsional stiffness to resist the effect of wind forces. Whereas the aerodynamically "streamlined" cross section used on the Severn suspension bridge in England (Fig. 12.21) is designed to minimize the excitation force and motion which cause aerodynamic instability. As a result, the streamlined cross section seeks to eliminate the cause rather than totally resist the effect. A secondary improvement to some designers is the aesthetics of the structure.

Leonhardt first reported this concept of streamlining the cross section for cable-stayed bridges in 1968,[36] although aerodynamic tests were conducted in 1959 for a monocable suspension structure that was unsuccessful in the Tagus River Bridge competition at Lisbon. This same concept was again proposed unsuccessfully for the Rheinbrücke-Emmerich bridge.

The first modern structure to use the aerodynamically shaped cross section in its deck structure was the Severn Bridge in England, designed by

**FIGURE 12.21.   Severn Suspension Bridge, England.**

Freeman, Fox and Partners. Christen Ostenfeld of Copenhagen used the streamlined concept on the Lillebelt Bridge in Denmark, although the original design was made with the conventional stiffening girder concept. Freeman, Fox and Partners have also used the streamlined cross section in the Bosporus Bridge. Therefore, aerodynamic stability can be attained even for extremely long spans with a sufficiently wide aerodynamically-shaped girder that is continuous at the pylons. Although the previous discussion has indicated a relatively favorable response for cable-stayed structures compared to the conventional suspension structure, disturbing aerodynamic oscillations can and do occur in cable-stayed bridges.

It has been illustrated that the cable-stayed system is not as sensitive to wind oscillations as a suspension bridge because of the inherent system stiffness and damping. Further, it has been shown that vortex shedding, bending and torsional oscillations can be minimized with streamlined deck cross sections. However, the reader is cautioned that these conclusions are generalizations based on data available from limited tests conducted on relatively few structures. It is suggested that wind tunnel tests be performed for any major cable-stayed structure.

## 12.9   STABILITY DURING ERECTION

It is most important to note that the validation of stability of the completed structure for expected wind speeds at the site is mandatory. However, this does not necessarily imply that the most critical stabiltiy condition of the structure occurs when the structure is fully completed. A more dangerous condition may occur during erection, when the joints have not been fully connected and, therefore, full stiffness of the structure has not yet been realized. In the erection stage, the frequencies are lower than in the final condition and the ratio of torsional frequency to flexural frequency may approach unity. Various stages of the partially erected structure may be more critical than the completed bridge. The use of welded components in towers has contributed to their susceptibility to vibration during erection.

The erection method used in the Severn Bridge[37] (Fig. 12.21) was to hoist 60 ft long segments of the deck structure from barges on the river to their connection to the suspender ropes. With only a moderate structural connection between segments, the critical flutter speed was established at less than 50 mph. By introducing a more effective torsional connection, the critical speed was raised to 100 mph for all stages of erection (Fig. 12.22). Scruton also noted that while the individual components were being erected, they were subject to violent yawing and pitching oscillations at low wind speeds unless a system of check ropes was operative.[37]

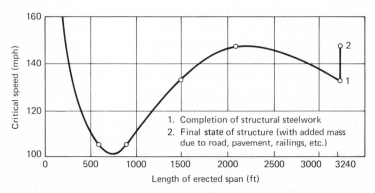

**FIGURE 12.22.   Variation of critical wind speed with length of erected center span for the Severn Suspension Bridge. (Ref. 37.)**

Contractors must assure themselves that the structure will be aerodynamically stable during erection.

Component parts of the structure are also in themselves susceptible to wind excitation, either during erection or in the final condition. The cables of cable-stayed bridges, the hangers of suspension and arch bridges, and the towers of suspension and cable-stayed[38] bridges have been known to exhibit oscillations, usually from vortex excited vibrations.

## 12.10   WIND TUNNEL INVESTIGATIONS

Because no other analytical procedures are yet available, wind tunnel tests are used to evaluate the aerodynamic characteristics of the cross section of an existing or proposed bridge deck, tower, or total bridge. More importantly, the wind tunnel tests may be used during the design process to evaluate the performance of a number of proposed cross sections for a particular project. In this manner, the wind tunnel investigations become a

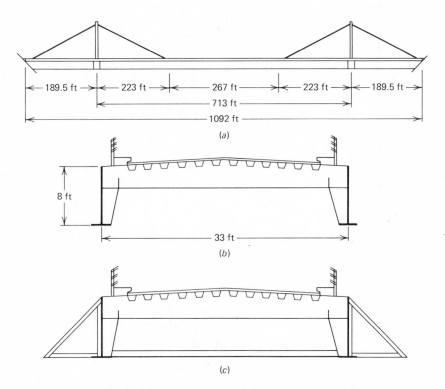

**FIGURE 12.23.   Long's Creek Bridge. (Ref. 13.)**

part of the design decision process and not a postconstruction corrective action. The following discussion of specific structures that have been investigated by wind tunnel tests has been extracted from reference 13 unless otherwise noted.

### 12.10.1  Long's Creek Bridge, Canada

The Long's Creek Bridge in New Brunswick, Canada, has a center span of 713 ft, with two vertical cable planes in cross section and only one pair of

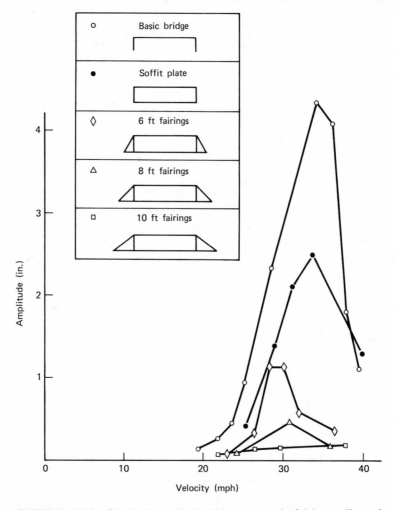

**FIGURE 12.24.  Vertical amplitude with asymmetric fairings, effect of fairing width (bridge height = 15 ft), Long's Creek Bridge. (Ref. 13.)**

radiating stays from each pylon, Fig. 12.23*a*. At wind speeds of 30 mph, the structure was observed to have a vibration frequency of 0.6 cps and an amplitude reaching 4 in., or 8 in. when snow fills the railing openings,[12,26] (Fig. 12.15). The bluff cross section produces an unfavorable aerodynamic section, Fig. 12.23*b*. The single pair of stays in each plane has a natural low resonant oscillation. If three to five stays had been used, the system damping as a whole would have been increased.[35] As a result of wind tunnel tests, a soffit, or bottom plate, was installed to produce a closed box section, thus increasing torsional stiffness, Fig. 12.23*c*. The addition of the soffit plate decreased the amplitude of motion by as much as 40%, Fig. 12.24.

Additional streamlining of the cross section by the addition of triangular edge fairings further reduced amplitude. The values indicated in Fig. 12.24

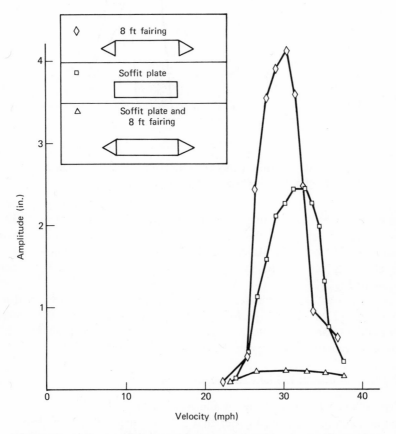

**FIGURE 12.25.   Vertical amplitude with soffit plate and 8 ft symmetric edge fairings (bridge height = 100 ft), Long's Creek Bridge. (Ref. 13.)**

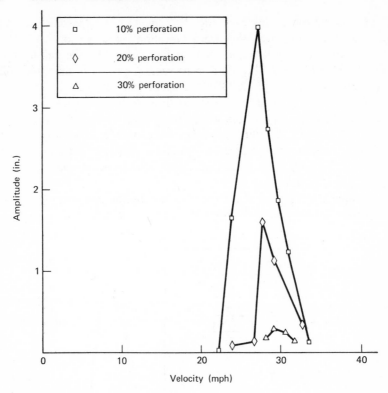

**FIGURE 12.26.    Vertical amplitude with girder web perforations (bridge height = 100 ft), Long's Creek Bridge. (Ref. 13.)**

are for a height of structure 15 ft above water. The structure was 100 ft above water level before the reservoir in the valley below the bridge was filled. Further tests established that had the structure been maintained at an elevation of 100 ft, symmetrical edge fairings would have been required, Fig. 12.25.

A study was also conducted to determine the effect of girder perforations, Fig. 12.26. A 30% perforation would have been required to reduce the motion to an acceptable level. However, this condition would have had serious effects on other aspects of the design.

## 12.10.2    Papineau-Leblanc Bridge, Canada

The Papineau Leblanc Bridge is illustrated in Fig. 12.27. The cross section is a central spine box girder that is aerodynamically bluff. However, the deck overhang provides a degree of streamlining such that model tests

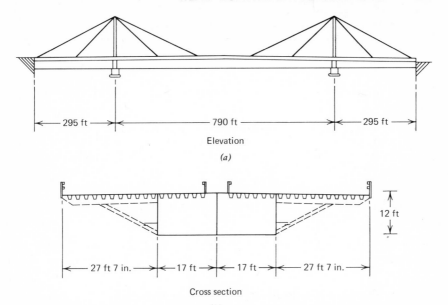

FIGURE 12.27.   Papineau-Leblanc Bridge: (*a*) elevation; (*b*) cross section. (Ref. 13.)

indicated only a small torsional eddy shedding response at wind velocities which are not expected to occur at the structure site.

### 12.10.3   Kniebrucke Bridge, Germany

The usual wind tunnel tests of a section model of the Kniebrücke Bridge in Düsseldorf indicated instability at high wind speeds. The instability was thought to be due to the sensitivity to torsional oscillation attributed to the omission of the bottom plate, which would have formed a closed box section.[3] Because of the width of the structure, this bottom plate was deliberately omitted as an economy measure, Fig. 12.28a. The reason the bridge has not shown any signs of disturbing oscillations has been attributed to its low elevation above the water level and to turbulence caused by the urban terrain.

Further studies were performed to determine whether remedial measures could be undertaken should undesirable wind oscillations develop in the prototype bridge. These studies indicated that an additional lining outside of the main girders would be sufficient to produce stability, Fig. 12.28b. As a consequence, Leonhardt[3,35] has proposed that a cross sectional configuration with triangular boxes at the edges and an open bottom deck be used on subsequent structures. He has further suggested that the slope

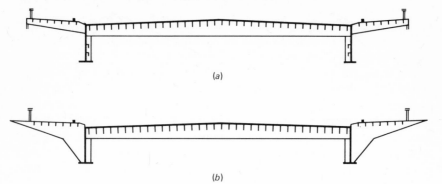

(a)

(b)

**FIGURE 12.28.**   (*a*) and (*b*) Cross sections, Kniebrücke, Düsseldorf. (Ref. 35.)

of the soffit plate of the outside edge boxes should not exceed an angle of 35 degrees.[3]

### 12.10.4   Proposed New Burrard Inlet Crossing, Canada

Wind tunnel tests were conducted at the National Physical Laboratory in Ottawa[13,39,40,41] on a proposed structure of approximately 2500 ft main span in Vancouver, B.C., on Canada's Pacific Coast. The deck of this bridge is to have an elevation of 208 ft above mean sea level at the towers. An artist's rendering of the cable-stayed bridge proposal is shown in Fig. 12.29.

Section model tests were conducted on six basic bridge configurations: (1) plate girder suspension bridge, (2) stiffening truss suspension bridge, (3) box girder suspension bridge, (4) trapezoidal box girder cable stay,

**FIGURE 12.29.   Artist's rendering of the proposed cable-stayed New Burrard Inlet Crossing Bridge. (Ref. 35.)**

(5) twin edge triangular box girder cable stay, and (6) box section cable stay. A number of variations were also studied that considered edge geometry, deck perforations, and girder perforations.

The trapezoidal box section was found to exhibit a degree of instability at low angles of attack that made it unacceptable, Fig. 12.30a. Edge modifications provided a high degree of stability, Fig. 12.30b. Water tunnel flow tests at an angle of attack of + 4 degrees show a large separated flow with no reattachment and a wide wake for the initial section, compared with the reattachment at the upper and lower surfaces and a narrower wake for the modified section, Fig. 12.31. These comparison features were evident at angles of attack as high as ± 10 degrees.

Wind tunnel tests were conducted for a twin triangular edge box girder section to determine the necessity for torsional stiffness and requirements for extremely wide bridges with a streamlined cross section, Fig. 12.31. These tests indicated that aerodynamic stability could be achieved on the deck cross section without the enclosing bottom plate and without con-

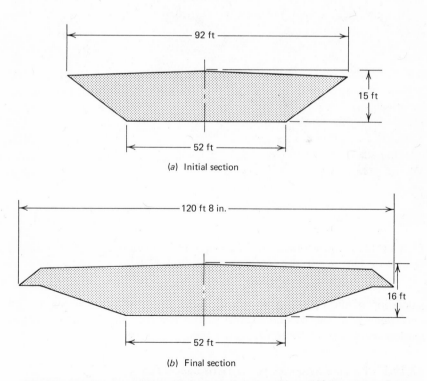

(a) Initial section

(b) Final section

FIGURE 12.30.    The initial and final road deck sections for the proposed new Burrard Inlet Crossing Bridge: (a) initial section; (b) final section. (Ref. 13.)

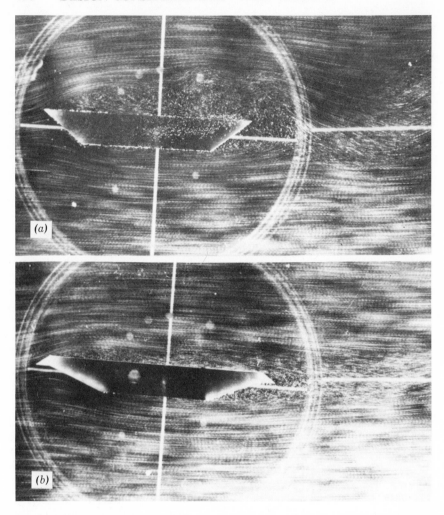

**FIGURE 12.31.** Flow visualization, New Burrard Inlet Crossing Bridge: (*a*) original section; (*b*) improved section (edge extension). (Ref. 13.)

sideration of the favorable system damping provided by cable stays. The performance of the section is due to the incorporation of a downward sloping wind nose at the top outside edges.

### 12.10.5 Pasco-Kennewick Intercity Bridge, U.S.A.

The Pasco-Kennewick Bridge is a prestressed segmental concrete super-structure that is supported by cable stays, Fig. 12.32. The concrete girder

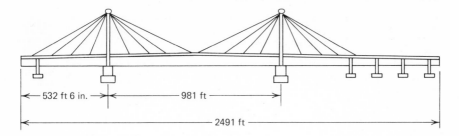

**FIGURE 12.32.** Pasco-Kennewick Intercity Bridge. (Ref. 13.)

has a much higher mass than that of other structures considered in the design studies. The superstructure cross section is a twin triangular edge box configuration with a depth to width ratio of 11. The deck streamlining provided a highly favorable aerodynamic response despite the low value of the ratio of torsional frequency to vertical bending frequency of $(N_\theta/N_y) = 1.4$.

A number of damping coefficients and edge configurations were investigated, and a typical result for a bridge height above water of 57 ft is shown in Fig. 12.33. As indicated, flutter and vortex excitation oscillations were found to exist at large angles of attack and velocities of the wind that are above those assumed for design at the site.

### 12.10.6 The Narrows Bridge, Canada

The Narrows Bridge, Halifax, Canada, is a suspension bridge with a center span of 1400 ft, end spans of 513 ft 10 in., and a vertical navigation clearance of 165 ft at midspan. This structure is important because it is one of a few bridges that has been tested using not only a section model but also a full model test in both uniform flow and turbulent boundary layer flow,[32,42] Fig. 12.17.

Comparisons of test results were made in this study for a 1:40 and 1:320 scale section model and a 1:320 full bridge model in both uniform flow and turbulent boundary layer flow. Based on wind normal to the longitudinal axis of the bridge, the following observations were made: (1) the section models exhibited a coupled vertical-torsional oscillatory

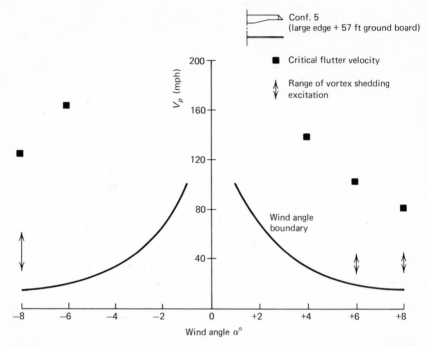

**FIGURE 12.33.   Critical flutter velocity and range of vortex shedding excitation. (Ref. 13.)**

instability at critical wind velocity; (2) a divergent instability was noted for the full model in a uniform flow at velocities well above the critical velocity observed for the sectional model; (3) for the full model in turbulent flow, random vertical oscillations were observed to increase in amplitude with increased wind velocity and turbulence but no instability or torsional motion was recorded.

The results are very dissimilar, especially between the section model and full model tests in uniform flow. They are also contrary to previous conclusions derived from the work of Farquharson et al.[1] and Frazer and Scruton[43] whose data formed the basis for the validity of the procedure for section model testing. At first it was felt that the difference in scale of 1:40 for the section model compared to the scale of 1:320 for the full model accounted for the dissimilarity of results. However, comparison of the results for the 1:40 and 1:320 sectional models were in fairly good agreement, which would appear to discount the scaling problem explanation.

Davenport[32] suggests the following explanation for the variation of results: (1) The orthotropic deck structure is relatively light compared to more conventional structures that have been tested. Therefore, aerodynamic and mechanical parameters, including participation of the towers and cables in responding to the wind, which are generally ignored in section model tests were exerting their influence. (2) The possibility of static deformations interfering with the instability mechanisms which are observed in the full model but not in the section model. Leonhardt has also commented on the system stability of the total structure, with special reference to the effect of the cables in a cable-stayed structure, (Section 12.7).

## 12.11   MOTION TOLERANCE

Consideration of an acceptable level of motion falls into two categories: (1) structurally damaging motion, and (2) human response motion. The first relates to violent motion that may be catastrophic or motion that over a period of time may lead to related fatigue failures. The second relates to motion that may not be structurally damaging but may be objectionable from the standpoint of user acceptance, such as vibrations that are noticeable to pedestrians or occupants of standing or moving vehicles.

The critical wind velocity level at which structural damage begins to occur is defined as that velocity which causes violent flutter motions. Velocities below the flutter speed will not produce structural damage. Very little information on the subject of flutter of bridges is presented in the literature; however, Buckland and Wardlaw[26] have suggested a design approach which is presented below.

A bridge is designed for static wind loads at a certain wind speed. This is not a collapse condition, but usually produces an allowable overstress. From this design condition the wind speed at which failure or yield would be expected to occur can be calculated.

With a flutter type of motion, the onset of oscillations is sudden and violent; and occurs at the critical flutter speed. It is suggested that as long as the flutter speed is greater than that calculated to cause failure, the same degree of safety against failure is obtained as for other design conditions. This is best demonstrated by an example:

Assume a bridge designed for a maximum wind gust speed of 90 mph, for which the basic allowable stress is $f$. Assume also that:

Dead Load + Live Load produce                1.00 $f$
Dead Load Only produces                      0.80 $f$
Dead Load + 90 mph wind load produces        1.25 $f$
Failure occurs at                            1.70 $f$

Because the wind load is proportional to the square of the velocity:

the stress due to a 90 mph wind $= K \times 90^2 = (1.25 - .8) f$

the stress due to wind at failure $= K \times V^2 = (1.7 - .8) f$

where $V$ is the wind velocity that could cause failure and $K$ is a constant of proportionality.

Therefore,

$$V^2 = (90)^2 \left( \frac{1.7 \ - .8}{1.25 - .8} \right)$$

and $V = 127$ mph. In this case we would expect a gust of 127 mph to be the maximum that the bridge can withstand. It follows then that if the "flutter speed" is greater than 127 mph, flutter is less likely to cause destruction of the bridge than what structural engineers refer to as "wind pressure." This is correct, but still unduly conservative.

It must be remembered that "wind pressure" is based upon a gust of wind—the greatest gust velocity which can be expected to occur at least once during a period of several years. But a gust is not of sufficient duration or spatial extent to build up flutter, so we should also calculate the mean wind velocity which corresponds to a gust of 127 mph. The National Building Code[44] of Canada suggests a formula:

$$V_{mean} = \frac{V_{gust} - 5.8}{1.29}$$

The result is that wind with a mean velocity of 94 mph could be expected to gust to 127 mph, which could destroy the bridge. Consequently, if the flutter speed is greater than 94 mph, flutter should not be a cause of failure.

Some vibrations that do not damage a structure may be unacceptable to the user of the bridge. Little data is available with respect to peoples' reaction to motion while standing, walking, or in automobiles. This subject is also discussed by Buckland and Wardlaw[26] and their suggested tentative criteria for motions affecting people is presented below:

1. For wind speeds up to 30 mph, 2% of $g$.
2. For wind speeds from 30 to 70 mph, 5% of $g$.
3. Over 70 mph, effects on observers may be disregarded in design considerations.

In the above $g$ is the acceleration of gravity in ft per $\sec^2$.

## REFERENCES

1. Farquharson, F. B., Smith, F. C. and Vincent, G. S., "Aerodynamic Stability of Suspension Bridges," Univ. of Washington Bull. 116, parts I through V, 1949–1954.
2. Bleich, F., McCullough, C. B., Rosecrans, R. and Vincent, G. S., "The Mathematical Theory of Vibration in Suspension Bridges," Bureau of Public Roads, U.S. Department of Commerce, Government Printing Office, Washington, D.C., 1950.

3. Leonhardt, F. and Zellner, W., "Cable-Stayed Bridges: Report on Latest Developments," Canadian Structural Engineering Conference, 1970, Canadian Steel Industries Construction Council, Toronto, Ontario, Canada.

4. Jakkula, A. A., "A History of Suspension Bridges in Bibliographical Form," *Texas A. & M. College of Engineering Experimental Station Bulletin* No. 57, July 1, 1941.

5. Steinman, D. B., "Design of Bridges Against Wind," *Civil Engineering*, ASCE, October, November, and December 1945 and January and February 1946.

6. Reid, W., "A Short Account of the Failure of a Part of the Brighton Chain Pier in the Gale of the 30th of November, 1836," *Professional Papers of the Corps of Royal Engineers*, Vol. 1, 1844, p. 99.

7. Provis, W. A., "Observations on the Effect of Wind on the Suspension Bridge over Menai Straits," *Minutes of Proceedings of the Institution of Civil Engineers*; Vol. 1, 1837–41, Session of 1841, p. 74.

8. Vincent, G. S., "Golden Gate Bridge Vibration Studies," *Journal of the Structural Division*, ASCE, Proc. Paper 1817, October 1958.

9. Steinman, D. B., "Aerodynamics of Suspension Bridges," Proceedings of the Thirty-Second Annual Roads School, Purdue University, January 1946.

10. Anon., "Stays and Brakes Check Oscillation of Whitestone Bridge," *Engineering News-Record*, December 5, 1940.

11. Ammann, O. H., "Additional Stiffening of Bronx-Whitestone Bridge," *Civil Engineering*, ASCE, Vol. 16, No. 3, March 1946.

12. Wardlaw, R. L. and Ponder, C. A., "Wind Tunnel Investigation on the Aerodynamic Stability of Bridges," Canadian Structural Engineering Conference, 1970, Canadian Steel Industries Construction Council, Toronto, Ontario, Canada.

13. Wardlaw, R. L., "A Review of the Aerodynamics of Bridge Road Decks and the Role of Wind Tunnel Investigation," U.S. Department of Transportation, Federal Highway Administration, Report No. FHWA-RD-73-76.

14. Taylor, P. R., "Cable-Stayed Bridges and Their Potential in Canada," *Engineering Journal* (Canada), Vol. 52, No. 11 (November, 1969).

15. Davenport, A. G., "Rationale for Determining Design Wind Velocities," *Journal of the Structural Division*, ASCE, paper 2476, Vol. 86, No. ST 5, May 1960.

16. Davenport, A. G., ed., "New Approaches to Design Against Wind Action," Course notes, ASCE Wind Seminar, Cleveland, Ohio, April 1972, Boundary Layer Wind Tunnel, University of Western Ontario, London, Canada.

17. Davenport, A. G., "The Application of Statistical Concepts to the Wind Loading of Structures," *Proceedings of the Institute of Civil Engineers*, Vol. 19, Paper No. 6480, August 1961.

18. Hirai, A. and Okubo, T., "On the Design Criteria Against Wind Effects for Proposed Honshu-Shikoku bridges," Paper No. 10, Symposium on Suspension Bridges, Lisbon, November 1966.

19. Ishizaki, H. and Mitsuda, Y., "On the Extent of Gusts and the Gust Factors of Strong Wind (in Japanese), Ann. Disas. Prev. Res. Inst. Kyoto Univ. No. 5A, 1962.

20. Sherlock, R. H., "Variations of Wind Velocity and Gusts with Height," *Proceedings* ASCE, Vol. 78, No. 126, 1952.

21. Deacon, E. L., "Gust Variations with Height up to 150 m," *Quarterly Journal of the Royal Meteorological Society*, Vol. 81, Ab. 350, 1955.

22. Roberts, G., "Severn Bridge—Design and Contract Arrangements," *Proceedings of the Institute Civil Engineers*, Vol. 41, September 1958.

23. Steinman, D. B., *Suspension Bridges: The Aerodynamic Problem and Its Solution*, International Association for Bridge and Structural Engineering, Volume 14, 1954.

24. Steinman, D. B., "Problems of Aerodynamic and Hydrodynamic Stability," *Proceedings Third Hydraulics Conference*, June 10–12, 1946, University of Iowa, Studies in Engineering, Bulletin 31.

25. Scanlan, R. H., "Aeroelastic Stability of Long-Span Bridges," U.S. Department of Transportation, Federal Highway Administration, Report No. FHWA-RD-73-75.

26. Buckland, P. G. and Wardlaw, R. L., "Some Aerodynamic Considerations in Bridge Design," *Engineering Journal* (Canada), Engineering Institute of Canada, April 1972.

27. Vincent, G. S., "A Summary of Laboratory and Field Studies in the United States on Wind Effects on Suspension Bridges," *Proceedings of Symposium No. 16, Wind Effects on Buildings and Structures*, National Physical Lab., England, 1963.

28. Steinman, D. B., "Aerodynamic Theory of Bridge Oscillation," *Transactions*, ASCE, Vol. 115, 1950.

29. Bleich, F., "Dynamic Instability of Truss-Stiffened Suspension Bridges Under Wind Action," ASCE, *Proceedings* Vol. 74, October 1948.

30. Kloppel, K. and Thiele, F., "Modellversuche im Windkanal zur Bemessung von Brücken gegen die Gefahr winderregter Schwingungen," *Der Stahlbau*, No. 12, December, 1967.

31. Scanlan, R. H., "Studies of Suspension Bridge Deck Flutter Instability," American Institute of Aeronautics and Astronautics, Paper No. 69.744, July 1969.

32. Davenport, A. G. et al., "A Study of Wind Action on a Suspension Bridge During Erection and on Completion," Research Report of University of Western Ontario, Canada, BLWT, 3, 69.

33. Walshe, D. E. J., "The Use of Models to Predict the Oscillatory Behavior of Suspension Bridges in Wind," National Physical Laboratory,. Teddington, England, Proc. Symposium No. 16, "Wind Effects on Buildings and Structures," 1963.

34. Walshe, D. E. J. and Rayner, D. V., "A Further Aerodynamic Investigation for the Proposed River Severn Bridge," National Physical Laboratory, England, NPL Aero Report 1010, March 1962.

35. Leonhardt, F. and Zellner, W., "Vergleiche zwichen Hängebrücken und Schrägkabel brücken für Spannweiten über 600 m," *International Association for Bridge and Structural Engineering*, Vol. 32, 1972.

36. Leonhardt, F., "Zur Entwicklung aerodynamisch stabiler Hängebrücken," *Die Bautechnik* 45 (1968), Nos. 10 and 11.

37. Scruton, C., "Aerodynamics of Structures," *Wind Effects on Buildings and Structures, Proceedings of the International Research Seminar* held at the National Research Council, Ottawa, Canada on 11–15 September 1967, University of Toronto Press.

38. Gade, R. H., "Status of the Investigation of Aerodynamic Behavior of the Sitka Harbor Cable-Stayed Bridge," presented at the Symposium on Full Scale Measurements of Wind Effects on Tall Buildings and Other Structures, London, Ontario, Canada, June 1974.

39. Wardlaw, R. L., "A Preliminary Wind Tunnel Study of the Aerodynamic Stability of Four Bridge Sections for Proposed New Burrard Inlet Crossing," National Research Council of Canada, National Aeronautical Establishment, Tech. Report LTR-LA-31, July 14, 1969.

40. Wardlaw, R. L., "Some Approaches for Improving the Aerodynamic Stability of Bridge Road Decks," *Proceedings of the Third International Conference on Wind Effects on Buildings and Structures*, Tokyo, 1971.

41. Wardlaw, R. L., "Static Force Measurements of Six Deck Sections for the Proposed New Burrard Inlet Crossing," National Research Council of Canada, National Aeronautical Establishment, Tech. Report LTR-LA-53, June 4, 1970.

42. Davenport, A. G., Isyumov, N. and Miyata, T., "The Experimental Determination of the Response of Suspension Bridges to Turbulent Wind," *Proceedings of the Third International Conference on Wind Effects on Buildings and Structures*, Tokyo, 1971.
43. Frazer, R. A. and Scruton, C., "A Summarized Account of the Severn Bridge Aerodynamic Investigation," Report NPL/Aero/222, H.M.S.O., 1952.
44. National Building Code of Canada, 1965 Edition, Supplement No. 1, "Climatic Information for Building Design in Canada, 1965" NRC No. 8329.

# APPENDIX A

# Typical Pedestrian Bridge

From Standard Plans for Highway Bridges Volume V, Typical Pedestrian Bridges U. S. Department of Commerce Bureau of Public Roads, Washington, D. C., October, 1964 (now U. S. Department of Transportation Federal Highway Administration).

# APPENDIX B

# Chronological Bibliography

1. Fidler, T. C., "Straight Link Suspension Bridges," *Engineering* (London), Vol. 31 (March 25, 1881), pp. 297–300, 373–374, 584–585.
2. Mehrtens, *Eisenbrückenbau* [*Steel Bridge Construction*], Vol. I, (Verlag Engelmann, Leipzig, 1908).
3. "New Type Bridge Proposed by Germans," *Engineering News-Record* (September 2, 1948), pp. 86–87.
4. Dischinger, F., "Hängebrücken für schwerste Verkehrslasten," *Der Bauingenier* (March 1949), pp. 65, 107.
5. Wenk, H., "Die Strömsundbrücke [The Strömsund Bridge]," *Der Stahlbau*, Vol. 23, No. 4 (April 1954), pp. 73–76.
6. Beyer, E. and Tussing, F., "Nordbrücke Düsseldorf," *Der Stahlbau*, Vol. 24, No. 2, 3, and 4 (February, March, April, 1955), pp. 25–33, 63–67, 79–88.
7. Homberg, H., "Einflusslinien von Schrägseilbrücken," *Der Stahlbau*, Vol. 24, No. 2 (February 1955), pp. 40–44.
8. Klingenberg, W. and Plum, "Versuche an dem Drähten und Seilen der neuen Rheinbrücke in Rodenkirchen bei Köln," *Der Stahlbau*, Vol. 24, No. 12 (December 1955), pp. 265–272.
9. Sievers, H. and Görtz, W., "Der Wiederaufbau der Strassenbrücke über den Rhein zwischen Duisburg Ruhort and Homberg," *Der Stahlbau*, Vol. 25, No. 4 (April 1956), pp. 77–88.
10. Ernst, H. J., "Montage eines seilverspannten Balken im Gross-Brückenbau," *Der Stahlbau*, Vol. 25, No. 5 (May 1956), pp. 101–108.
11. Godfrey, G. B., "Post-War Development in Germany: Steel Bridges and Structures," *Structural Engineer*, Vol. 35, No. 2 (February 1957), pp. 53–68.
12. Godfrey, G. B., "Post-War Development in Germany Steel Bridges and Structures (Discussion)," *Structural Engineer*, Vol. 35, No. 10 (October 1957), pp. 390–398.
13. Kunz, R., Trappmann, H. and Tröndle, E., "Die Büchenauer Brücke, eine neue Schrägseilbrücke der Bundesstrasse 35 in Bruchsal," *Der Stahlbau*, Vol. 26, No. 4 (April 1957), pp. 98–102.
14. Beyer, E. and Ernst, H. J., "Erfahrungen und Seilversuche an einer seilunterspannten Verbundkonstruktion," *Der Stahlbau*, Vol. 26, No. 7 (July 1957), pp. 177–183.
15. Schüssler, K. and Braun, F., "Wettbewerb 1954 zum Bau einer Rheinbrücke oder eines Tunnels in Köln im Zuge Klappergasse-Gotenring," *Der Stahlbau*, Vol. 26, No. 8, 9, 10, and 11 (August, September, October, and November, 1957), pp. 205–217, 253–274, 294–312, and 326–348.
16. *Denkschrift über die Nordbrücke Düsseldorf* (Springer-Verlag, Berlin-Göttingen-Heidelberg, 1958).
17. Beyer, E., "Nordbrücke Düsseldorf I. Theil: Gesamtanlauge und Montage der neuen Rheinbrücke," *Der Stahlbau*, Vol. 27, No. 1 (January 1958), pp. 1–6.
18. Hadley, H. M., "Tied Cantilever Bridge—Pioneer Structure in U.S.," *Civil Engineering*, ASCE, January 1958.

19. Lewenton, G., "Die deutschen Pavillonbauten auf der Weltausstellung Brüssel 1958," *Der Stahlbau*, Vol. 27, No. 4 (April 1958), pp. 1–91.

20. Fuchs, D., "Der Fussgängersteg auf der Brüsseler Weltausstellung 1958," *Der Stahlbau*, Vol. 27, No. 4 (April 1958), pp. 91–97.

21. Wintergerst, L., "Nordbrücke Düsseldorf III. Theil: Statik und Konstruktion der Strombrücke," *Der Stahlbau*, Vol. 27, No. 6 (June 1958), pp. 147–158.

22. Schreier, G., "North Bridge at Düsseldorf: Analysis, Design, Fabrication and Erection of the Bridge Spanning the River," *Acier-Stahl-Steel* (English version), Part 1, No. 9 (September 1958), Part 2, No. 11 (November 1958).

23. Fischer, G., "Die Severinsbrücke in Köln" ["The Severin Bridge at Cologne"], *Acier-Stahl-Steel*, Vol. 25, No. 3 (March 1960), p. 101 (pp. 97–107, English version).

24. Vogel, G., "Erfahrungen mit geschweissten Montagestossen beim Bau der Severinsbrücke in Köln," *Schweissen ünd Schneiden*, Vol. 12, No. 5 (May 1960), pp. 189–194.

25. Hess, H., "Die Severinsbrücke Köln, Entwurf und Fertigung der Strömbrücke," *Der Stahlbau*, Vol. 29, No. 8 (August 1960), pp. 225–261.

26. Vogel, G., "Die Montage des Stahlüberbaues der Severinsbrücke Köln," *Der Stahlbau*, Vol. 29, No. 9 (September 1960), pp. 269–293.

27. Michalos, J. and Birnstiel, C., "Movements of a Cable due to Changes in Loading," *Journal of the Structural Division*, ASCE, Vol. 86, No. ST 12, Proc. Paper 2674 (December 1960), pp. 23–38.

28. Goschy, Be'la, "Dynamics of Cable-Stayed Pipe Bridges," *Acier-Stahl-Steel*, Vol. 26, No. 6 (June 1961), pp. 277–282.

29. Dotzauer, H. K. and Hess, H., "Belastungsprobe der Severinsbrücke Köln," *Der Stahlbau*, Vol. 30, No. 10 (October 1961), pp. 303–311.

30. Leonhardt, F. and Andrä, W., "Fussgängersteg über die Schillerstrasse im Stuttgart," *Die Bautechnik*, Vol. 39, No. 4 (April 1962), pp. 110–116.

31. Appelbaum, G., and Rokicki, K., "Wesentliche Merkmale der Vorfertigung und Montage der Autobahn-Norderelbbrücke bei Hamburg," *Schweissen und Schneiden*, Vol. 14, No. 6 (June 1962), pp. 255–258.

32. Havemann, H. K., "Die Seilverspannung der Autobahnbrücke über die Norderelbe-Bericht über Versuche zur Daverfestigkeit der Drahtseile," *Der Stahlbau*, Vol. 31, No. 8 (August 1962), pp. 225–232.

33. Jennings, A., "The Free Cable," *The Engineer* (December 28, 1962), pp. 1111–1112.

34. *Design Manual for Orthotropic Steel Plate Deck Bridges* (New York: American Institute of Steel Construction, 1963), pp. 7–10.

35. "The Bridge Spanning Lake Maracaibo in Venezuela, The General Rafael Urdaneta Bridge" (Bauverlang GmbH Weisbaden, Berlin, 1963).

36. Klöppel and Weber, "Teilmodellversuche zur Beurteilung des aerodynamischen Verhaltens von Brücken," *Der Stahlbau*, Vol. 32, No. 3 and 4 (March and April, 1963), pp. 65–79, 113–121.

37. Poskitt, T. J. and Livesley, R. H., "Structural Analysis of Guyed Masts," *Proceedings of the Institution of Civil Engineers*, Vol. 24 (March 1963), pp. 373–386.

38. Brotton, D. M., Williamson, N. W. W. and Millar, M., "The Solution of Suspension Bridge Problems by Digital Computers—Part I," *Structural Engineer*, Vol. 41, No. 4 (April 1963).

39. Poskitt, T. J., "The Application of Elastic Catenary Functions to the Analysis of Suspended Cable Structures," *Structural Engineer*, Vol. 41, No. 5 (May 1963).

40. Reimers, K., "Fussgängerbrücke über die Glacischaussee in Hamburg für die Internationale Gartenbau-Ausstellung 1963," *Schweissen und Schneiden*, Vol. 15, No. 6 (June 1963), pp. 262–264.

41. Braun, F. and Moors, J., "Wettbewerb zum Bau einer Rheinbrücke im Zuge der Inneren Kanalstrasse in Köln (Zoobrücke), *Der Stahlbau*, Vol. 32, No. 6, 7, and 8 (June, July, and August, 1963), pp. 174–183, 204–213, 248–254.

42. Havemann, H. K., "Die Brücke über die Norderelbe im Zuge der Bundesautobahn Südliche Umgehung Hamburg, Teil I: Ideen-und Bauwettbewerb," *Der Stahlbau*, Vol. 32, No. 7 (July 1963), pp. 193–198.

43. Brotton, D. M. and Arnold, G., "The Solution of Suspension Bridge Problems by Digital Computers—Part II," *Structural Engineer*, Vol. 41, No. 7 (July 1963).

44. Aschenberg, H. and Freudenberg, G., "Die Brücke über die Norderelbe im Zuge der Bundesautobahn Südliche Umgehung Hamburg, Teil II: Konstruktion des Brückenüberbaus," *Der Stahlbau*, Vol. 32, No. 8 (August 1963), pp. 240–248.

45. Aschenberg, H. and Freudenberg, G., "Die Brücke über die Norderelbe im Zuge der Bundesautobahn Südliche Umgehung Hamburg, Teil III: Statische Berechnung des Brückenüberbaus," *Der Stahlbau*, Vol. 32, No. 9 (September 1963), pp. 281–287.

46. Poskitt, T. J. and Livesley, R. H., "Structural Analysis of Guyed Masts (Discussion)," *Proceedings of the Institution of Civil Engineers*, Vol. 26 (September 1963), pp. 185–186.

47. Havemann, H. K. and Freudenberg, G., "Die Brücke über die Norderelbe im Zuge der Bundesautobahn Südliche Umgehung Hamburg, Teil IV: Bauausführung der stählernen Uberbauten," *Der Stahlbau*, Vol. 32, No. 10 (October 1963), pp. 310–317.

48. "Norderelbe Bridge K6: A Welded Steel Motorway Bridge," *Acier-Stahl-Steel*, Vol. 28, No. 11 (November 1963), pp. 499–500.

49. *Stahlbau, Ein Handbuch für Studium und Praxis* (Vol. II, 2nd ed.; Cologne: Stahlbauverlag GmbH, 1964), "Seilverspannte Balken," p. 584.

50. Leonhardt, F., "Aerodynamisch stabile Hängebrücke für grosse Spannweiten," International Association for Bridge and Structural Engineering, Preliminary Publication, Seventh Congress, Rio de Janeiro (1964), pp. 155–167.

51. Leonhardt, F., "Kabel mit hoher Ermüdungsfestigkeit für Hängebrücken," International Association for Bridge and Structural Engineering, Preliminary Publication, Seventh Congress, Rio de Janeiro (1964), pp. 1055–1060.

52. Homberg, H., "Fortschritt im deutschen Stahlbrückenbau [Progress in German Steel Bridge Construction]," *Report on Steel Congress 1964 of the High Authority of the European Economic Community*.

53. Feige, A., "Steel Motorway Bridge Construction in Germany," *Acier-Stahl-Steel*, Vol. 29, No. 3 (March 1964), pp. 113–126.

54. Shaw, F. S., "Some Notes on Cable Suspension Roof Structures," *Journal of the Institution of Engineers* (Australia) (April–May 1964), pp. 105–113.

55. O'Brien, W. T. and Francis, A. J., "Cable Movements under Two Dimensional Loads," *Journal of the Structural Division*, ASCE, Vol. 90, No. ST 3 (June 1964), pp. 89–123.

56. Havemann, H. K., "Spannungs-und Schwingungsmessungen an der Brücke über die Norderelbe im Zuge der Bundesautobahn Südliche Umgehung Hamburg," *Der Stahlbau*, Vol. 33, No. 10 (October 1964), pp. 289–297.

57. Lohmer, G., "Brückenbaukunst," *Der Stahlbau*, Vol. 33, No. 11 (November 1964).

58. Beyer, E. and Ernst, H. J., "Brücke Jülicher Strasse in Düsseldorf," *Der Bauingenieur*, Vol. 39, No. 12 (December 1964), pp. 469–477.

59. Ernst, H. J., "Der E-Modul von Seilen unter Berücksichtigung des Durchhanges," *Der Bauingenieur*, Vol. 40, No. 2 (February 1965), pp. 52–55.

60. Daniel, H., "Die Bundesautobahnbrücke über den Rhein bei Leverkusen. Planung, Wettbewerb und seine Ergebnisse," *Der Stahlbau*, Vol. 34, No. 2, 3, 4, 5, and 12

(February, March, April, May, and December 1965), pp. 33–36, 83–86, 115–119, 153–158, 362–368.

61. "Montreal Hosts a Double Bridge Spectacular in the St. Lawrence," *Engineering News Record* (August 5, 1965), pp. 24–27, 31.

62. Brown, C. D., "Design and Construction of the George Street Bridge over the River Usk, at Newport, Monmouthshire," *Proceedings of the Institution of Civil Engineers*, Vol. 32 (September 1965), pp. 31–52.

63. Francis, A. J., "Single Cables Subjected to Loads," *Civil Engineering Transactions*, Institution of Engineers, Australia (October 1965), pp. 173–180.

64. Klöppel, K., Esslinger, M. and Kollmeier, H., "Die Berechnung eingespannter und fest mit dem Kabel verbundener Hängebrückenpylonen bei Beanspruhnung in Brückenlängs-richtung," *Der Stahlbau*, Vol. 34, No. 12 (December 1965), pp. 358–361.

65. Poskitt, T. J., "The Structural Analysis of Suspension Bridges," *Journal of the Structural Division*, ASCE, Vol. 92, No. ST 1, Proc. Paper 4664 (February 1966), pp. 49–73.

66. Brown, C. D., "Design and Construction of the George Street Bridge over the River Usk, at Newport, Monmouthshire (Discussion)," *Proceedings of the Institution of Civil Engineers*, Vol. 33 (March 1966), pp. 552–561.

67. Thul, H., "Stählerne Strassenbrücken in der Bundesrepublik," *Der Bauingenieur*, Vol. 41, No. 5 (May 1966), pp. 169–189.

68. Daniel, H. and Urban, J., "Die Bundesautobahnbrücke über den Rhein bei Leverkusen," *Der Stahlbau*, Vol. 35, No. 7 (July 1966).

69. Protte, W. and Tross, W., "Simulation als Vorgehensweise bei der Berechnung von Schragseilbrucken," *Der Stahlbau*, Vol. 35, No. 7 (July 1966).

70. Freudenberg, G. and Ratka, O., "Die Zoobrücke über den Rhein in Köln," *Der Stahlbau*, Vol. 38, No. 8, 9, and 11 (August, September, and November 1966).

71. Thul, H., "Cable Stayed Bridges in Germany," *Proceedings of the Conference on Structural Steelwork* held at the Institution of Civil Engineers (September 26–28, 1966), The British Constructional Steelwork Association, Ltd., London, pp. 69–81.

72. "The Slender Severn Suspension Bridge," *Engineering*, Vol. 202, No. 5238 (September 9, 1966), pp. 449–456.

73. "Continuing the Severn Crossing: The Wye Viaduct," *Engineering*, Vol. 202, No. 5239 (September 16, 1966).

74. Gimsing, N. J., "Anchored and Partially Anchored Stayed Bridges," Symposium on Suspension Bridges, Paper No. 30, Lisbon, Laboratorio Nacional De Engenharia Civil, November, 1966.

75. Feige, A., "The Evolution of German-Stayed Bridges: An Overall Survey," *Acier-Stahl-Steel* (English version), Vol. 31, No. 12 (December 1966), pp. 523–532.

76. *Steel Footbridges*, British Constructional Steelwork Association, Ltd., 16м/623/1266.

77. Goschy, B., "The Torsion of Skew-Cable Suspension Bridges," *Space Structures*, edited by R. M. Davies (Oxford and Edinburgh: Blackwell Scientific Publications, 1967), pp. 213–220.

78. Okaucki, I., Yabe, A. and Ando, K., "Studies on the Characteristics of a Cable-Stayed Bridge," *Bulletin of the Faculty of Science and Engineering, Chuo University*, Vol. 10 (1967).

79. "Bridge May Hang Like a Roof," *Engineering News-Record* (January 26, 1967), p. 56.

80. Heeb, A., Gerold, W. and Dreher, W., "Die Stahlkonstruktion der Neckarbrücke Untertürkheim," *Der Stahlbau*, Vol. 36, No. 2 (February 1967), pp. 33–38.

81. "Great Belt Bridge Award Winner," *Consulting Engineer* (England) (March 1967).

82. "Another Cable-Stayed Bridge Conquers the Rhine," *Engineering News-Record* (May 25, 1967), pp. 102–103, 107.

83. Smith, B. S., "The Single Plane Cable-Stayed Girder Bridge: A Method of Analysis Suitable for Computer Use," *Proceedings of the Institution of Civil Engineers*, Vol. 37 (May 1967), pp. 183–194.

84. Feige, A., "The Evolution of German Cable-Stayed Bridges: An Overall Survey," *Engineering Journal*, American Institute of Steel Construction (July 1967), pp. 113–122.

85. Daniel, H. and Schumann, H., "Die Bundesautobahnbrücke über den Rhein bei Leverkussen," *Der Stahlbau*, Vol. 36, No. 8 (August 1967), pp. 225–236.

86. Klöppel, K. and Thiele, F., "Modellversuche im Windkanal zur Bemessung von Brücken gegen die Gefahr winderregter Schwingungen," *Der Stahlbau*, Vol. 36, No. 12 (December 1967), pp. 353–365.

87. Payne, R. J., "The Structural Requirements of the Batman Bridge as They Affect Fabrication of the Steelwork," *Journal of the Institution of Engineers* (Australia), Vol. 39, No. 12 (December 1967).

88. Bresler, B., Lin, T. Y. and Scalzi, J. B., *Design of Steel Structures*, Chapter 15, Gillespie, J. W., McDermott, J. F. and Podolny, W., Jr., "Special Structures" (2nd ed.; New York: John Wiley & Sons, Inc., 1968), pp. 752–754.

89. Heckel, R., "The Use of Orthotropic Steel Decks in Austria," *Proceedings of the Conference on Steel Bridges*, The British Constructional Steelwork Association, Ltd., London (1968), pp. 143–150.

90. Schor, R. J., "Steel Bridges in Holland," *Proceedings of the Conference on Steel Bridges*, The British Constructional Steelwork Association, Ltd., London (1968), pp. 161–168.

91. Allen, J. S., Leeson, J. and Upstone, M. P., "River Severn Pipe Bridge and Road Crossing for the South Staffordshire Waterworks Company," *Proceedings of the Conference on Steel Bridges*, The British Constructional Steelwork Association, Ltd., London (1968), pp. 169–176.

92. Elliott, P., "Can Steel Bridges Become More Competitive?," *Proceedings of the Conference on Steel Bridges*, The British Constructional Steelwork Association, Ltd., London (1968), pp. 199–210.

93. Foucriat, J. and Sfintesco, M., "Steel Bridges in France," *Proceedings of the Conference on Steel Bridges*, The British Constructional Steelwork Association, Ltd., London (1968), pp. 217–227.

94. Troitsky, M. S., *Orthotropic Bridges Theory and Design* (Cleveland, Ohio: The James F. Lincoln Arc Welding Foundation, 1968), pp. 46–52.

95. *Suspended Structures*, British Constructional Steelwork Association, Ltd., 16M/842/68.

96. Tung, D. H. H. and Kudder, R. J., "Analysis of Cables as Equivalent Two-Force Members," *Engineering Journal*, American Institute of Steel Construction (January 1968), pp. 12–19.

97. Smith, B. S., "A Linear Method of Analysis for Double-Plane Cable-Stayed Girder Bridges," *Proceedings of the Institution of Civil Engineers*, Vol. 39 (January 1968), pp. 85–94.

98. Schottgen, J. and Wintergerst, L., "Die Strassenbrücke über den Rhein bei Maxau," *Der Stahlbau*, Vol. 37, No. 1 (January 1968), pp. 1–9.

99. Morandi, R., "Ill viadotto—dell' Ansa della Magliana—per la Autostrada Roma—Aeroporto di Fiumicino," *L'Industria Italiana del Cemento*, No. 38 (March 1968), pp. 147–162.

100. Feige, A., "Fussgängerbrücken aus Stahl," *Merkblatt 251*, Beratungsstelle für Stahl-verwendung, Düsseldorf, 3 Auflage 1968.

101. "The Expo Bridge: Study in Steel Quality (Canada)," *Acier-Stahl-Steel*, Vol. 33, No. 5 (May 1968), pp. 238–240.

102. "Opening Batman Bridge 18th May 1968," Department of Public Works, Tasmania, Australia.

103. Demers, J. G. and Marquis, P., "Le Pont a Haubans de la Riviere-des-Prairies," *L'Ingenieur*, Vol. 54, No. 231 (June 1968), pp. 24–28.

104. Tesár, A., "Das Projekt der neuen Strassenbrücke über die Donau in Bratislava/ CSSR," *Der Bauingenieur*, Vol. 43, No. 6 (June 1968), pp. 189–198.

105. Klingenberg, W. and Thul, H., "Ideenwettbewerb für einen Brückenschlag über den Grossen Belt," *Der Stahlbau*, Vol. 37, No. 8 (August 1968).

106. Moser, K., "Der Einfluss des zeitabhängigen Verhaltens bei Hänge und Schrägseil-brückensystemen," *International Association for Bridge and Structural Engineering*, *Final Report*, Eighth Congress, New York (September 9–14, 1968), pp. 119–129.

107. Murakami, E. and Okubo, T., "Wind Resistant Design of a Cable-Stayed Girder Bridge," *International Association for Bridge and Structural Engineering*, *Final Report*, Eighth Congress, New York (September 9–14, 1968), pp. 1263–1274.

108. Leonhardt, F., "Zur Entwicklung aerodynamisch stabiler Hängebrücken," *Die Bau-technik*, Vol. 45, No. 10 and 11 (1968), pp. 1–21.

109. Morandi, R., "Some Types of Tied Bridges in Prestressed Concrete," *First International Symposium*, *Concrete Bridge Design*, ACI Publication SP-23, Paper SP 23–25 (1969).

110. Tamms and Beyer, "Kniebrücke Düsseldorf," Beton-Verlag GmbH, Düsseldorf, 1969.

111. Balbachevsky, G. N., "Study Tour of the A.F.P.C.," *Acier-Stahl-Steel*, Vol. 34, No. 2 (February 1969), pp. 73–82.

112. Pflüger, A., "Schwingungsverhalten der Schwebebahnbrücke Alter Markt Wupper-tal," *Der Stahlbau*, Vol. 38, No. 5 (May 1969), pp. 140–144.

113. "Erskine Bridge," *Building with Steel*, Vol. 5, No. 4 (June 1969), pp. 28–32.

114. Wardlaw, R. L., "A Preliminary Wind Tunnel Study of the Aerodynamic Stability of Four Bridge Sections for Proposed New Burrard Inlet Crossing," National Research Council of Canada, National Aeronautical Establishment, Tech. Report LTR-LA-31 (July 14, 1969).

115. Andrä, W. and Zellner, W., "Zugglieder aus Paralleldrahtbündeln und ihre Veran-kerung bei hoher Dauerschwellbelastung," *Die Bautechnik*, Vol. 46, No. 8 and 9 (1969), pp. 1–12.

116. Rooke, W. G., "Papineau Bridge Steel Erected in Record Time," *Heavy Construction News* (September 1, 1969).

117. Tschemmernegg, F., "Über die Aerodynamik und Statik von Monokabelhänge-brücken," *Der Bauingenieur*, Vol. 44, No. 10 (October 1969), pp. 353–362.

118. Scalzi, J. B., Podolny, W., Jr., and Teng, W. C., "Design Fundamentals of Cable Roof Structures," United States Steel Corporation, ADUSS 55-3580-01 (October 1969).

119. Taylor, P. R., "Cable Stayed Bridges and Their Potential in Canada," *Engineering Journal* (Canada), Vol. 52, No. 11 (November 1969), pp. 15–21.

120. Simpson, C. V. J., "Modern Long Span Steel Bridge Construction in Western Europe," *Proceedings of the Institution of Civil Engineers* (1970), Supplement (ii).

121. Leonhardt, F. and Zellner, W., "Cable-Stayed Bridges: Report on Latest Develop-

ments," Canadian Structural Engineering Conference, 1970, Canadian Steel Industries Construction Council, Toronto, Ontario, Canada.

122. Wardlaw, R. L. and Ponder, C. A., "Wind Tunnel Investigation on the Aerodynamic Stability of Bridges," Canadian Structural Engineering Conference, 1970, Canadian Steel Industries Construction Council, Toronto, Ontario, Canada.

123. "Der Fussgängersteg Raxstrasse in Wien 10," *Stahlbau Rundschau*, 34 (February 1970), pp. 2-3.

124. "Japanese Try Shop Fabricated Bridge Cables," *Engineering News-Record* (February 26, 1970), p. 14.

125. Bachelart, H., "Pont de la Bourse, Footbridge over the Bassin du Commerce Le Havre (France)," *Acier-Stahl-Steel*, Vol. 35, No. 4 (April 1970), pp. 167-169.

126. Tamhankar, M. G., "Design of Cable-Stayed Girder-Bridges," *Journal of the Indian Roads Congress*, Vol. XXXIII-1 (May 1970).

127. "The Paris-Masséna Bridge: A Cable-Stayed Structure," *Acier-Stahl-Steel*, Vol. 35, No. 6 (June 1970), pp. 278-284.

128. Wardlaw, R. L., "Static Force Measurements of Six Deck Sections for the Proposed New Burrard Inlet Crossing," National Research Council of Canada, National Aeronautical Establishment, Tech. Report LTR-LA-53 (June 4, 1970).

129. "Record All-Welded, Cable-Stayed Span Hangs from Pylons," *Engineering News-Record* (September 3, 1970), pp. 20-21.

130. Freudenberg, G., "Die Stahlhochstrasse über den neuen Hauptbahnhof in Ludwigshafen/Rhein," *Der Stahlbau*, Vol. 39, No. 9 (September 1970), pp. 257-267.

131. Wardlaw, R. L., "Some Approaches for Improving the Aerodynamic Stability of Bridge Road Decks," *Proceedings of the Third International Conference on Wind Effects on Buildings and Structures*, Tokyo (1971).

132. O'Conner, Colin, *Design of Bridge Superstructures* (New York: Wiley-Interscience, John Wiley & Sons, Inc., 1971).

133. "Bridge Bidder Leaves $21 Million on the Table," *Engineering News-Record* (January 7, 1971), p. 11.

134. Podolny, W., Jr. and Fleming, J. F., "Cable-Stayed Bridges—A State of the Art," Preprint Paper 1346, ASCE National Water Resources Engineering Meeting, Phoenix, Arizona (January 11-15, 1971).

135. Seim, C., Larsen, S. and Dang, A., "Design of the Southern Crossing Cable Stayed Girder," Preprint Paper 1352, ASCE National Water Resources Engineering Meeting, Phoenix, Arizona (January 11-15, 1971).

136. Troitsky, M. S. and Lazar, B. E., "Model Analysis and Design of Cable-Stayed Bridges," *Proceedings of the Institution of Civil Engineers* (March 1971).

137. Seim, C., Larsen, S. and Dang, A., "Analysis of Southern Crossing Cable-Stayed Girder," Preprint Paper 1402, ASCE National Structural Engineering Meeting, Baltimore, Maryland (April 19-23, 1971).

138. Tang, Man-Chung, "Analysis of Cable-Stayed Girder Bridges," *Journal of the Structural Division*, ASCE, Vol. 97, No. ST 5 (May 1971), pp. 1481-1496.

139. Feige, A. and Idelberger, K., "Long-Span Steel Highway Bridges Today and Tomorrow," *Acier-Stahl-Steel* (English version), No. 5 (May 1971).

140. Donnelly, J. A., "Beauty of Steel Footbridges," *Acier-Stahl-Steel* (English version), No. 6 (June 1971).

141. Baron, F. and Lien, S. Y., "Analytical Studies of the Southern Crossing Cable-Stayed Girder Bridge," Report No. UC SESM 71-10, Vols. I and II, Department of Civil Engineering, University of California, Berkeley, California (June 1971).

142. Podolny, W., Jr., "Static Analysis of Cable-Stayed Bridges," Ph.D. thesis, University of Pittsburgh, 1971.
143. "River Foyle Bridge," *Civil Engineering and Public Works Review*, Vol. 66, No. 780 (July 1971), p. 753.
144. "Longest Concrete Cable-Stayed Span Cantilevered over Tough Terrain," *Engineering News-Record* (July 15, 1971), pp. 28–29.
145. "Feasibility Study of Mississippi River Crossings Interstate Route 410," Report to Louisiana Department of Highways in Cooperation with Federal Highway Administration, Modjeski and Masters, Consulting Engineers, Harrisburg, Pennsylvania (July 1971).
146. Demers, J. G. and Simonsen, O. F., "Montreal Boasts Cable-Stayed Bridge," *Civil Engineering*, ASCE (August 1971).
147. Lazar, B. E., "Analysis of Cable-Stayed Girder Bridges (Discussion)," by Man-Chung Tang, *Journal of the Structural Division*, ASCE, Vol. 97, No. ST 10 (October 1971), pp. 2631–2632.
148. Gee, A. F., "Cable-Stayed Concrete Bridges," *Developments in Bridge Design and Construction*, edited by Rockey, Bannister, and Evans (London: Crosby Lockwood & Son, Ltd., October 1971).
149. Scalzi, J. B. and McGrath, W. K., "Mechanical Properties of Structural Cables," *Journal of the Structural Division*, ASCE, Vol. 97, No. ST 12 December 1971), pp. 2837–2844.
150. "Der Bau der 2. Mainbrücke der Farbverke Hoechst AG." New York: Dickerhoff and Widemann, Inc., 1972).
151. Leonhardt, F. and Zellner, W., "Vergleiche zwischen Hängebrücken und Schrägseilbrücken für Spannweiten über 600 m [Comparative Investigations Between Suspension Bridges and Cable-Stayed Bridges for Spans Exceeding 600 m], *IABSE Publication 32-I*, 1972.
152. "Dywidag-Spannverfhren, Paralleldrahtseil," Bericht Nr. 14, Herausgegeben von der Aubteilung für Entwicklung (New York: Dickerhoff and Wideman, Inc., June 1972).
153. Taylor, P. R. and Demers, J. G., "Design, Fabrication and Erection of the Papineau-Leblanc Bridge," Canadian Structural Engineering Conference, 1972, Canadian Steel Industries Construction Council, Toronto, Ontario, Canada.
154. Dubrova, E., "On Economic Effectiveness of Application of Precast Reinforced Concrete and Steel for Large Bridges (USSR)," *IABSE Bulletin*, 28 (1972).
155. Daniel, H., "Die Rheinbrücke Duisburg-Neuenhamp," *Der Stahlbau*, Vol. 41, No. 1 (January 1972), pp. 7–14.
156. "Die Erskine-Brücke-eine 1321 m langes Schrägseilbrücke in Schottland," *Der Stahlbau*, Vol. 41, No. 1 (January 1972), pp. 26–29.
157. Daniel, H., "Die Rheinbrücke Duisburg-Neuenkamp," *Der Stahlbau*, Vol. 41, No. 3 (March 1972), pp. 73–78.
158. Jonatowski, J., "Analysis of Cable-Stayed Girder Bridges (Discussion)," by Man-Chung Tang, *Journal of the Structural Division*, ASCE, Vol. 98, No. ST 3 (March 1972), pp. 770–774.
159. Lin, T. Y. and Kulka, F., "Basic Design Concepts for Long Span Structures," Preprint Paper 1727, ASCE National Structural Engineering Meeting, Cleveland, Ohio (April 24–28, 1972).
160. Kerensky, O. A., Henderson, W. and Brown, W. C., "The Erskine Bridge," *Structural Engineer*, Vol. 50, No. 4 (April 1972).
161. Buckland, P. G. and Wardlaw, R. L., "Some Aerodynamic Considerations in Bridge Design," *Engineering Journal* (Canada), Engineering Institute of Canada (April 1972).

162. Finsterwalder, U. and Finsterwalder, K., "Neue Entwicklung von Parallel-drahtseil für Schrägseil und Spannbandbrücken," Preliminary Report, Ninth IABSE Congress, Amsterdam (May 1972), pp. 877–883.

163. ERRATA, "Analysis of Cable-Stayed Girder Bridges," by Man-Chung Tang, *Journal of the Structural Division*, ASCE, Vol. 98, No. ST 5 (May 1972), p. 1191.

164. Schreier, G., "Bridge over the Rhine at Düsseldorf: Design, Calculation, Fabrication and Erection," *Acier-Stahl-Steel* (English version), No. 5 (May 1972).

165. Leonhardt, F., "Seilkonstrucktonen und seil verspannte Konstruktionen," Introductory Report, Ninth IABSE Congress, Amsterdam (May 1972).

166. "Cable-Stayed Bridges with Bolted Galvanized Joints," *Civil Engineering*, ASCE (May 1972), p. 98.

167. Thul, H., "Schrägseilbrücken," Preliminary Report, Ninth IABSE Congress, Amsterdam (May 1972), pp. 249–258.

168. "First U.S. Stayed Girder Span Is a Slim, Economical Crossing," *Engineering News-Record* (June 29, 1972), pp. 14–15.

169. Thul, H., "Entwicklungen im Deutschen Schrägseilbrückenbau," *Der Stahlbau*, Vol. 41, No. 6 (June 1972), pp. 161–171.

170. Kondo, K., Komatsu, S., Inoue, H. and Matsukawa, A., "Design and Construction of Toyosato-Ohhashi Bridge," *Der Stahlbau*, Vol. 41, No. 6 (June 1972), pp. 181–189.

171. "Seattle Plans Second U.S. Stayed-Girder Bridge," *Engineering News-Record* (July 27, 1972), p. 10.

172. Lazar, B. E., "Stiffness Analysis of Cable-Stayed Bridges," *Journal of the Structural Division*, ASCE, Vol. 98, No. ST 7 (July 1972), pp. 1605–1612.

173. Weisskopf, F., "World's Longest-Span Cable-Stayed Girder Bridge over the Rhine near Duisburg (Germany)," *Acier-Stahl-Steel* (English version) (July–August, 1972).

174. "Record Stayed-Girder Span Goes Up Amid Controversy," *Engineering News-Record* (August 3, 1972), p. 13.

175. "Repair Work Gives New Lift to Old Bridge," *Engineering News-Record* (August 24, 1972), p. 17.

176. Lazar, B. E., Troitsky, M. S. and Douglass, M. McC., "Load Balancing Analysis of Cable Stayed Bridges," *Journal of the Structural Division*, ASCE, Vol. 98, No. ST 8 (August 1972), pp. 1725–1740.

177. Tang, M. C., "Design of Cable-Stayed Girder Bridges," *Journal of the Structural Division*, ASCE, Vol. 98, No. ST 8 (August 1972), pp. 1789–1802.

178. Birdsall, Blair, "Mechanical Properties of Structural Cables (Discussion)," by Scalzi and McGrath, *Journal of the Structural Division*, ASCE, Vol. 98, No. ST 8 (August 1972), pp. 1883–1884.

179. Borelly, W., "Nordbrücke Mannheim-Ludwigshafen," *Der Bauingenieur*, Heft 8 u. 9 (August and September 1972).

180. Podolny, W., Jr., and Fleming, J. F., "Historical Development of Cable-Stayed Bridges," *Journal of the Structural Division*, ASCE, Vol. 98, No. ST 9 (September 1972), pp. 2079–2095.

181. "Concrete Cable-Stayed Bridge over River Main in Germany," *Civil Engineering and Public Works Review*, Vol. 67, No. 796 (November 1972).

182. Kajita, T. and Cheung, Y. K., "Finite Element Analysis of Cable-Stayed Bridges," *IABSE Publication 33-II*, 1973.

183. Baron, F. and Lien, S. Y., "Analytical Studies of a Cable-Stayed Girder Bridge," *Computers & Structures*, Vol. 3 (New York: Pergamon Press, 1973).

184. Gray, N., "Chaco/Corrientes Bridge in Argentina," *Municipal Engineers Journal*, Paper No. 380, Vol. 59 (Fourth Quarter, 1973).

185. "Schräseilbrücke in Kanada," *Der Stahlbau*, Heft 1 (January 1973).

186. Podolny, W., Jr., "Cable-Stayed Bridges of Prestressed Concrete," *PCI Journal*, Vol. 18, No. 1 (January–February 1973).

187. Podolny, W., Jr., "Cable Connections in Stayed Girder Bridges," Meeting Preprint 1933, ASCE National Structural Engineering Meeting, April 9–13, 1973, San Francisco, California.

188. Gute, W. L., "Design and Construction of the Sitka Harbor Bridge," Meeting Preprint 1957, ASCE National Structural Engineering Meeting, April 9–13, 1973, San Francisco, California.

189. Kealey, R. T., "Feasibility Study of Mississippi River Crossings—Interstate 410," Meeting Preprint 2003, ASCE National Structural Engineering Meeting, San Francisco, California (April 9–13, 1973).

190. Woods, S. W., "Historical Development of Cable-Stayed Bridges (Discussion)," by Podolny and Fleming, *Journal of the Structural Division*, ASCE, Vol. 99, No. ST 4 (April 1973).

191. Volke, E., "Die Strombrücke im Zuge der Nordbrücke Mannheim-Ludwigshafen (Kurt-Schumacher Brücke)," *Der Stahlbau*, Heft 4 u. 5 (April and May 1973).

192. Rademacher, C. H., "Die Strombrücke im Zuge der Nordbrücke Mannheim-Ludwigshafen (Kurt-Schumacher Brücke), Teil II: Werkstattfertig und Montage," *Der Stahlbau*, Heft 6 (June 1973).

193. Kavanagh, T. C., Discussion to "Historical Development of Cable-Stayed Bridges," by Podolny and Fleming, *Journal of the Structural Division*, ASCE, Vol. 99, No. ST 7 (July 1973).

194. Podolny, W., Jr. and Fleming, J. F., "Cable-Stayed Bridges—Single Plane Static Analysis," *Highway Focus*, Vol. 5, No. 2 (August 1973), U.S. Dept. of Transportation, Federal Highway Administration, Washington, D. C.

195. Podolny, W., Jr., "Economic Comparisons of Stayed Girder Bridges," *Highway Focus*, Vol. 5, No. 2 (August 1973), U.S. Dept. of Transportation, Federal Highway Administration, Washington, D. C.

196. Podolny, W., Jr., "Cable-Stayed Bridges and Wind Effects," *Highway Focus*, Vol. 5, No. 2 (August 1973), U.S. Dept. of Transportation, Federal Highway Administration, Washington, D. C.

197. Podolny, W., Jr., Readers Comment, "Cable-Stayed Bridges of Prestressed Concrete," *PCI Journal*, September–October, 1973.

198. Naruoka, M. and Sakamoto, T., "Cable-Stayed Bridges in Japan," *Acier-Stahl-Steel* (English version), No. 10 (October 1973).

199. "Seattle Plans Cable-Stayed Bridge," *Engineering News-Record* (November 29, 1973), p. 13.

200. Gute, W. L., "First Vehicular Cable-Stayed Bridge in the U.S.," *Civil Engineering*, ASCE, Vol. 43, No. 11 (November 1973).

201. Burgholzer, L., Garn, E., Schimetta, O., "Die 2. Donaubrücke Linz," *Der Stahlbau*, Heft 11 (November 1973).

202. Podolny, W., Jr., "Cable-Stayed Bridges," *Engineering Journal*, American Institute of Steel Construction (First Quarter 1974).

203. "Cable-Stayed Span Will Set U.S. Record," *Engineering News-Record* (January 3, 1974), p. 10.

204. "Cable-Stayed Bridge Will Have Record 981 ft. Span," *Engineering News-Record* (March 21, 1974), p. 39.

205. "First Stayed Swing Span Cuts Costs," *Engineering News-Record* (March 28, 1974), p. 11.

206. Rothman, H. B. and Chang, F. K., "Longest Precast-Concrete Box-Girder Bridge in Western Hemisphere," *Civil Engineering*, ASCE (March 1974).

207. "Chaco/Corrientes, Latin America's Longest Cable-Supported Bridge," *World Construction* (English version), Vol. 27, No. 3 (March 1974).

208. Burns, C. A. and Fotheringham, W. D., "Deck Panels for West Gate Bridge (Australia)," *Acier-Stahl-Steel* (English version) (June 1974).

209. Mohsen, H., "Trends in the Construction of Steel Highway Bridges," *Acier-Stahl-Steel* (English version) (June 1974).

210. Gade, R. H., "Status of the Investigation of Aerodynamic Behavior of the Sitka Harbor Cable-Stayed Bridge," presented at the Symposium on Full Scale Measurements of Wind Effects on Tall Buildings and Other Structures, London, Ontario, Canada (June 1974).

211. "Rigid Stays Slim Box Girder Bridge and Reduce Deflection," *Engineering News-Record* (June 20, 1974), p. 58.

212. "Cable-Stayed Crossing Bids Top Estimate by 40%," *Engineering News-Record* (July 4, 1974), p. 13.

213. Andrä, W. and Saul, R., "Versuche mit Bundeln aus parallelen Drähten und Litzen für die Nordbrücke Mannheim-Ludwigshafen und das Zeltdach München," *Die Bautechnik*, Heft 9 u. 10 (September and October 1974).

214. "Site Conditions Turn Bridge Tower Upside Down," *Engineering News-Record* (October 17, 1974), p. 25–26.

215. Podolny, W., Jr., "Design Considerations in Cable-Stayed Bridges," *Proceedings of the Specialty Conference on Metal Bridges*, ASCE, St. Louis, Missouri (November 12–13, 1974).

216. Podolny, W., Jr., "Cable Connections in Stayed Girder Bridges," *Engineering Journal*, American Institute of Steel Construction (Fourth Quarter, 1974).

217. Beyer, E. and Ramberger, G., "Die Franklinbrücke in Düsseldorf," *Der Stahlbau*, Heft 5 (May 1975).

218. Schwab, R., "Die Köhlbrandreuzung, Überbrückung einer Seeschiffahrtsstrasse im Hamburger Hafen," *Strasse und Autobahn* (May 1975).

219. Schwab, R. and Homann, H., "Der Bau der Köhlbrandbrücke," *Die Bautechnik*, Heft 5 (May 1975).

220. Warolus, L., "Cable-Stayed Bridge over the Meuse at Heer-Agimont (Belgium)," *Acier-Stahl-Steel* (English version) (May 1975).

221. Boué, P. and Höhne, K. J., "Der Stromüberbau der Köhlbrandbrücke," *Der Stahlbau*, Heft 6 (June 1975).

222. Rabe, J. and Baumer, H., "Die Gründungen und Pfeiler der Köhlbrandbrücke," *Die Bautechnik*, Heft 6 (June 1975).

# Index of Bridges

# Index of Personal Names

499

500    **INDEX OF PERSONAL NAMES**

# Index of Firms and Organizations

501

# Index of Subjects